Spring Cloud Alibaba
大型微服务架构项目实战

下册

十三 / 著

电子工业出版社
Publishing House of Electronics Industry
北京·BEIJING

内 容 简 介

本书重在引导读者体验真实的项目开发，围绕Spring Cloud Alibaba技术栈全面展开，兼顾相关技术的知识拓展，由浅入深，步步为营，对一个单体API项目进行拆解和微服务化，并从零到一落地一个功能完整、流程完善的微服务架构项目。本书的目标是让读者拥有完整且高质量的学习体验，远离"Hello World"项目，为技术深度的挖掘和薪水、职位的提升提供保障。

本书分为13章。第1章主要介绍大型微服务架构项目设计与实战，包括它的主要功能模块、从单体模式到前后端分离模式再到微服务架构模式的开发历程、微服务架构项目改造前的拆分思路、微服务架构项目的启动等注意事项。第2章至第8章主要介绍微服务架构项目的开发过程，介绍开发步骤、微服务模块的编码过程、微服务组件的整合，涉及的主要微服务组件包括Nacos、OpenFeign和Spring Cloud LoadBalancer。第9章至第13章对实战的微服务架构项目进行补充，将Spring Cloud Gateway、Sentinel、Seata、Sleuth、Zipkin等微服务组件整合到实战项目中，并对该过程中遇到的问题进行复盘及处理。本书实战项目整合热门的微服务组件，手把手地教读者如何在实战中运用这些知识点，让读者掌握高阶的使用技巧，并且能够将其运用到实际生产项目中。

本书的内容丰富，案例通俗易懂，几乎涵盖了目前Spring Cloud Alibaba的全部热门组件，特别适合想要了解Spring Cloud Alibaba热门组件及想搭建微服务架构系统的读者阅读。

未经许可，不得以任何方式复制或抄袭本书之部分或全部内容。
版权所有，侵权必究。

图书在版编目（CIP）数据

Spring Cloud Alibaba 大型微服务架构项目实战. 下册 / 十三著. -- 北京 : 电子工业出版社, 2024. 9.
ISBN 978-7-121-48663-0

Ⅰ．TP368.5

中国国家版本馆 CIP 数据核字第 2024UN8657 号

责任编辑：石　悦
文字编辑：戴　新
印　　刷：三河市君旺印务有限公司
装　　订：三河市君旺印务有限公司
出版发行：电子工业出版社
　　　　　北京市海淀区万寿路 173 信箱　　邮编：100036
开　　本：787×980　1/16　印张：20.75　字数：455 千字
版　　次：2024 年 9 月第 1 版
印　　次：2024 年 9 月第 1 次印刷
定　　价：99.00 元

凡所购买电子工业出版社图书有缺损问题，请向购买书店调换。若书店售缺，请与本社发行部联系，联系及邮购电话：(010) 88254888，88258888。
质量投诉请发邮件至 zlts@phei.com.cn，盗版侵权举报请发邮件至 dbqq@phei.com.cn。
本书咨询联系方式：faq@phei.com.cn。

前　　言

大家好，我是十三。

非常感谢你们阅读本书，在技术的道路上，我们从此不再独行。

写作背景

2017 年 2 月 24 日，笔者正式开启技术写作之路，也开始在 GitHub 网站上做开源项目，由于一直坚持更新文章和开源项目，慢慢地被越来越多的人所熟悉。2018 年 6 月 7 日，电子工业出版社的陈林编辑通过电子邮件联系笔者并邀请笔者出书。从此，笔者与电子工业出版社结缘。2018 年，笔者也被不同的平台邀请制作付费专栏和课程。从 2018 年 9 月开始，笔者陆续在 CSDN 图文课、实验楼、蓝桥云课、掘金小册、极客时间等平台上线了多个付费专栏和多门课程。2020 年，笔者与电子工业出版社的陈林编辑联系并沟通了写作事宜，之后签订了图书出版合同，第一本书在 2021 年正式出版。

笔者写作的初衷是把自己对技术的理解及实战项目开发的经验分享给读者。笔者把过去几年的经历整理成一张图，如下图所示。"免费文章→付费专栏→付费视频→实体图书"，从无到有，是一步一步走过来的。这些是本书的写作背景。

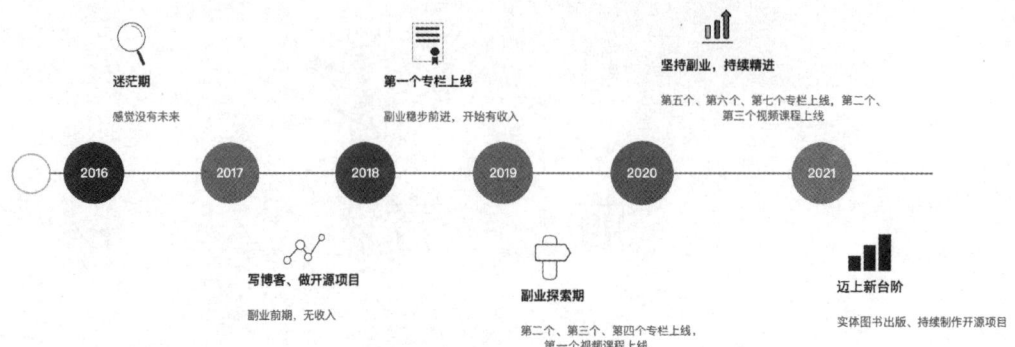

笔者将付费专栏和本书中所用到的实战项目开源到 GitHub 和 Gitee 两个开源代码平台上，本书中基于 Spring Cloud Alibaba 的微服务架构项目 newbee-mall-cloud 是笔者开发的一个开源项目。

随着越来越完善的微服务技术栈的发布，以及越来越多的微服务架构项目落地和上线，使用 Java 技术栈的企业应该都在尝试或已经落地了各自的微服务架构项目。通过招聘网站的信息和每次面试的反馈，Java 开发人员能够清晰地认识到：**微服务技术已经成为 Java 开发者必须掌握的一门技术。**

读者可能对微服务架构有所耳闻，也能够看出它是未来的一种流行架构，进而非常希望能够了解微服务技术体系，甚至动手实践完成微服务架构项目的开发与维护，掌握微服务技术这个非常宝贵的技能。不过，在掌握这个技能时，可能会遇到如下几个问题：

- 微服务技术的体系复杂，从何学起？学习哪些知识点？有没有简捷而有效的学习路径？
- 微服务架构中的组件和中间件很多，如何选择一套合适且可落地的方案？
- 在搭建与开发微服务架构项目时，会遇到哪些问题？该如何解决这些问题？
- 想要自己动手开发一个大型微服务架构项目，有没有适合的源代码？有没有可以借鉴的经验？

针对这些问题，笔者结合自己的开发经验和一个可操作的大型微服务架构项目，从复杂的微服务技术体系中梳理一条明确而有效的学习路径，让读者可以成体系地学习微服务架构，本书的知识点规划和学习路径如下图所示。

前言

Spring Cloud Alibaba 大型微服务架构项目实战

1. 梳理微服务架构
- 讲解微服务的概念，介绍它的"前世今生"
- 介绍与Spring Cloud Alibaba相关的技术和微服务组件

2. 拆解出微服务架构搭建的步骤
- 基础的开发环境准备
- 化繁为简，微服务组件+编码实践
- 确认微服务组件的技术选型

3. 搭建并整合微服务组件
- Spring Cloud Alibaba 基础模板项目
- 服务治理——Nacos
- 服务通信——OpenFeign
- 负载均衡器——Spring Cloud LoadBalancer
- 服务网关——Spring Cloud Gateway
- 分布式事务——Seata
- 服务容错——Sentinel
- 链路追踪——Sleuth+Zipkin
- 日志中心——ElasticSearch+Logstash+Kibana(ELK)

4. 从零到一开发大型的微服务架构项目
- 项目规划及功能确认
- 服务边界确认与项目拆分
- 项目编码并整合各微服务组件

以上就是笔者为读者整理的微服务架构项目的学习路径和实战步骤：梳理微服务架构、拆解出微服务架构搭建的步骤、搭建并整合微服务组件、从零到一开发大型的微服务架构项目。

你会学到什么

《Spring Cloud Alibaba 大型微服务架构项目实战》（上册）已经对概念性的知识点进行了介绍，包括微服务的技术选型、对比技术栈、确定实战项目所选择的微服务组件，并对这些组件进行了讲解，包括组件的作用、搭建和优化，主要介绍了 Nacos、Spring Cloud Gateway、Sentinel、Seata 等组件的搭建和整合，完成了微服务架构项目里中间件搭建和整合的工作。除基础的整合外，也对重点技术栈的源代码进行了详细的剖析，让读者能够"知其然，知其所以然"。

作为对《Spring Cloud Alibaba 大型微服务架构项目实战》（上册）的补充，本书对一个大型的商城项目进行拆解和微服务化，并从零到一落地一个功能完整、流程完善的微服务架构项目，由浅入深，逐一击破微服务架构项目中的难点。通过本书的讲解和提供的完整代码，读者可以更深入地掌握 Spring Cloud Alibaba 技术栈中的组件、知识点，并且能够将其应用到自己所开发的项目中，在实战中深入理解微服务技术，掌握微服务架构项目开发的核心知识点。

本书的代码基于 Spring Boot 2.6.3 版本和 Spring Cloud Alibaba 2021.0.1.0 版本。需要注意的是，本书从书稿整理完成至正式出版耗时近一年，在这段时间里，Spring Boot、Spring Cloud Alibaba 及相关技术栈有一些版本升级，比如 Spring Boot 3.x 发版、Spring Cloud Alibaba 2022.x 发版。对于这些情况，笔者会在本书实战项目的开源仓库中创建不同的代码分支，保持实战项目的源代码更新，保证读者不会学习过时的知识点。

读者将通过本书学到以下内容。

- Spring Cloud Alibaba 微服务组件的整合与使用。
- 服务治理之服务注册与服务发现。
- 服务间的通信方式。
- 负载均衡器的原理与实践。
- 服务网关的搭建与使用。
- 分布式事务的处理。
- 服务容错之限流及熔断。

- 微服务间的链路追踪。
- ELK 日志中心的搭建与使用。
- 针对各个知识点的实战源代码和一套可执行的微服务架构项目源代码。

适宜人群

- 从事 JavaWeb 开发的技术人员。
- 希望进阶高级开发的后端开发人员。
- 对微服务架构感兴趣、想要了解 Spring Cloud Alibaba 热门组件的开发人员。
- 希望将微服务架构及相关技术实际运用到项目中的开发人员。
- 想要独立完成一个微服务架构项目的开发人员。

源代码

本书的每个实战章节都有对应的源代码，读者可以在本书封底扫码获取。

最终的实战项目是笔者的开源项目 newbee-mall-cloud，在开源网站 GitHub 和 Gitee 上都能搜索并下载到最新的源代码。

- https://www.hxedu.com.cn/Resource/OS/github.htm
- https://www.hxedu.com.cn/Resource/OS/gitee.htm

致谢

感谢电子工业出版社的陈林、石悦、李玲和其他老师，本书能够顺利出版离不开你们的奉献，感谢你们辛苦、严谨的工作。

感谢 newbee-mall 系列开源仓库的各位用户及笔者专栏文章的所有读者。他们提供了非常多的修改和优化意见，使这个微服务架构项目变得更加完善，为笔者提供了持续写作的动力。

感谢掘金社区、运营负责人优弧和运营人员 Captain。本书部分内容是基于掘金小册《Spring Cloud Alibaba 大型微服务项目实战》中的章节扩展而来的，本书能顺利出版得到了掘金社区的大力支持。

特别感谢一下家人，没有他们的默默付出和巨大的支持，笔者不可能有如此多的时间和精力专注于本书的写作。

感谢每一位没有提及名字，但是曾经帮助过笔者的贵人。

韩帅
2024 年 6 月 1 日 于杭州

目 录

第 1 章 大型微服务架构项目设计与实战 ·· 1

1.1 微服务架构项目详解 ·· 1
    1.1.1 实战项目介绍 ··· 2
    1.1.2 新蜂商城项目的开源历程 ·· 3
    1.1.3 新蜂商城项目的功能及数据库设计 ·· 4
1.2 从单体项目到微服务架构项目的拆分思路 ·· 6
1.3 微服务架构项目源代码获取和项目启动 ··· 8
    1.3.1 基础环境准备及微服务组件安装和配置 ·· 8
    1.3.2 下载微服务架构项目的源代码 ·· 8
    1.3.3 微服务架构项目的目录结构讲解 ·· 10
    1.3.4 启动并验证微服务实例 ·· 13
1.4 微服务架构项目的功能演示 ··· 21
    1.4.1 商城用户的注册与登录演示 ··· 21
    1.4.2 添加商品到购物车的功能演示 ·· 25
    1.4.3 下单流程演示 ··· 28
    1.4.4 后台管理系统的部分功能演示 ·· 32
1.5 微服务架构项目中接口的参数处理及统一结果响应 ··· 38
1.6 微服务架构项目打包和部署的注意事项 ··· 41

第 2 章 实战项目基础构建及公用模块引入 48

- 2.1 编码前的准备 48
- 2.2 搭建项目骨架 49
  - 2.2.1 构建项目并整理依赖关系 49
  - 2.2.2 编写测试代码 54
- 2.3 用户微服务编码 58
  - 2.3.1 引入业务依赖 58
  - 2.3.2 商城用户模块中的接口改造 60
  - 2.3.3 用户微服务改造过程中遇到的问题 62
- 2.4 引入公用模块 64
- 2.5 用户微服务模块改造 66
- 2.6 OpenFeign 编码暴露远程接口 71
- 2.7 远程调用 OpenFeign 应该如何设置 72

第 3 章 用户微服务编码实践及功能讲解 74

- 3.1 登录流程介绍 74
  - 3.1.1 什么是登录 74
  - 3.1.2 用户登录状态 75
  - 3.1.3 登录流程设计 75
- 3.2 登录功能的源代码介绍 78
- 3.3 token 值处理及鉴权源代码介绍 80
- 3.4 用户微服务代码改造 83
  - 3.4.1 引入 Redis 进行鉴权改造 83
  - 3.4.2 用户微服务中登录代码及鉴权代码修改 86
- 3.5 网关层鉴权 88
  - 3.5.1 在网关层引入 Redis 88
  - 3.5.2 鉴权的全局过滤器编码实现 89
  - 3.5.3 功能测试 91

第 4 章 商品微服务编码实践及功能讲解 ·········· 95

4.1 商品微服务介绍 ·········· 95
4.1.1 商品分类管理模块介绍 ·········· 95
4.1.2 商品管理模块介绍 ·········· 99

4.2 创建商品微服务编码 ·········· 103

4.3 商品微服务与用户微服务通信 ·········· 108
4.3.1 为什么需要调用用户微服务 ·········· 109
4.3.2 商品微服务调用用户微服务编码实践 ·········· 110
4.3.3 功能测试 ·········· 113

4.4 商品微服务编码 ·········· 117
4.4.1 商品微服务代码改造 ·········· 117
4.4.2 OpenFeign 编码暴露远程接口 ·········· 120
4.4.3 功能测试 ·········· 121

4.5 改造过程中遇到的问题总结 ·········· 124
4.5.1 问题 1：循环依赖 ·········· 124
4.5.2 问题 2：缺少 LoadBalancer 依赖 ·········· 125

第 5 章 推荐微服务编码实践及功能讲解 ·········· 126

5.1 推荐微服务主要功能模块介绍 ·········· 126
5.1.1 轮播图管理模块介绍 ·········· 126
5.1.2 商品推荐管理模块介绍 ·········· 129
5.1.3 表结构设计 ·········· 130

5.2 创建推荐微服务编码 ·········· 133

5.3 推荐微服务编码 ·········· 139
5.3.1 推荐微服务调用用户微服务编码实践 ·········· 139
5.3.2 推荐微服务编码 ·········· 141
5.3.3 推荐微服务远程调用商品微服务编码实践 ·········· 144
5.3.4 功能测试 ·········· 147

第 6 章 用户微服务及商品微服务功能完善 ……………………………………… 151

6.1 增加商城用户的相关功能 ……………………………………………………… 151
6.1.1 商城用户模块介绍 ……………………………………………………… 151
6.1.2 商城用户功能模块编码 ………………………………………………… 152
6.1.3 商城用户模块代码完善 ………………………………………………… 154
6.1.4 OpenFeign 编码暴露远程接口 ………………………………………… 158
6.1.5 商城用户鉴权功能测试 ………………………………………………… 160

6.2 新增商城端网关模块 …………………………………………………………… 164
6.2.1 创建商城端网关 newbee-mall-cloud-gateway-mall ………………… 164
6.2.2 商城端网关功能测试 …………………………………………………… 168

6.3 商城首页数据的接口实现 ……………………………………………………… 170
6.3.1 首页的排版设计 ………………………………………………………… 171
6.3.2 首页接口的响应结果设计 ……………………………………………… 172
6.3.3 业务层代码的实现 ……………………………………………………… 174
6.3.4 调用商品微服务进行数据的查询与封装 ……………………………… 175
6.3.5 首页接口控制层代码的实现 …………………………………………… 177
6.3.6 首页接口网关配置 ……………………………………………………… 179

6.4 商城分类页面的接口实现 ……………………………………………………… 180
6.4.1 分类页面的接口响应数据 ……………………………………………… 180
6.4.2 业务层代码的实现 ……………………………………………………… 182
6.4.3 分类页面数据接口控制层代码的实现 ………………………………… 185
6.4.4 分类接口网关配置 ……………………………………………………… 187

6.5 商品列表和商品详情页面的接口实现 ………………………………………… 187
6.5.1 接口传参解析及返回字段定义 ………………………………………… 188
6.5.2 业务层代码的实现 ……………………………………………………… 191
6.5.3 控制层代码的实现 ……………………………………………………… 194
6.5.4 商品接口网关配置 ……………………………………………………… 197

6.6 商城端部分接口的功能测试 …………………………………………………… 198
6.6.1 获取首页数据的接口测试 ……………………………………………… 198
6.6.2 获取分类页面的数据接口测试 ………………………………………… 201

第 7 章 购物车微服务编码实践及功能讲解 ··205

7.1 购物车微服务主要功能介绍 ···205
7.1.1 购物车模块介绍 ··205
7.1.2 购物车模块的表结构设计 ···206
7.2 创建购物车微服务模块 ···207
7.3 远程调用用户微服务及其他注意事项 ··································212
7.4 购物车微服务编码 ··215
7.4.1 购物车微服务代码改造 ··216
7.4.2 网关模块配置 ···218
7.5 购物车微服务远程调用商品微服务编码实践 ························219
7.6 购物车微服务功能测试 ···222
7.7 OpenFeign 编码暴露远程接口 ··226

第 8 章 订单微服务编码实践及功能讲解 ··228

8.1 订单微服务主要功能模块介绍 ···228
8.1.1 订单模块介绍 ···228
8.1.2 订单模块的表结构设计 ··229
8.1.3 订单模块中的主要功能分析 ·······································234
8.1.4 订单处理流程及订单状态的介绍 ································239
8.2 创建订单微服务模块 ··242
8.3 订单微服务编码 ···243
8.4 订单微服务远程调用商品微服务和购物车微服务编码实践 ·····246
8.5 订单微服务功能测试 ··252
8.5.1 添加收货地址接口演示 ··253
8.5.2 生成订单接口演示 ··255
8.5.3 订单列表接口演示 ··256

第 9 章 Spring Cloud Gateway 聚合 Swagger 接口文档 ························259

9.1 为什么要聚合 Swagger 接口文档 ··259

9.2 网关层聚合 Swagger 接口文档的实现思路 ·· 260

9.3 网关层聚合 Swagger 接口文档编码 ··· 263

第 10 章 微服务架构项目中整合 Seata ··· 268

10.1 实战项目中整合 Seata 编码实践 ··· 268

10.2 "分支事务不回滚"问题的复盘 ·· 271

    10.2.1 发现问题 ··· 271

    10.2.2 尝试解决问题 ·· 271

    10.2.3 分析问题产生的原因 ·· 272

    10.2.4 查看源代码并确定问题所在 ·· 273

    10.2.5 解决问题 ··· 275

第 11 章 微服务架构项目中整合 Sentinel ··· 278

11.1 实战项目中整合 Sentinel 编码实践 ··· 278

11.2 "Sentinel 控制台页面中的微服务数据空白"问题的处理 ······················· 280

    11.2.1 错误的解决思路 ·· 280

    11.2.2 正确的解决思路 ·· 282

第 12 章 微服务架构项目中整合 Seluth、Zipkin ···································· 291

12.1 整合 Sleuth 编码实践 ··· 291

12.2 在全局异常处理类中增加日志 ·· 305

12.3 整合 Zipkin 实践 ··· 307

第 13 章 微服务架构项目中整合 ELK 日志中心 ······································ 311

13.1 微服务架构项目中的日志输出配置 ·· 311

13.2 通过 Kibana 查询日志 ·· 314

    13.2.1 查看日志 ··· 314

    13.2.2 日志定时刷新 ·· 315

    13.2.3 常用的日志搜索条件 ·· 316

    13.2.4 根据 traceId 搜索日志 ·· 317

第 1 章

大型微服务架构项目设计与实战

本章将介绍微服务架构项目中的主要功能模块,从单体模式到前后端分离模式,再到微服务架构模式的开发历程,微服务架构项目的启动、打包和部署过程中的注意事项,微服务架构项目改造前的拆分思路,并通过 Swagger 接口文档工具、调用接口带领读者熟悉新蜂商城项目的主要功能。

作者不是简简单单地介绍这个项目,而是讲解这个项目的演进过程、功能设计、服务拆分过程等背后的原理,还是那句话:知其然,也要知其所以然。当能够完整地了解和掌握这个项目时,读者对微服务架构肯定会有更好的理解,在实际工作或面试时,面对"微服务架构"这个技术点就会更加得心应手。

1.1 微服务架构项目详解

本章内容与《Spring Cloud Alibaba 大型微服务架构项目实战》(上册)中第 14 章的内容基本一致,因为这部分内容比较重要,部分读者如果只购买了本书,则可能会在学习过程中遇到一些问题,因此在本书中加入了本章。

本来并不打算将微服务架构项目的开发过程单独整理成书,不过,在与出版社的编辑们沟通后,决定将本书的内容整理得更加丰富、完整。在上册的基础上,整理了实战项目的开发过程,其实是对整个开发过程的现场还原,重点是十几份开发过程的步骤源代码,每份代码都可以独立运行,一个步骤接着一个步骤,把整个项目开发完毕。读者学习后,如果能够自行还原整个过程并把项目独立开发出来,说明已经完全掌握了微服务开发的相关知识点。这个过程对笔者来说肯定不复杂,但是很多读者可能没有机会接触微服务架构项目的开发,这部分读者迫切需要这样的开发经验。开发过程比较详细,讲解了每个模块和微服务的开发过程,这种详尽的开发过程讲解一般不用文字的方式呈现,而是使用视频教程,所以编写本书也是一次全新的尝试。

1.1.1　实战项目介绍

　　newbee-mall 是一套电商系统,包括基础版本(Spring Boot+Thymeleaf)、前后端分离版本(Spring Boot+Vue 3+Element Plus+Vue Router 4+Pinia+Vant 4)、秒杀版本、Go 语言版本、微服务版本(Spring Cloud Alibaba+Nacos+Sentinel+Seata+Spring Cloud Gateway+OpenFeign+ElasticSearch+Logstash+Kibana)。商城端包括首页门户、商品分类、新品上线、首页轮播、商品推荐、商品搜索、商品展示、购物车、订单结算、订单流程、个人订单管理、会员中心、帮助中心等模块。商城后台管理系统包括数据面板、轮播图管理、商品管理、订单管理、会员管理、分类管理、设置等模块。

　　该项目包括商城端和商城后台管理系统。对应的用户包括商城会员和商城后台管理员。商城端是所有用户都可以浏览使用的系统,商城会员在这里可以浏览、搜索、购买商品。商城后台管理员在商城后台管理系统中管理商品信息、订单信息、会员信息等,具体包括商城基本信息的录入和更改、商品信息的添加和编辑、处理订单的拣货和出库,以及商城会员信息的管理。

　　该项目的具体特点如下。

　　(1) newbee-mall 对开发人员十分友好,无须复杂的操作步骤,仅需 2 秒就可以启动完整的商城项目。

　　(2) newbee-mall 是一个企业级别的 Spring Boot 大型项目,对各个阶段的 Java 开发人员都是极佳的选择。

　　(3) 开发人员可以把 newbee-mall 作为 Spring Boot 技术栈的综合实践项目,其在技术上符合要求,并且代码开源、功能完备、流程完整、页面美观、交互顺畅。

　　(4) newbee-mall 涉及的技术栈新颖、知识点丰富,有助于读者理解和掌握相关知识,可以进一步提升开发人员的职场竞争力。

1.1.2　新蜂商城项目的开源历程

笔者在 2019 年 8 月 12 日写下了新蜂商城项目的第一行代码，经过近两个月的开发和测试，新蜂商城项目于 2019 年 10 月 9 日正式开源在 GitHub 网站上，当时的提交记录如图 1-1 所示。

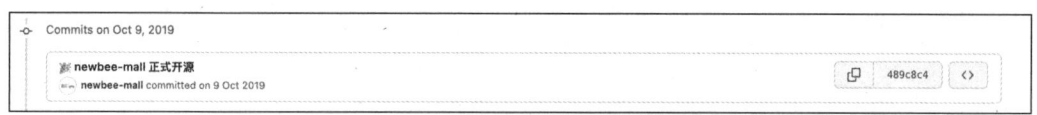

图 1-1　新蜂商城开源代码提交记录

由于避开了其他商城开源项目的不足之处，并且学习和使用的成本不高，因此新蜂商城项目在开源的第一年就取得了不错的成绩，获得近 6000 个 Star 和 1500 个 Fork，成为一个比较受欢迎的开源项目。

最让笔者感到欣慰的一点是新蜂商城开源项目帮助了很多技术人员和学生。在项目开源之后，笔者经常收到留言和邮件，得知读者在学习和使用该商城开源项目后，对 Spring Boot 技术栈有了更深刻的认识，并且拥有了项目实战经验，可以顺利地完成工作或学业，甚至在找心仪工作的过程中起到了关键作用。

这些反馈不仅让笔者欣慰，还让笔者更加有动力不断地完善新蜂商城开源项目。为了让新蜂商城开源项目保持长久的生命力，并且帮助更多的朋友，笔者一直优化和升级。截至 2023 年 2 月，新蜂商城已经发布了 7 个重要的版本。

（1）新蜂商城 v1 版本，于 2019 年 10 月 9 日开源，主要技术栈为 Spring Boot + MyBatis + Thymeleaf。

（2）新蜂商城 Vue 2 版本，于 2020 年 5 月 30 日开源，主要技术栈为 Vue 2.6。

（3）新蜂商城 Vue 3 版本，于 2020 年 10 月 28 日开源，主要技术栈为 Vue 3。

（4）新蜂商城后台管理系统 Vue 3 版本，于 2021 年 3 月 29 日开源，主要技术栈为 Vue 3 + Element Plus。

（5）新蜂商城升级版本，于 2021 年 6 月 2 日开源，增加了秒杀、优惠券等功能。

（6）新蜂商城 Go 语言版本，于 2022 年 4 月开源，主要技术栈为 Go + Gin。

（7）新蜂商城微服务版本，于 2022 年 6 月开源，整合了 Spring Cloud Alibaba 及相关的微服务组件。

软件的需求是不断变化的，技术的更新迭代越来越快，新蜂商城系统会一步一步跟上技术演进的脚步，未来会不断地更新和完善。

由于篇幅有限，不可能将新蜂商城所有版本的开发讲解都写在一本书中。本书主要介绍微服务版本，技术栈为 Spring Cloud Alibaba、Nacos、Sentinel、OpenFeign、Seata 等。

关于新蜂商城的版本迭代记录，笔者整理了重要版本的时间轴，如图 1-2 所示，今后会一直完善和迭代新蜂商城项目。

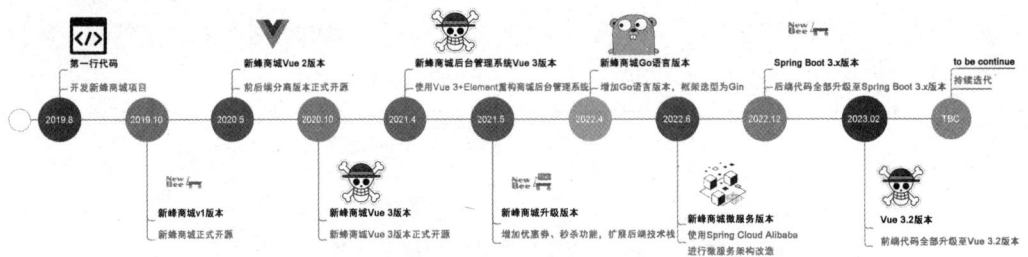

图 1-2　新蜂商城重要版本的时间轴

新蜂商城由最初的单体项目，逐步过渡到前后端分离和微服务架构项目，到现在已经"开枝散叶"，成长为一系列项目的集合。想要一个开源作品保持长久而健康的生命，这是一个非常不错的办法。由基础项目慢慢优化，不断地增加技术栈，在让读者学习越来越多知识点的同时，对开源作者的技术提升也有很大的帮助。开源作者和读者通过这个开源项目都能够学习到很多，达到在技术层面"共同富裕"的目的。

1.1.3　新蜂商城项目的功能及数据库设计

新蜂商城商城端功能汇总如图 1-3 所示。其主要功能包括会员、商城首页、商品搜索、商品展示、购物车、订单和支付。

新蜂商城后台管理系统功能汇总如图 1-4 所示。其主要功能包括系统管理员、轮播图管理、热销商品配置、新品上线配置、推荐商品配置、分类管理、商品管理、会员管理和订单管理。后台管理系统中的功能模块主要是让商城后台管理员管理运营数据及用户交易数据。这里通常就是基本的增、删、改、查功能。

在数据库方面，第一个版本中共有 9 张表，分别是商品分类表、商品表、轮播图表、首页推荐表、购物车表、订单表、订单项表、商城用户表和商城管理员表。

第二个版本主要把单体项目重构为前后端分离版，技术栈是 Spring Boot 和 Vue，功能与第一个版本的功能并没有太多的差异。在数据库方面，第二个版本中共有 13 张表，比第一个版本多了 4 张表，其中两张表是与 token 相关的表，另外两张表是与收货地址相关的表。

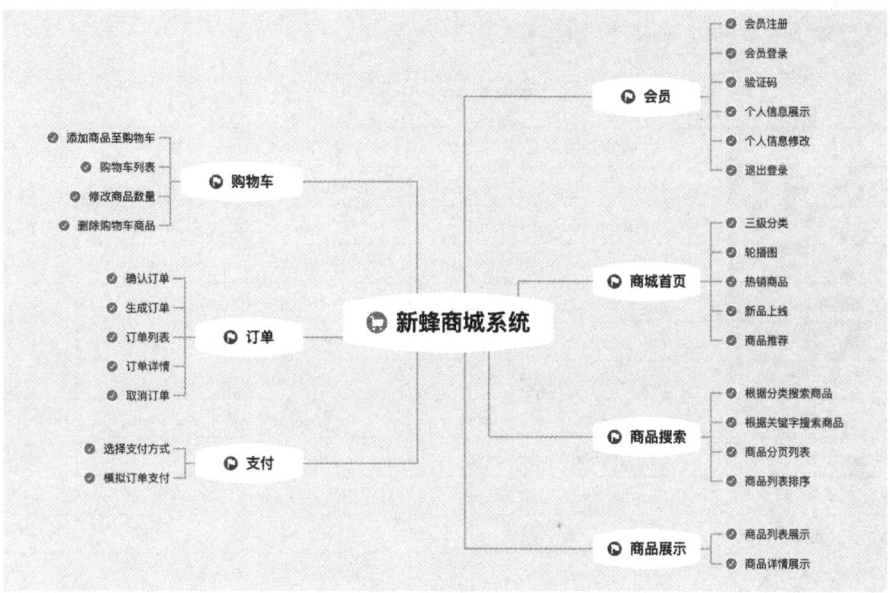

图 1-3 商城端功能汇总

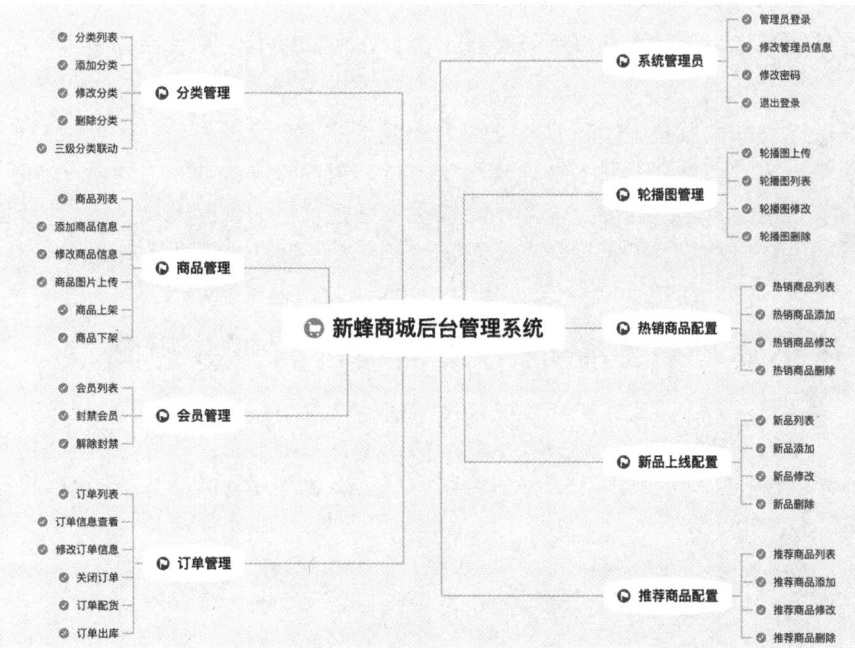

图 1-4 后台管理系统功能汇总

单体版与前后端分离版的表设计对比见表1-1。

表1-1 单体版与前后端分离版的表设计对比

表名	新蜂商城单体版	新蜂商城前后端分离版	备注
商品分类表	tb_newbee_mall_goods_category	tb_newbee_mall_goods_category	字段相同
商品表	tb_newbee_mall_goods_info	tb_newbee_mall_goods_info	字段相同
轮播图表	tb_newbee_mall_carousel	tb_newbee_mall_carousel	字段相同
首页推荐表	tb_newbee_mall_index_config	tb_newbee_mall_index_config	字段相同
购物车表	tb_newbee_mall_shopping_cart_item	tb_newbee_mall_shopping_cart_item	字段相同
订单表	tb_newbee_mall_order	tb_newbee_mall_order	部分调整
订单项表	tb_newbee_mall_order_item	tb_newbee_mall_order_item	字段相同
商城用户表	tb_newbee_mall_user	tb_newbee_mall_user	部分调整
商城管理员表	tb_newbee_mall_admin_user	tb_newbee_mall_admin_user	字段相同
商城用户token表	无	tb_newbee_mall_user_token	新增表
商城管理员token表	无	tb_newbee_mall_admin_user_token	新增表
收货地址表	无	tb_newbee_mall_user_address	新增表
订单-收货地址关联表	无	tb_newbee_mall_order_address	新增表

与单体版相比，前后端分离版只是做了部分字段的调整，并且完善了用户收货地址模块。

本书所讲解的微服务架构项目是在前两个版本的基础上开发的，因此表结构、功能基本上都是一致的。想要更深入地理解这个项目，笔者建议读者先去体验一下新蜂商城单体版和前后端分离版的功能。笔者已经提供了这些项目的体验网站，读者可以在开源仓库中看到。在实际体验之后再学习微服务版本的源代码和功能设计，会更加顺畅一些。

1.2 从单体项目到微服务架构项目的拆分思路

在进行微服务架构改造前，要对系统的功能进行归纳和总结，确定拆分出哪些微服务，这样才能进行后续的服务化拆分和编码测试。笔者在开发微服务架构版本时的拆分思路如下。

首先，拆分的粒度不能太细。比如，项目里有10张表，拆分成10个微服务，这种做法既不合理，也没有必要，完全是为了拆分而拆分。

其次，拆分的粒度不能太粗。比如，项目里有10张表，拆分成2个微服务，拿新蜂商城项目来说，把商城用户表和商城管理员表拆分成一个用户微服务，把剩余的表拆分成商品订单微服务。这种做法也不是很好，有些糊弄的意味。用这种粗粒度拆分的方式，拆分后与拆分前没有什么区别，与微服务架构的初衷就背道而驰了。

新蜂商城项目中的商城用户模块、商品模块、订单模块间的功能边界是非常清晰的，所以这三个模块分别拆分成三个单独的微服务是没有问题的。以此为基础，把商城用户表、商城管理员表划分到用户微服务中，把商品分类表、商品表划分到商品微服务中，把订单表、订单项表划分到订单微服务中。

以上是最基本的划分，接下来分析剩余的表和功能模块。

轮播图管理模块与上述三个微服务没有强关联性，可以不划进任何一个微服务中。商品推荐管理模块与商品模块是有关联性的，可以放到商品微服务中。不过笔者觉得商品推荐管理模块可以和轮播图管理模块重新组合成一个推荐微服务，所以就把它和轮播图管理模块放在一起了。

购物车模块与商品模块有一定的关联性，与订单模块的关联性也很强，所以将其放到订单微服务中也是可以的。笔者在设计时考虑到分布式事务的问题，为了更好地演示和处理分布式事务，就将购物车模块单独作为一个微服务。

通过数据库表和功能模块的总结，最终的拆分方案见表1-2。

表1-2 拆分方案

微服务	功能模块	涉及的表
用户微服务	管理员模块、商城用户模块	tb_newbee_mall_user、tb_newbee_mall_admin_user
商品微服务	商品分类模块、商品模块	tb_newbee_mall_goods_category、tb_newbee_mall_goods_info
推荐微服务	轮播图管理模块、商品推荐管理模块	tb_newbee_mall_carousel、tb_newbee_mall_index_config
购物车微服务	购物车模块	tb_newbee_mall_shopping_cart_item
订单微服务	订单模块、订单项模块、收货地址模块	tb_newbee_mall_order、tb_newbee_mall_order_item、tb_newbee_mall_user_address、tb_newbee_mall_order_address

当然，笔者的拆分思路及最终实战项目的源代码只是一种实现思路，读者也可以不按照上述思路进行拆分，在熟悉和掌握本书所讲解的微服务相关知识点后，再自行实现另外一套拆分思路和源代码也是完全可以的。比如，把用户微服务拆分得更细一些，拆分为商城用户微服务和管理员微服务；不单独拆分出购物车微服务，而是把购物车模块放到订单微服务中；把商品推荐管理模块放到商品微服务中，而不是放到推荐微服务中；把收货地址模块放到用户微服务中，而不是放在订单微服务中。

以上就是新蜂商城微服务版本的拆分思路。通过以上内容的讲解，读者应该对本书的最终实战项目有了更清晰的认识。当然，看完本节的拆分思路和拓展思路后，读者也可以确定自己的拆分思路，并且尝试着用编码去实现自己的想法。

1.3 微服务架构项目源代码获取和项目启动

本节主要介绍新蜂商城微服务版本的源代码下载和项目启动，包括准备基础环境、安装和配置微服务组件、下载源代码、介绍源代码的目录结构、准备数据库、配置项目启动和注意事项，让读者能够顺利启动最终的微服务架构项目并进行个性化修改。

1.3.1 基础环境准备及微服务组件安装和配置

把 JDK、Maven、IDEA、Lombok、MySQL、Redis 这类基础环境都安装和配置完成，以便进行后续的项目启动工作。

目前已经集成和改造的微服务组件整理如下。

（1）微服务框架——Spring Cloud Alibaba。

（2）服务中心——Nacos。

（3）通信服务——OpenFeign。

（4）网关服务——Spring Cloud Gateway。

（5）负载均衡器——LoadBalancer。

（6）分布式事务处理——Seata。

（7）流控组件——Sentinel。

（8）链路追踪——Sleuth+Zipkin。

（9）日志中心——ElasticSearch+Logstash+Kibana。

对这些微服务组件，读者应该都不陌生，在启动项目前，依次参考对应章节中的讲解完成组件的搭建和启动即可。

1.3.2 下载微服务架构项目的源代码

在部署项目之前，需要把项目的源代码下载到本地，最终的微服务架构项目在 GitHub 和 Gitee 平台上都创建了代码仓库。由于国内访问 GitHub 网站的速度可能缓慢，因此笔者在 Gitee 上创建了一个同名的代码仓库，两个代码仓库会保持同步更新，它们

的网址分别如下:

```
https://g**hub.com/newbee-ltd/newbee-mall-cloud
```

```
https://g**ee.com/newbee-ltd/newbee-mall-cloud
```

读者可以直接在浏览器中输入上述链接到对应的代码仓库中查看源代码及相关文件。

1. 使用 clone 命令下载源代码

如果在本地计算机中安装了 Git 环境,就可以直接在命令行中使用 clone 命令把代码仓库中的文件全部下载到本地。

通过 GitHub 下载源代码,执行如下命令:

```
git clone https://g**hub.com/newbee-ltd/newbee-mall-cloud.git
```

通过 Gitee 下载源代码,执行如下命令:

```
git clone https://g**ee.com/newbee-ltd/newbee-mall-cloud.git
```

打开 cmd 命令行,切换到对应的目录。比如,下载到 D 盘的 java-dev 目录,先执行 cd 命令切换到该目录下,再执行 clone 命令。

等待文件下载,全部下载完成后就能够在 java-dev 目录下看到新蜂商城项目微服务版的所有源代码了。

2. 通过开源网站下载源代码

除通过命令行下载外,读者也可以选择更直接的下载方式。GitHub 和 Gitee 两个开源平台都提供了对应的下载功能,读者可以在代码仓库中直接单击对应的"下载"按钮进行源代码下载。

在 GitHub 网站上直接下载源代码,需要进入 newbee-mall-cloud 在 GitHub 网站中的代码仓库主页。

在 newbee-mall-cloud 代码仓库页面上有一个带有下载图标的绿色"Code"按钮,单击该按钮,再单击"Download Zip"按钮就可以下载 newbee-mall-cloud 源代码的压缩包文件,下载完成后解压,导入 IDEA 或 Eclipse 编辑器进行开发或修改。

在 Gitee 网站上下载源代码更快一些。在 Gitee 网站上直接下载源代码,需要进入 newbee-mall-cloud 在 Gitee 网站上的代码仓库页面,在代码仓库页面上有一个"克隆/下载"按钮,单击该按钮,再单击"下载 Zip"按钮。在 Gitee 网站上下载源代码多了一步验证操作,单击"下载 Zip"按钮后会跳转到验证页面,输入正确的验证码就可以下载源代码的压缩包文件,下载完成后解压缩,导入 IDEA 或 Eclipse 编辑器进行开发或修改。

3. 通过本书提供的源代码地址下载源代码

除在开源网站中下载完整的源代码外，本书还提供了对应章节的源代码，按照笔者提供的下载地址就可以进行源代码的下载。

1.3.3 微服务架构项目的目录结构讲解

下载源代码并解压缩后，在代码编辑器中打开项目，这是一个标准的 Maven 多模块项目。笔者使用的开发工具是 IDEA，导入之后 newbee-mall-cloud 源代码的目录结构如图 1-5 所示。

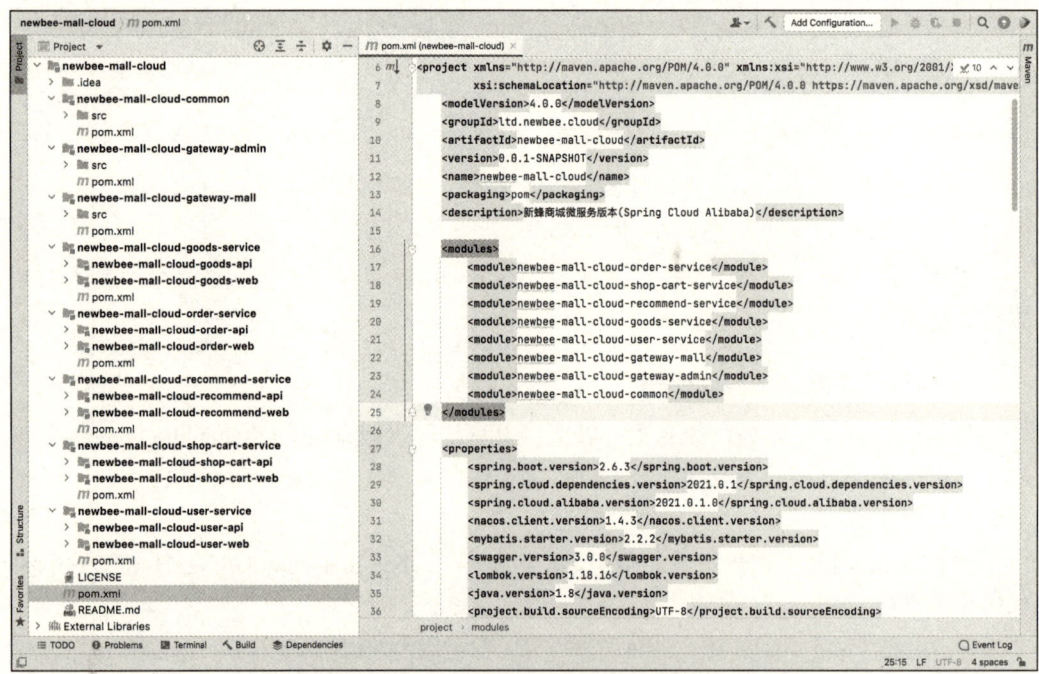

图 1-5 newbee-mall-cloud 源代码的目录结构

下面介绍一下目录的内容和作用，整理如下。

```
newbee-mall-cloud
    ├── newbee-mall-cloud-common              // 1.公用模块，存放与项目相关的工具
                                              //   类、DTO 对象等
    ├── newbee-mall-cloud-gateway-admin       // 2.后台管理系统的网关服务
        newbee-mall-cloud-gateway-mall        // 3.商城端的网关服务
```

```
├── newbee-mall-cloud-goods-service              // 4.商品微服务
│   ├── newbee-mall-cloud-goods-api              // 存放商品模块中暴露的用于远程
│   │                                                调用的FeignClient类
│   └── newbee-mall-cloud-goods-web              // 商品模块API的代码及逻辑
├── newbee-mall-cloud-order-service              // 5.订单微服务
│   ├── newbee-mall-cloud-order-api              // 存放订单模块中暴露的用于远程
│   │                                                调用的FeignClient类
│   └── newbee-mall-cloud-order-web              // 订单模块API的代码及逻辑
├── newbee-mall-cloud-recommend-service          // 6.推荐微服务
│   ├── newbee-mall-cloud-recommend-api          // 存放商品推荐管理模块中暴露的用于
│   │                                                远程调用的FeignClient类
│   └── newbee-mall-cloud-recommend-web          // 商品推荐管理模块API的代码及逻辑
├── newbee-mall-cloud-shop-cart-service          // 7.购物车微服务
│   ├── newbee-mall-cloud-shop-cart-api          // 存放购物车模块中暴露的用于远程
│   │                                                调用的FeignClient类
│   └── newbee-mall-cloud-shop-cart-web          // 购物车模块API的代码及逻辑
├── newbee-mall-cloud-user-service               // 8.用户微服务
│   ├── newbee-mall-cloud-user-api               // 存放商城用户模块中暴露的用于
│   │                                                调用的FeignClient类
│   └── newbee-mall-cloud-user-web               // 商城用户模块API的代码及逻辑
└── pom.xml    // root节点的Maven配置文件
```

以上是项目结构的整体概览，具体到某一个 Maven 模块中，有些内容依然需要讲解。笔者以 newbee-mall-cloud-gateway-mall 模块和 newbee-mall-cloud-goods-service 模块为例，介绍子模块中的详细目录结构。

图 1-6 是商城端的网关服务的目录结构，这是一个标准的 Maven 项目。

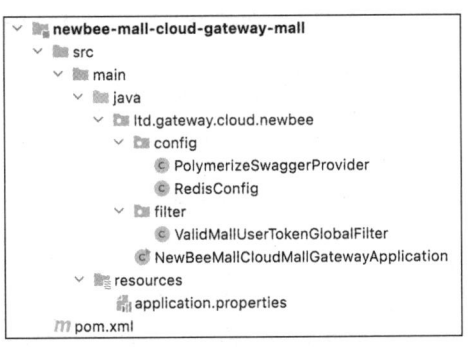

图 1-6 商城端的网关服务的目录结构

在 newbee-mall-cloud-gateway-mall 模块中，代码目录的内容和作用整理如下。

```
newbee-mall-cloud-gateway-mall
```

```
├── src/main/java
│    └── ltd.gateway.cloud.newbee
│         ├── config   // 存放配置类,如 Swagger 整合、Redis 配置
│         ├── filter   // 存放网关过滤器
│         └── NewBeeMallCloudMallGatewayApplication  // Spring Boot 项目主类
├── src/main/resources
│    └── application.properties  // 项目配置文件
└── pom.xml  // Maven 配置文件
```

图 1-7 是商品微服务的目录结构,这也是一个多模块的 Maven 项目,包括三个 Maven 配置文件,分别是商品微服务的主配置文件,以及 api 和 web 两个子配置文件。当然,这三个 Maven 配置文件都依赖 root 节点的 Maven 配置文件。

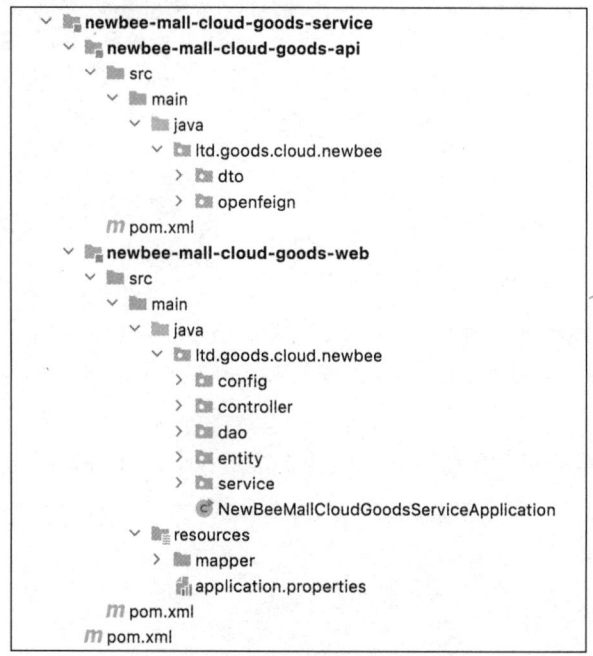

图 1-7 商品微服务的目录结构

在 newbee-mall-cloud-goods-service 模块中,代码目录的内容和作用整理如下。

```
newbee-mall-cloud-goods-service
    ├── newbee-mall-cloud-goods-api     // 存放商品模块中暴露的用于远程调用
    │                                       的 FeignClient 类
    │    ├── src/main/java
    │    │    └── ltd.goods.cloud.newbee
    │    │         ├── dto              // FeignClient 类中所需的 Java Bean
```

```
            └── openfeign                    // 商品模块中暴露的 FeignClient 类
        └── pom.xml                          // Maven 配置文件
    └── newbee-mall-cloud-goods-web          // 商品模块 API 的代码及逻辑
        ├── src/main/java
        │   └── ltd.goods.cloud.newbee
        │       ├── config                   // 存放 Web 配置类
        │       ├── controller               // 存放与商品微服务相关的控制类，包括商城端和后台
        │       │                               管理系统中所需的 Controller 类
        │       ├── dao                      // 存放与商品微服务相关的数据层接口
        │       ├── entity                   // 存放与商品微服务相关的实体类
        │       ├── service                  // 存放与商品微服务相关的业务层方法
        │       └── NewBeeMallCloudGoodsServiceApplication
        │                                    // Spring Boot 项目主类
        ├── src/main/resources
        │       ├── mapper                   // 存放 MyBatis 的通用 Mapper 文件
        │       └── application.properties   // 商品微服务项目的配置文件
        └── pom.xml                          // Maven 配置文件
```

除基本目录中的源代码外，笔者在 static-files 目录中也上传了数据库文件和与本项目相关的一些图片文件。

1.3.4 启动并验证微服务实例

下面讲解微服务架构项目的启动和启动前的准备工作。

1. 数据库准备

在最终的微服务架构项目中，共有五个微服务需要连接 MySQL 数据库，分别是用户微服务、购物车微服务、商品微服务、订单微服务和推荐微服务。因此，需要分别创建五个数据库，并导入对应的建表语句。

打开 MySQL 软件，新建五个数据库，SQL 语句如下：

```
# 创建用户微服务所需数据
CREATE DATABASE /*!32312 IF NOT EXISTS*/'newbee_mall_cloud_user_db' /*!40100
DEFAULT CHARACTER SET utf8 */;

# 创建购物车微服务所需数据
CREATE DATABASE /*!32312 IF NOT EXISTS*/'newbee_mall_cloud_cart_db' /*!40100
DEFAULT CHARACTER SET utf8 */;
```

```
# 创建商品微服务所需数据
CREATE DATABASE /*!32312 IF NOT EXISTS*/'newbee_mall_cloud_goods_db'
/*!40100 DEFAULT CHARACTER SET utf8 */;

# 创建订单微服务所需数据
CREATE DATABASE /*!32312 IF NOT EXISTS*/'newbee_mall_cloud_order_db'
/*!40100 DEFAULT CHARACTER SET utf8 */;

# 创建推荐微服务所需数据
CREATE DATABASE /*!32312 IF NOT EXISTS*/'newbee_mall_cloud_recommend_db'
/*!40100 DEFAULT CHARACTER SET utf8 */;
```

当然，读者在实际操作时也可以使用其他数据库名称，如自行定义的数据库名称。在数据库创建完成后就可以将五份数据库建表语句和初始化数据文件导入各个数据库，导入成功后可以看到数据库的表结构，如图1-8所示。

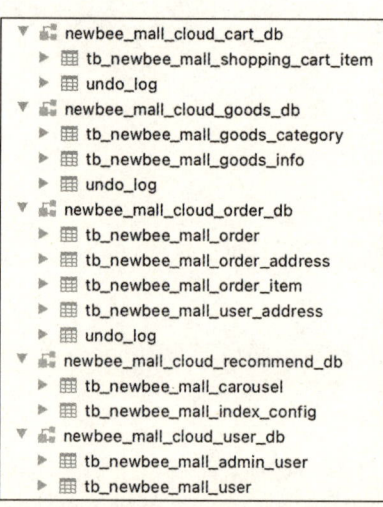

图1-8　newbee-mall-cloud 项目所需的数据库表结构

这五个数据库可以在不同的 MySQL 实例中，如果只是测试，则可以放在同一个 MySQL 实例中，读者可以自行决定。

2. 修改数据库连接配置

数据库准备完毕，接下来修改数据库连接配置。分别在 newbee-mall-cloud-goods-web 模块、newbee-mall-cloud-order-web 模块、newbee-mall-cloud-recommend-web 模块、newbee-mall-cloud-shop-cart-web 模块、newbee-mall-cloud-user-web 模块中打开 resources 目录下的 application.properties 配置文件，修改数据库连接的相关信息。代码中默认的数

据库配置如下。

```
spring.datasource.url=jdbc:mysql://localhost:3306/newbee_mall_cloud_good
s_db?useUnicode=true&serverTimezone=Asia/Shanghai&characterEncoding=utf8
&autoReconnect=true&useSSL=false&allowMultiQueries=true
spring.datasource.username=root
spring.datasource.password=123456
```

需要修改的配置如下。

（1）数据库地址和数据库名称为 localhost:3306/newbee_mall_cloud_goods_db。

（2）数据库登录账户名称为 root。

（3）账户密码为 123456。

根据开发人员所安装的数据库地址和账号信息进行修改。这五个微服务的配置文件中都有数据库连接的默认配置，如果与默认配置文件中的数据库名称不同，则需要将数据库连接中的数据库名称进行修改。数据库地址、登录账户名称、账户密码也需要修改为开发人员自己的配置内容。

3. 修改 Nacos 连接配置

分别在 newbee-mall-cloud-gateway-admin 模块、newbee-mall-cloud-gateway-mall 模块、newbee-mall-cloud-goods-web 模块、newbee-mall-cloud-order-web 模块、newbee-mall-cloud-recommend-web 模块、newbee-mall-cloud-shop-cart-web 模块和 newbee-mall-cloud-user-web 模块中打开 resources 目录下的 application.properties 配置文件，修改服务中心 Nacos 连接的相关信息。代码中默认的 Nacos 配置如下。

```
# Nacos 连接地址
spring.cloud.nacos.discovery.server-addr=127.0.0.1:8848
# Nacos 登录用户名(默认为 nacos，在生产环境中一定要修改)
spring.cloud.nacos.username=nacos
# Nacos 登录密码(默认为 nacos，在生产环境中一定要修改)
spring.cloud.nacos.password=nacos
```

需要修改的配置如下。

（1）Nacos 连接地址。

（2）Nacos 登录用户名。

（3）Nacos 登录密码。

根据开发人员所安装和配置的 Nacos 组件进行修改。这七个微服务的配置文件中都有服务中心连接的默认配置，如果与默认配置文件中的内容不同，则需要自行修改。当然，这七个微服务的配置文件中的 Nacos 配置都是一致的，所有服务都必须注册到同一个服务中心。

4. 修改 Redis 连接配置

分别在 newbee-mall-cloud-gateway-admin 模块、newbee-mall-cloud-gateway-mall 模块和 newbee-mall-cloud-user-web 模块中打开 resources 目录下的 application.properties 配置文件，修改 Redis 连接的相关信息。代码中默认的 Redis 配置如下。

```
##Redis 配置
# Redis 数据库索引（默认为 0）
spring.redis.database=13
# Redis 服务器连接地址
spring.redis.host=127.0.0.1
# Redis 服务器连接端口
spring.redis.port=6379
# Redis 服务器连接密码
spring.redis.password=
```

需要修改的配置如下。

（1）Redis 服务器连接地址。

（2）Redis 服务器连接端口。

（3）Redis 服务器连接密码。

根据开发人员所安装和配置的 Redis 数据库进行修改。项目中只有这三个模块使用 Redis，主要用于同步用户的登录信息和用户鉴权操作。三个模块的配置文件中都有 Redis 连接的默认配置，如果与默认配置文件中的内容不同，则需要自行修改。当然，这三个模块的配置文件中的 Redis 配置都是一致的，连接的是相同的 Redis 实例和相同的数据库。

5. 启动所有的微服务实例

做完以上工作后，就可以启动整个项目了，过程如图 1-9 所示。

依次启动 NewBeeMallCloudGoodsServiceApplication 类、NewBeeMallCloudOrderServiceApplication 类、NewBeeMallCloudRecommendServiceApplication 类、NewBeeMallCloudShopCartServiceApplication 类、NewBeeMallCloudUserServiceApplication 类、NewBeeMallCloudMallGatewayApplication 类、NewBeeMallCloudAdminGatewayApplication 类，共七个微服务实例。其中，前五个类分别是商品微服务、订单微服务、推荐微服务、购物车微服务、用户微服务的启动主类，剩下的两个类是网关服务的启动主类。

之后，耐心等待所有实例启动即可。笔者选择的是通过运行 main()方法的方式启动 Spring Boot 项目，读者也可以选择其他方式。

如果未能成功启动,则需要查看控制台中的日志是否报错,并及时确认问题和修复。启动成功后进入 Nacos 控制台,单击"服务管理"→"服务列表"选项,可以看到列表中已经存在这七个微服务的服务信息,如图 1-10 所示。

至此,微服务实例的启动及微服务实例的注册就完成了。

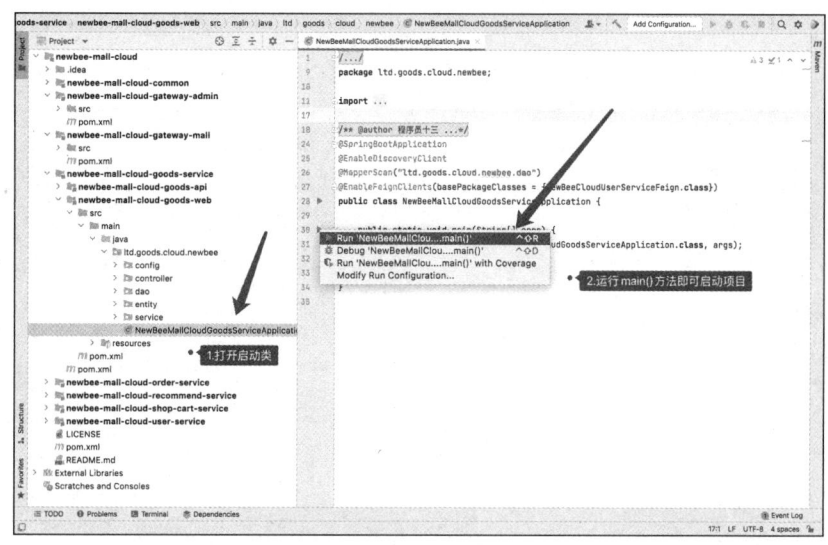

图 1-9　微服务实例启动过程

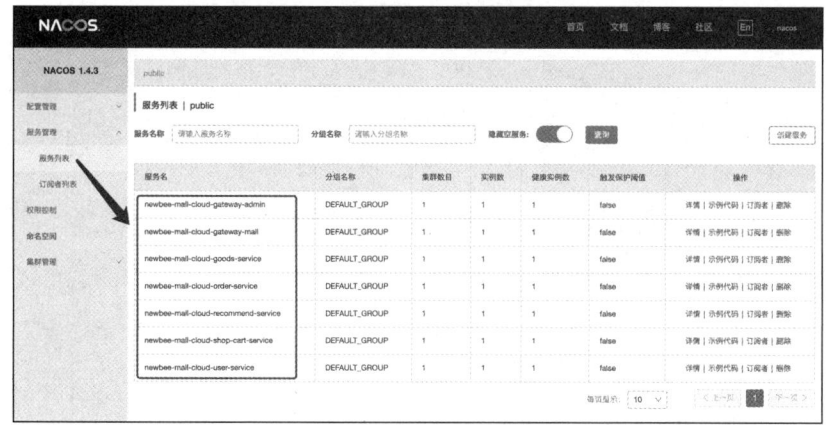

图 1-10　newbee-mall-cloud 项目中的所有微服务实例都完成了注册

6. 简单的功能验证

尽管微服务架构项目实战的启动和各微服务实例的注册都已经完成了,但是为了确认项目运行正常,还需要进行一些简单的验证。

因为各个微服务实例中的接口都使用 Swagger 接口文档工具,所以读者可以通过访问各个微服务实例的 Swagger 接口文档进行简单的测试。除微服务实例外,在网关服务层也做了 Swagger 文档的整合,方便开发人员做接口测试。各个微服务实例的接口文档网址见表 1-3。

表 1-3　各个微服务实例的接口文档网址

服务名称	Swagger 接口文档网址
用户微服务	http://localhost:29000/swagger-ui/index.html
商品微服务	http://localhost:29010/swagger-ui/index.html
推荐微服务	http://localhost:29020/swagger-ui/index.html
购物车微服务	http://localhost:29030/swagger-ui/index.html
订单微服务	http://localhost:29040/swagger-ui/index.html
后台管理系统的网关服务	http://localhost:29100/swagger-ui/index.html
商城端的网关服务	http://localhost:29110/swagger-ui/index.html

如果读者在启动时修改了项目的启动端口号,那么访问网址也需要进行对应的修改。有了 Swagger 接口文档工具,测试起来就方便多了,读者可以依次测试各个微服务实例的接口是否正常。

笔者在本小节的功能验证中,简单地测试一下首页接口是否正常,在地址栏中依次输入如下网址进行测试。

(1)推荐微服务:http://localhost:29020/swagger-ui/index.html。

(2)商城端的网关服务:http://localhost:29110/swagger-ui/index.html。

推荐微服务的 Swagger 接口文档页面如图 1-11 所示。

图 1-11　推荐微服务的 Swagger 接口文档页面

单击"新蜂商城首页接口"选项卡,可以看到"获取首页数据"的接口。打开该接口描述,单击页面中的"Try it out"按钮(单击后变为"Cancel"按钮)发送测试请求,之后单击页面上的"Execute"按钮,接口响应结果如图 1-12 所示。

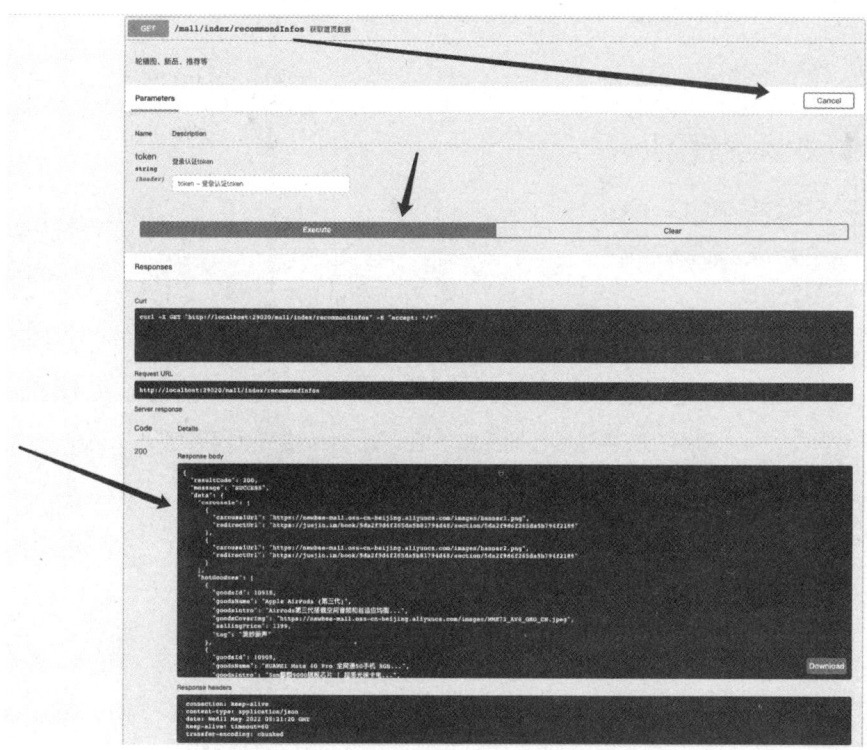

图 1-12 新蜂商城首页接口测试过程

测试通过!

接下来通过网关服务来访问首页接口。在一般情况下,微服务实例中的接口是不会对外暴露的,想要获取对应的数据,可以直接访问网关服务,由网关服务进行请求的转发和处理。

商城端的网关服务的 Swagger 接口文档页面如图 1-13 所示。

由于网关服务层是没有任何接口的,因此只能做下游微服务实例的 Swagger 接口文档聚合。单击页面右上方的"Select a definition"下拉列表框中的下拉按钮,选择"recommend-service-swagger-route"选项就能够看到推荐微服务的 Swagger 接口文档了。当然,通过网关服务访问与直接访问 URL 是不同的,通过网关服务实例的 Swagger 接口文档页面访问,由网关服务做一次转发。"获取首页数据"的接口访问结果如图 1-14 所示。

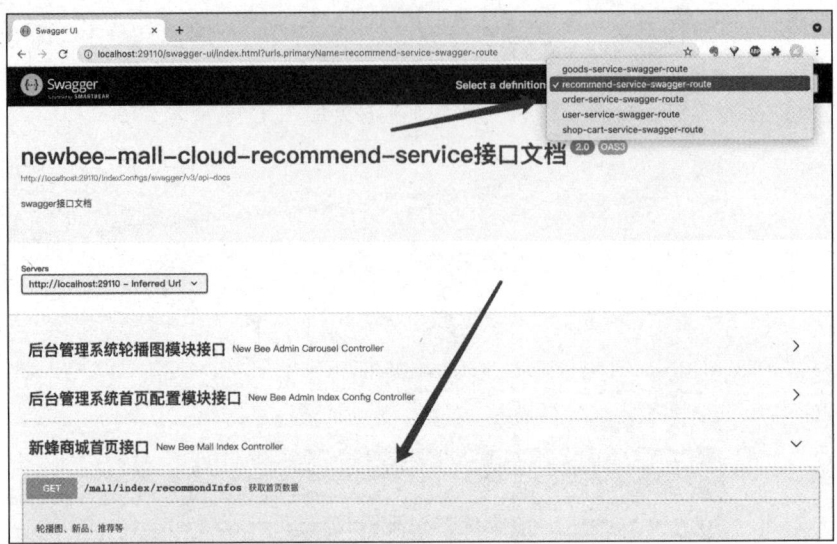

图 1-13　商城端的网关服务的 Swagger 接口文档页面

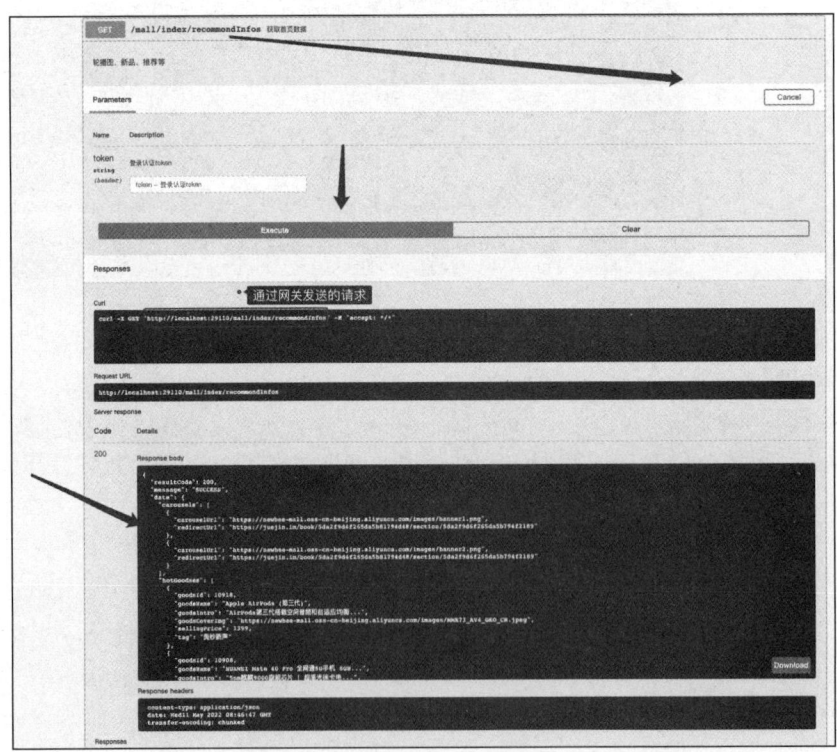

图 1-14　通过网关服务发起对新蜂商城首页接口测试

接口响应正常,结果与直接访问推荐微服务时的结果一样。读者在测试时需要注意,直接访问与通过网关服务访问,请求的 URL 是不同的,虽然结果相同,但是两种请求方式有着本质的区别。

对应到实际的项目页面中,是新蜂商城项目的首页,获取首页接口数据后的显示效果如图 1-15 所示。

图 1-15 新蜂商城首页的显示效果图

1.4 微服务架构项目的功能演示

本节将统一使用网关服务中的 Swagger 接口文档工具进行演示。

读者在测试时需要根据功能在商城端的网关服务的 Swagger 接口文档工具中和在后台管理系统的网关服务的 Swagger 接口文档工具中分别测试。因为二者的用户体系和身份验证不同,所以在通过商城端的网关服务调用后台管理系统的接口时会报错"无权限",在通过后台管理系统的网关服务调用商城端的接口时也会报错"无权限"。比如,在商城端的网关服务的 Swagger 接口文档工具中无法正常调用修改商品信息接口或添加轮播图接口;在后台管理系统的网关服务的 Swagger 接口文档工具中也无法正常调用购物车列表接口或生成订单接口。

1.4.1 商城用户的注册与登录演示

用户的注册与登录演示主要涉及用户注册接口和用户登录接口。

1. 用户注册接口演示

商城端的网关服务的 Swagger 接口文档页面如图 1-16 所示。单击页面右上方的"Select a definition"下拉列表框中的下拉按钮，选择"user-service-swagger-route"选项就能够看到用户微服务的 Swagger 接口文档了。

图 1-16　商城端的网关服务的 Swagger 接口文档页面

依次单击"用户注册""Try it out"按钮，在参数栏中输入 loginName 字段和 password 字段，之后单击"Execute"按钮就能够发送用户注册的请求了。比如，要注册一个用户名为 13700001234、密码为 123456 的用户，过程如图 1-17 所示。

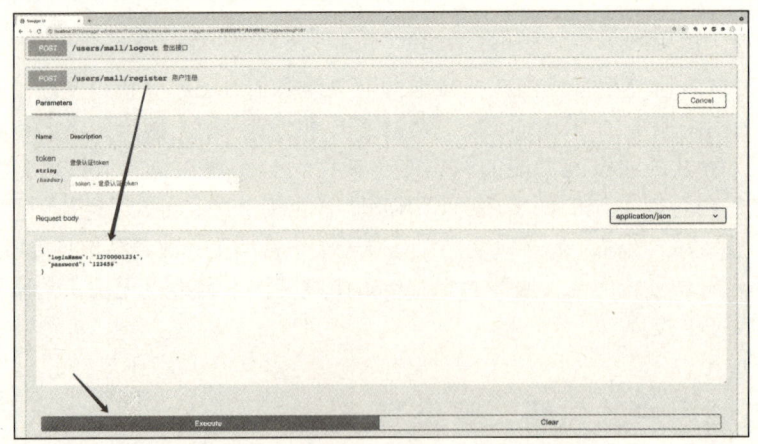

图 1-17　用户微服务中注册接口的测试过程

单击"Execute"按钮后，结果如图1-18所示。

图 1-18　用户微服务中注册接口的测试结果

如果参数信息都通过了基本的验证，就会得到注册成功的响应结果。此时，再去用户微服务的数据库中查看商城用户表中的数据，可以看到新增了一条用户名为13700001234、密码为123456的用户数据。

2. 用户登录接口演示

接下来测试用户登录接口。依次单击"登录接口""Try it out"按钮，在参数栏中输入loginName字段和password字段（注意，这里的password字段需要进行MD5加密），之后单击"Execute"按钮就能够发送用户登录的请求了。使用刚刚注册的用户名为13700001234、密码为123456的账号进行登录，过程如图1-19所示。

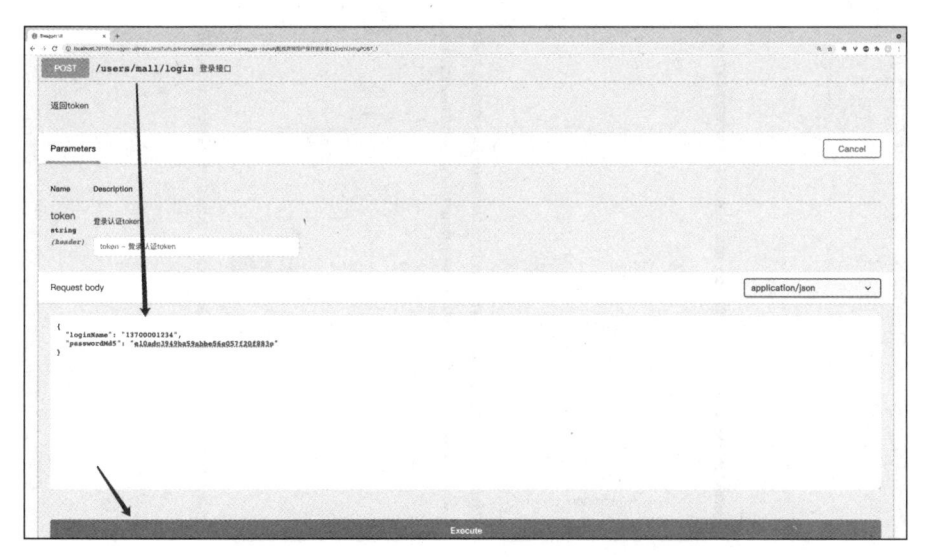

图 1-19　用户微服务中登录接口的测试过程

单击"Execute"按钮后，结果如图1-20所示。

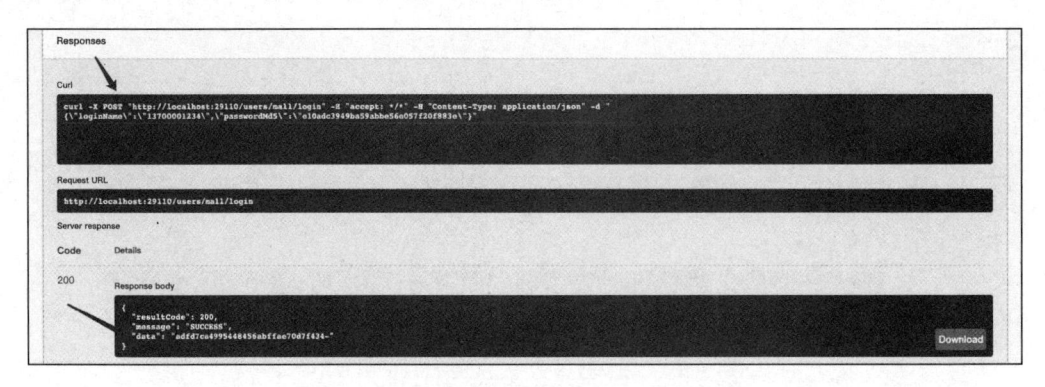

图1-20　用户微服务中登录接口的测试结果

如果登录信息都正确，就可以得到一个登录成功的token字段，该字段的值在响应对象Result的data字段中，用于身份认证。比如，当前登录接口的测试结果获取了值为"adfd7ca4995448456abffae70d7f434-"的token字段，之后就能够使用该token值访问项目中与商城用户相关的接口了。该token值是笔者在测试时生成的，读者在自行测试时生成的token值可能与此不同，不要弄混淆了。

这两个接口对应到实际的项目页面中，是新蜂商城项目的登录页面和注册页面，显示效果如图1-21所示。

图1-21　新蜂商城注册页面和登录页面的显示效果

至此，商城用户的注册功能和登录功能就测试完成了。读者在测试时可以关注 MySQL 数据库和 Redis 数据库中的相关记录。注册成功后会向用户表新增一条数据，用户登录成功后也会向 Redis 数据库中新增一条 token 记录，用于保存用户的登录信息。

1.4.2　添加商品到购物车的功能演示

在实际的项目中，添加商品到购物车需要在商品详情页面中操作。因此，这里的功能演示会涉及商品详情接口、添加商品到购物车接口和购物车列表接口。

1. 商品详情接口演示

单击页面右上方的"Select a definition"下拉列表框中的下拉按钮，选择"goods-service-swagger-route"选项就能够看到商品微服务的 Swagger 接口文档了。

比如，在前文的首页功能演示中，接口返回了一些商品信息并显示到首页，想要看到"华为 Mate 50 Pro 手机""iPhone 14 Pro 手机"两个商品的详情，就需要访问商品详情接口。

依次单击"商品详情接口""Try it out"按钮，在参数栏中输入商品 id 和 token 值，之后单击"Execute"按钮就能够发送获取商品详情接口的请求了，如图 1-22 所示。

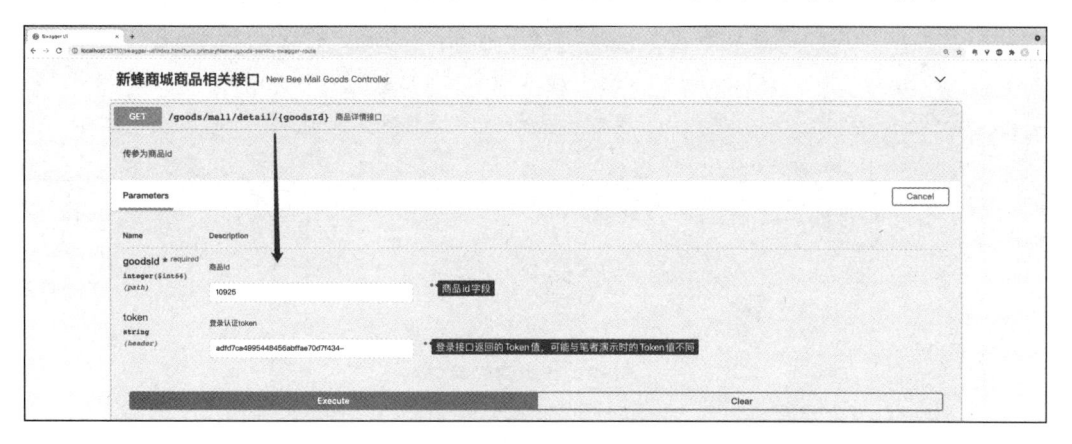

图 1-22　商品微服务中商品详情接口的测试过程

单击"Execute"按钮后，接口的测试结果如图 1-23 所示。

如果用户正常登录且商品 id 正确，就可以得到商品详情。

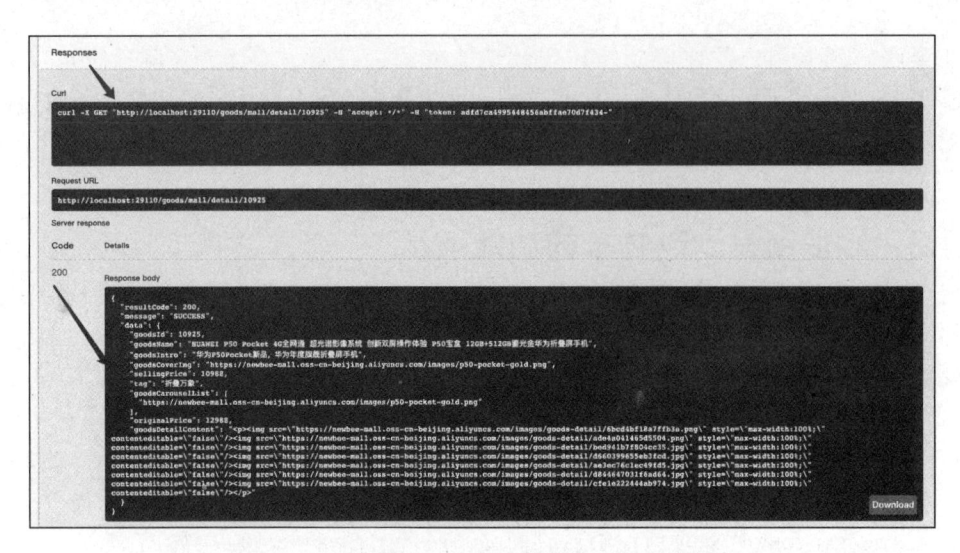

图 1-23　商品微服务中商品详情接口的测试结果

2. 添加商品到购物车接口演示

接下来将演示添加商品到购物车。单击页面右上方的"Select a definition"下拉列表框中的下拉按钮，选择"shop-cart-service-swagger-route"选项，就能够看到购物车微服务的 Swagger 接口文档了。

依次单击"添加商品到购物车接口""Try it out"按钮，在参数栏中输入商品 id 和添加数量，在登录认证 token 的输入框中输入登录接口返回的 token 值，如图 1-24 所示。

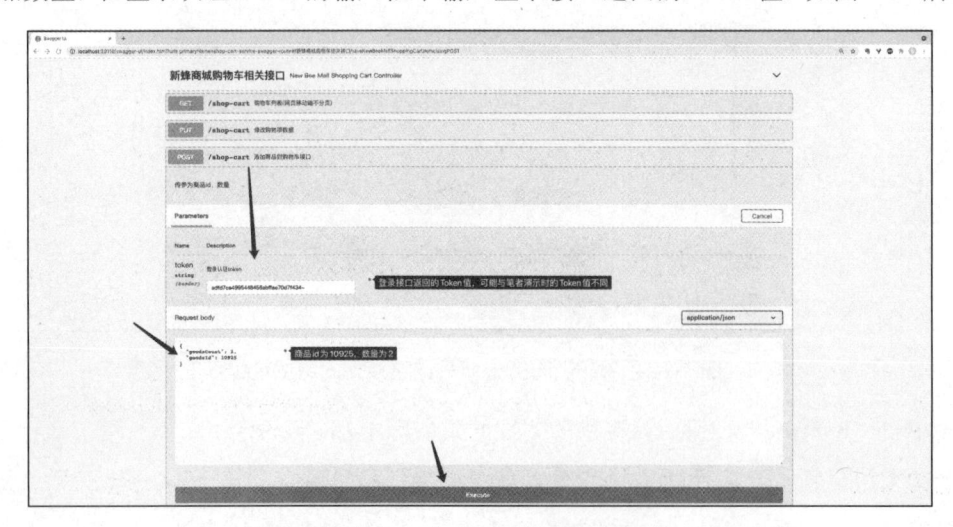

图 1-24　购物车微服务中添加商品到购物车接口的测试过程

单击"Execute"按钮,接口的测试结果如图1-25所示。

图1-25 购物车微服务中添加商品到购物车接口的测试结果

若后端接口的测试结果中有"SUCCESS",则表示添加商品成功。

笔者在测试时,输入的商品数量都是符合规范的,而且商品id是数据库中真实存在的。如果输入的商品数量过大,则会报错"超出单个商品的最大购买数量"。如果输入的商品id在数据库中并不存在,则会报错"商品不存在"。读者在测试时需要注意这一点。

3. 购物车列表接口演示

依次单击"购物车列表(网页移动端不分页)""Try it out"按钮,在登记认证token的输入框中输入刚刚获取的token值,单击"Execute"按钮发起测试请求,就能够看到此时的购物车列表数据了,测试过程和结果如图1-26所示。购物项id分别为11和12,后续在生成订单时需要用到。

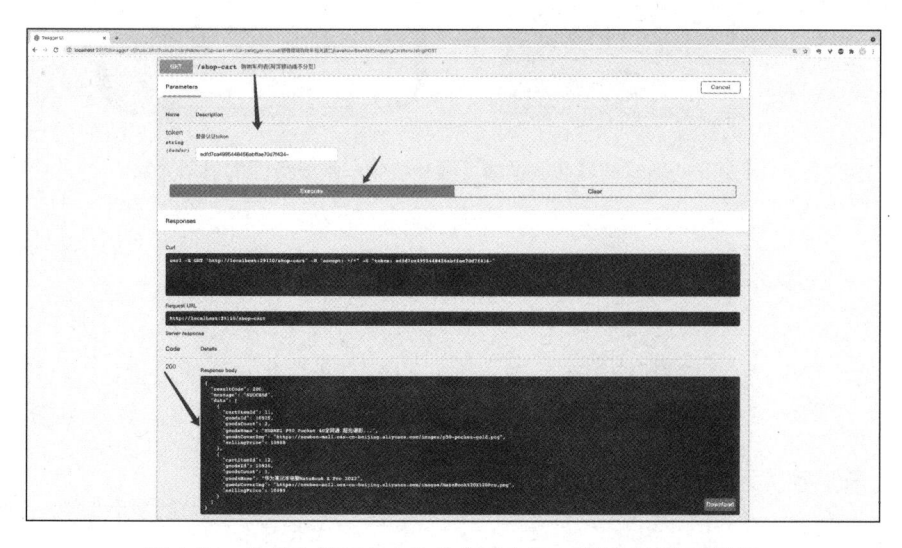

图1-26 购物车微服务中购物车列表接口的测试过程和结果

以上三个接口对应到实际的项目页面中，是新蜂商城项目的商品详情页面和购物车列表页面，显示效果如图1-27所示。

至此，添加商品到购物车并查看购物车列表的功能就演示完毕了，读者在测试时可以关注 MySQL 数据库中购物项表的变化。

图1-27 新蜂商城商品详情页面和购物车列表页面的显示效果

1.4.3 下单流程演示

在把心仪的商品添加到购物车并确定需要购买的商品和对应的数量后，就可以执行提交订单的操作。本节的功能演示涉及添加收货地址接口、生成订单接口、订单列表接口。

1. 添加收货地址接口演示

单击页面右上方的"Select a definition"下拉列表框中的下拉按钮，选择"order-service-swagger-route"选项，就能够看到订单微服务的 Swagger 接口文档了。

下单时需要用户的收货地址，否则无法正确地生成订单数据。依次单击"添加地址""Try it out"按钮，在参数栏中输入收货地址，在登录认证 token 的输入框中输入登录接口返回的 token 值，如图 1-28 所示。

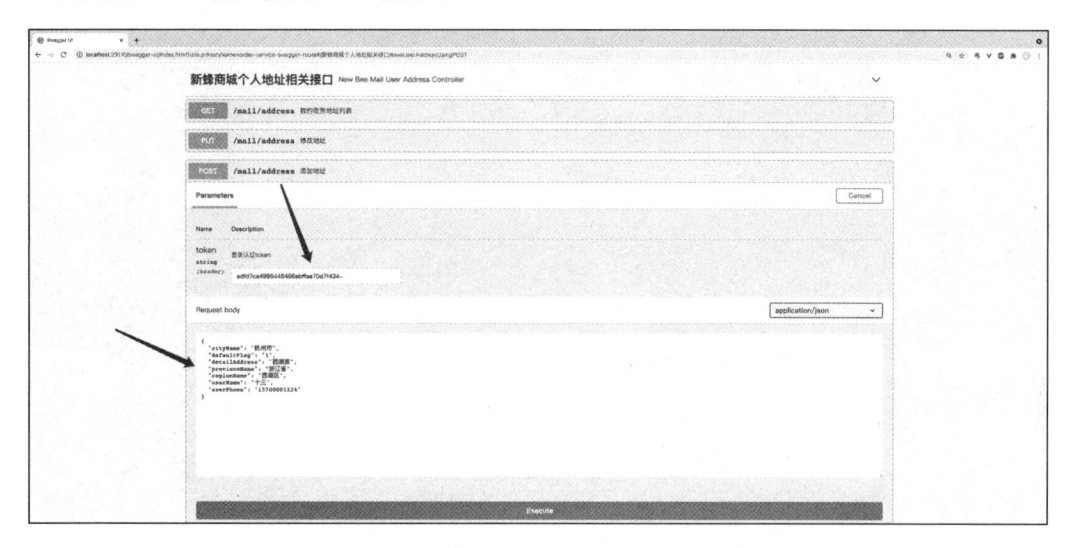

图 1-28　订单微服务中添加收货地址接口的测试过程

单击"Execute"按钮，接口的测试结果如图 1-29 所示。

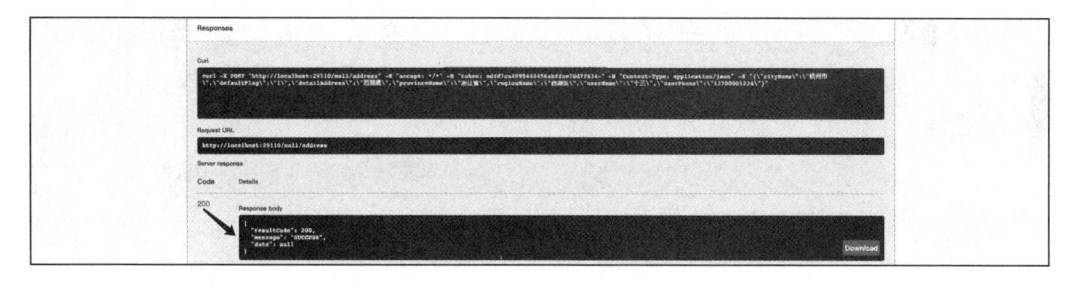

图 1-29　订单微服务中添加收货地址接口的测试结果

若后端接口的测试结果中有"SUCCESS"，则表示收货地址添加成功。此时，再去订单微服务的数据库中查看收货地址表中的数据，可以看到已经新增了一条地址信息，该数据的主键 id 为 1，后续生成订单时会用到。

2. 生成订单接口演示

依次单击"生成订单接口""Try it out"按钮，在参数栏中输入当前用户的地址 id 和需要结算的购物项 id 列表，这里输入的数据都是刚刚演示时生成的数据。在登录认证

token 的输入框中输入登录接口返回的 token 值，如图 1-30 所示。

单击"Execute"按钮，接口的测试结果如图 1-31 所示。

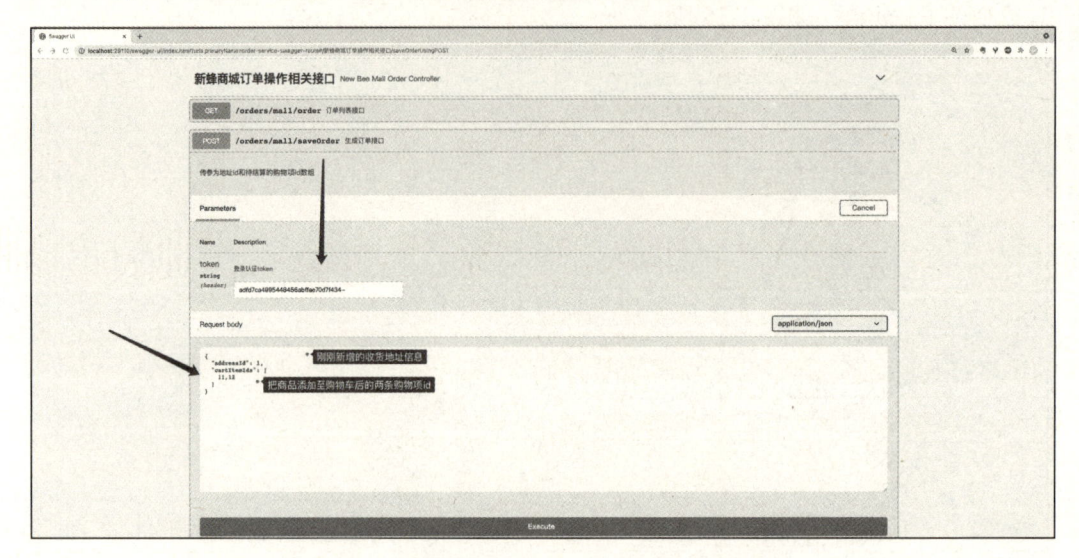

图 1-30　订单微服务中生成订单接口的测试过程

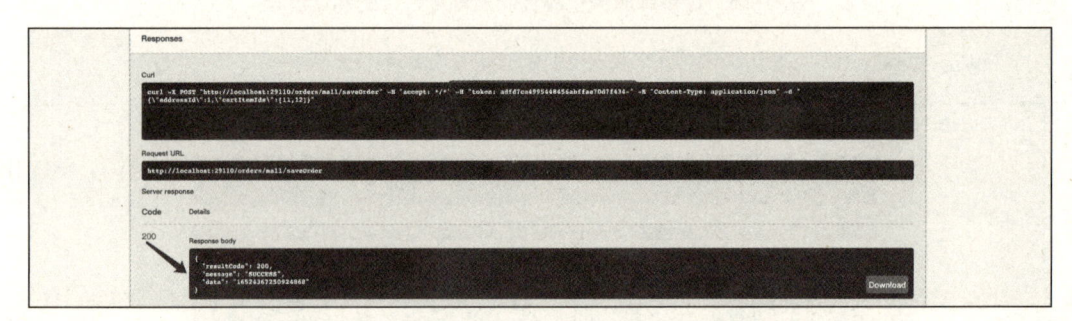

图 1-31　订单微服务中生成订单接口的测试结果

如果结算时提交的数据都正确，就可以得到一个订单生成后的订单号字段，该字段的值在响应对象 Result 的 data 字段中。比如，当前接口的测试结果获取了值为"16524367250924868"的订单号，之后就能够使用该订单号来测试取消订单、模拟支付、订单详情的接口了。

生成订单接口测试成功。

3. 订单列表接口演示

依次单击"订单列表接口""Try it out"按钮，在参数栏中输入页码和订单状态字段，在登录认证 token 的输入框中输入登录接口返回的 token 值，就可以查询当前用户的订单列表数据了，如图 1-32 所示。

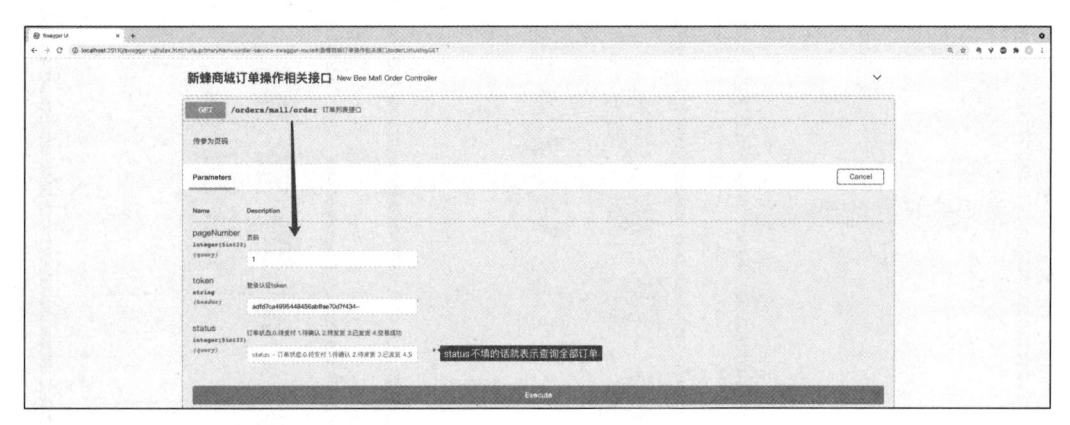

图 1-32　订单微服务中订单列表接口的测试过程

单击"Execute"按钮，接口的测试结果如图 1-33 所示。

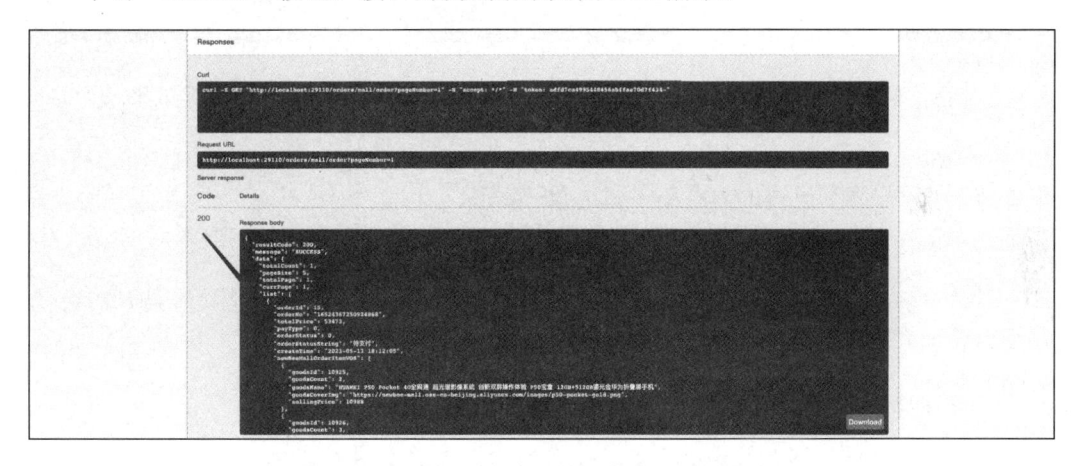

图 1-33　订单微服务中订单列表接口的测试结果

请求成功。订单列表所需的数据在 Result 类的 data 属性中，有分页信息、订单列表数据，每一条购物项中都包括订单号、订单状态、下单时间、订单中包含的商品等内容。

以上三个接口对应到实际的项目页面中，是新蜂商城项目的添加收货地址页面、生成订单页面和订单列表页面，显示效果如图 1-34 所示。

至此，下单流程中的部分功能就演示完毕了，读者在测试时可以关注 MySQL 数据库中购物项表的变化。

图1-34 新蜂商城添加收货地址页面、生成订单页面和订单列表页面的显示效果

1.4.4 后台管理系统的部分功能演示

后台管理系统的功能与商城端的功能不同，需要管理员用户的权限，在测试时会使用后台管理系统的网关服务。

1. 管理员用户登录接口演示

后台管理系统的网关服务的 Swagger 接口文档页面如图 1-35 所示。单击页面右上方的"Select a definition"下拉列表框中的下拉按钮，选择"user-service-swagger-route"选项，就能够看到用户微服务的 Swagger 接口文档了，管理员用户相关的接口也在用户微服务中。

先单击管理员用户操作相关接口中的"登录接口"按钮，再单击"Try it out"按钮，在参数栏中输入 userName 字段和 passwordMd5 字段（注意，密码字段需要进行 MD5 加密），之后单击"Execute"按钮就能够发送管理员用户登录的请求了。默认管理员用户名为 admin，密码为 123456，如果读者修改了数据库中的数据，就需要对应地修改这里的参数，如图 1-36 所示。

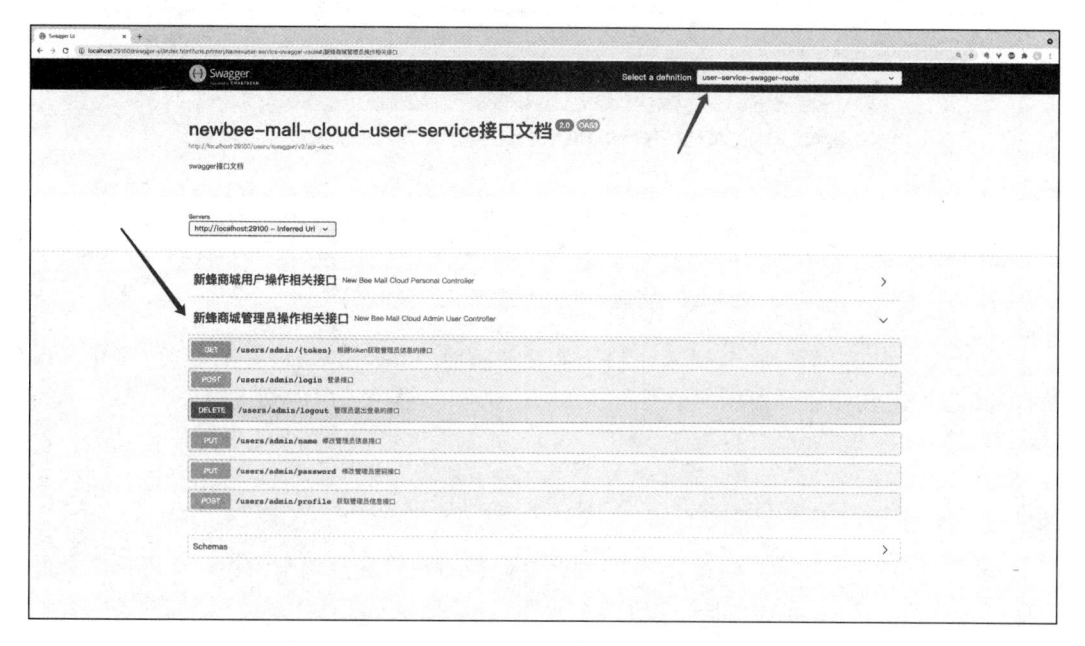

图 1-35　后台管理系统的网关服务的 Swagger 接口文档页面

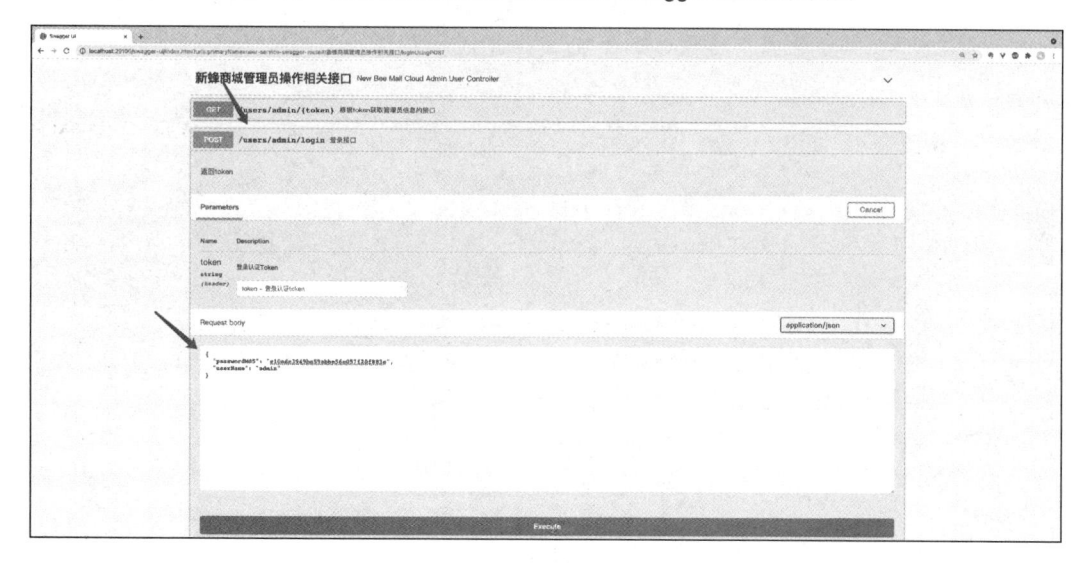

图 1-36　管理员用户登录接口的测试过程

单击"Execute"按钮后，接口的测试结果如图 1-37 所示。

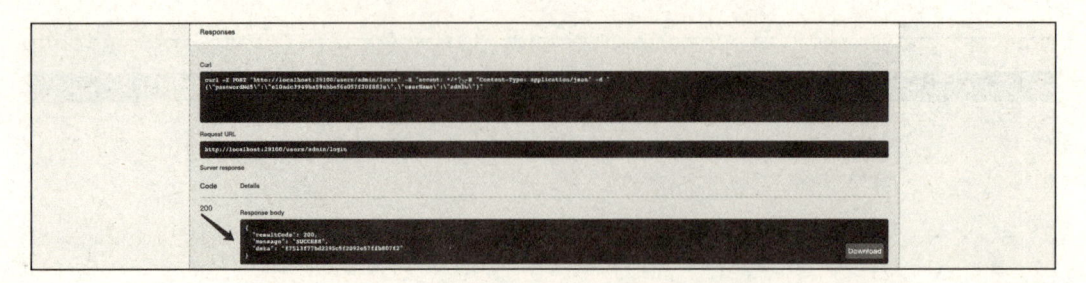

图 1-37 管理员用户登录接口的测试结果

如果登录信息都正确，就可以得到一个登录成功的 token 值，该值在响应对象 Result 的 data 字段中，用于管理员用户的身份认证。比如，当前登录接口的测试结果获取的 token 值为"f7513f77bd2395c5f2092e57ffb807f2"，之后就能够使用该 token 值访问项目中与管理员用户相关的接口了。该 token 值是笔者在测试时生成的，读者在自行测试时生成的值可能有所不同，不要弄混淆了。

该接口对应到实际的项目页面中，是新蜂商城后台管理系统中的管理员用户登录页面，显示效果如图 1-38 所示。

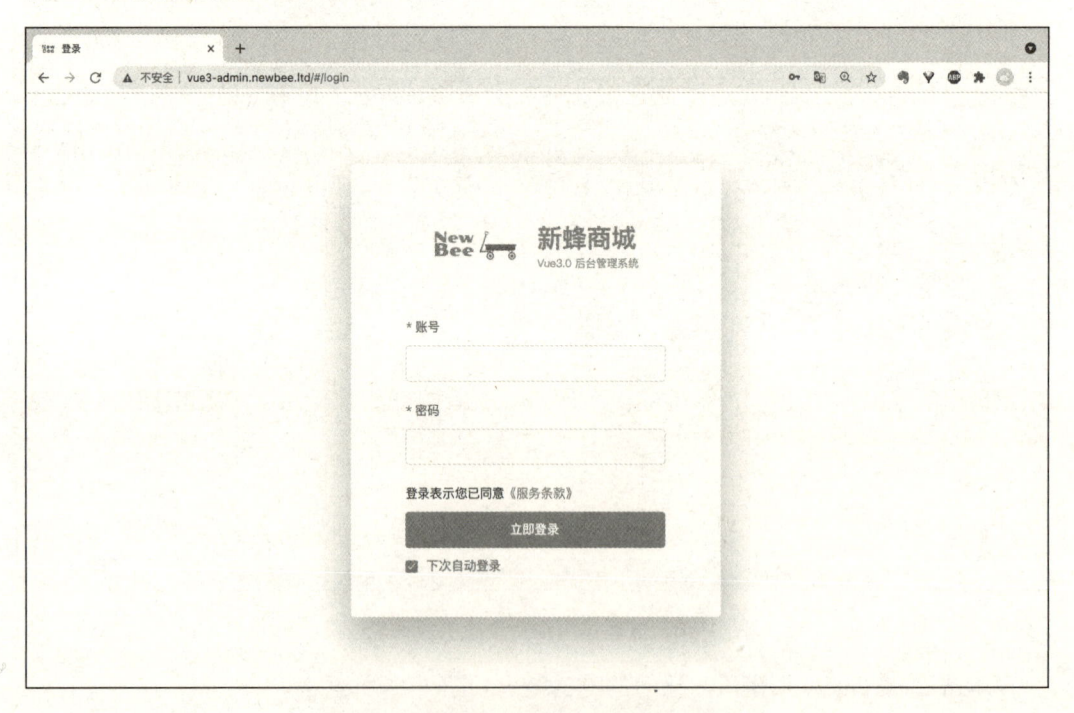

图 1-38 管理员用户登录页面的显示效果

2. 新增分类接口演示

接下来演示后台管理系统中的新增分类功能。单击页面右上方的"Select a definition"下拉列表框中的下拉按钮，选择"goods-service-swagger-route"选项，就能够看到商品微服务的 Swagger 接口文档了。

依次单击"新增分类""Try it out"按钮，在参数栏中输入分类名称、分类等级等字段，在登录认证 token 的输入框中输入管理员用户登录接口返回的 token 值，如图 1-39 所示。

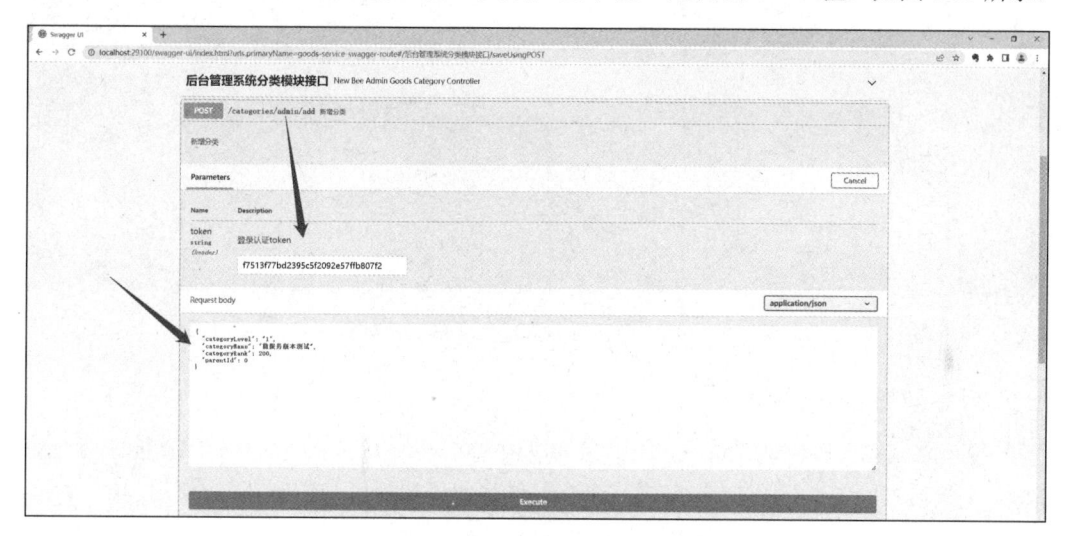

图 1-39 新增分类接口的测试过程

单击"Execute"按钮，接口的测试结果如图 1-40 所示。

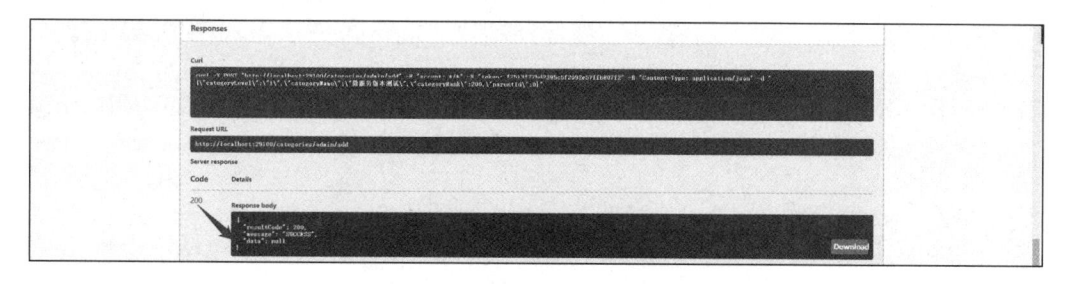

图 1-40 新增分类接口的测试结果

若后端接口的测试结果中有"SUCCESS"，则表示添加成功，查看商品微服务的数据库，分类表中已经新增了一条数据。笔者在测试时，输入的字段都是符合规范的。如果输入的字段没有通过基本的验证判断，就会报出对应的错误提示，读者在测试时需要注意这一点。

该接口对应到实际的项目页面中，是新蜂商城后台管理系统中的商品分类管理页面，显示效果如图 1-41 所示。

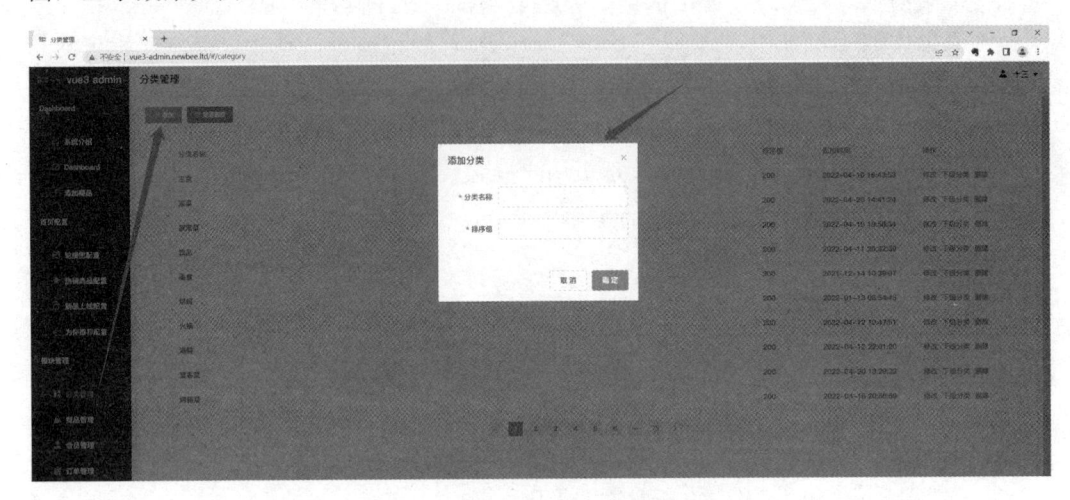

图 1-41　商品分类管理页面的显示效果

3. 下架商品接口演示

接下来演示在后台管理系统中把商品 id 为 10003 和 10005 的商品下架的功能。对应的接口在 Swagger 接口文档的"后台管理系统商品模块接口"选项卡中，如图 1-42 所示。

图 1-42　Swagger 接口文档的"后台管理系统商品模块接口"选项卡

依次单击"批量修改销售状态""Try it out"按钮，在参数栏中输入商品的销售状态（下架为 1，上架为 0）和需要下架的商品 id 数组，在登录认证 token 的输入框中输入管理员用户登录接口返回的 token 值，如图 1-43 所示。

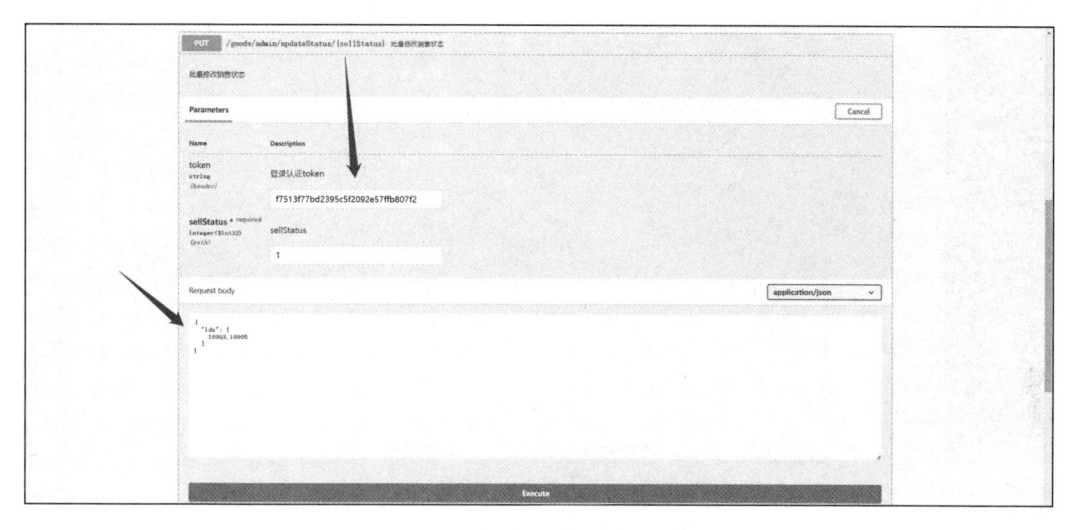

图 1-43　下架商品接口的测试过程

单击"Execute"按钮，接口的测试结果如图 1-44 所示。

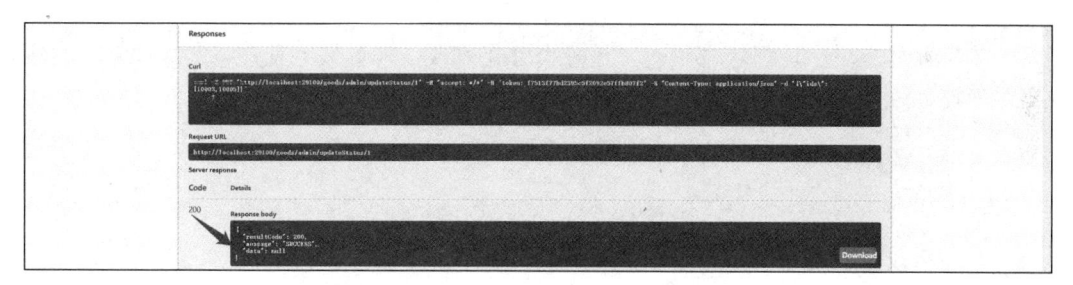

图 1-44　下架商品接口的测试结果

若后端接口的测试结果中有"SUCCESS"，则表示商品下架成功。此时查看数据库中对应的记录，可以看到这两个商品的 goods_sell_status 字段的值已经被修改为 1（下架状态）。

该接口对应到实际的项目页面中，是新蜂商城后台管理系统中的商品管理页面，显示效果如图 1-45 所示。

后台管理系统的功能很多，因篇幅有限，这里只演示了部分功能，主要是让读者进行参考。另外，在测试时一定要关注 MySQL 数据库和 Redis 数据库中的相关记录。

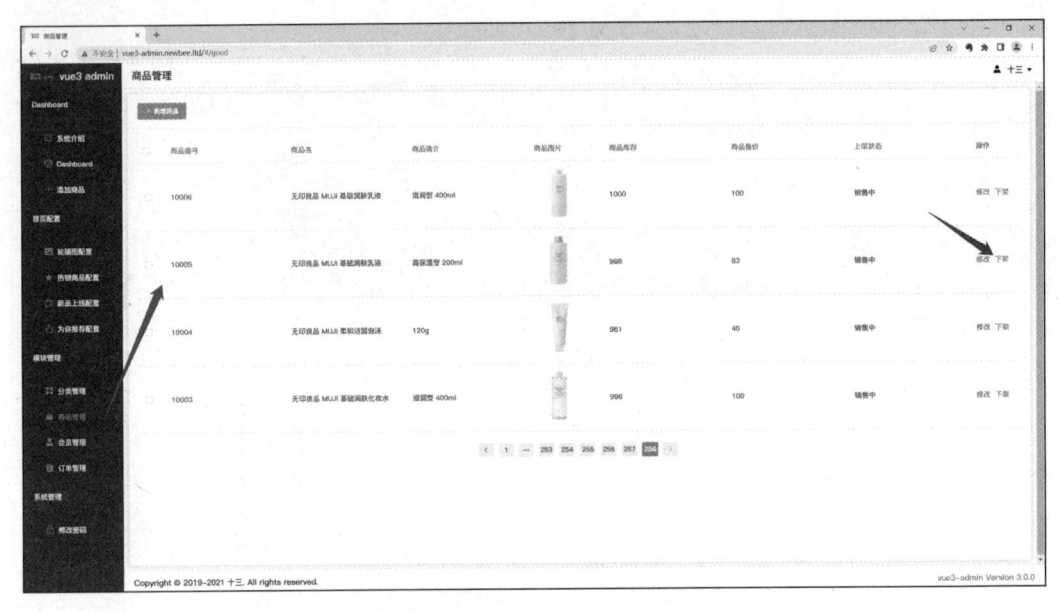

图1-45 商品管理页面的显示效果

1.5 微服务架构项目中接口的参数处理及统一结果响应

为了让读者更快地理解源代码并进行个性化修改，接下来笔者介绍实战项目中的接口是怎样处理参数接收和结果响应的。

1. 普通参数接收

读者应该比较熟悉普通参数接收方式，因为它是 GET 请求方式，所以传参时直接在路径后拼接参数和参数值即可。

```
@RequestMapping(value = "/list", method = RequestMethod.GET)
@ApiOperation(value = "商品列表", notes = "可根据名称和上架状态筛选")
public Result list(@RequestParam(required = false) @ApiParam(value = "页码") Integer pageNumber,
            @RequestParam(required = false) @ApiParam(value = "每页条数") Integer pageSize,
            @RequestParam(required = false) @ApiParam(value = "商品名称") String goodsName,
            @RequestParam(required = false) @ApiParam(value = "上架状态 0-上架 1-下架") Integer goodsSellStatus, @TokenToAdminUser LoginAdminUser adminUser) {
```

```
    省略部分代码
}
```

下面这行代码是商品列表接口的方法定义,格式如下:

```
?key1=value1&key2=value2
```

2. 路径参数接收

部分接口在设计时采用了将参数拼入路径的方式,当只需要一个参数时,可以考虑使用这种接口设计方式。路径参数接收与前文的普通参数接收没有很大的区别,也可以设计为普通参数接收的方式,主要是由开发者的开发习惯决定的。

```
@GetMapping("/detail/{goodsId}")
@ApiOperation(value = "商品详情接口", notes = "传参为商品id")
public Result<NewBeeMallGoodsDetailVO> goodsDetail(@ApiParam(value = "商品
id") @PathVariable("goodsId") Long goodsId, @TokenToMallUser MallUserToken
loginMallUserToken) {
    省略部分代码
}
```

上面这段代码是商品详情接口的方法定义,如果想要查询订单号为 10011 的商品信息,则直接请求/detail/10011 路径即可,代码中使用@PathVariable 注解来进行接收。

3. 对象参数接收

项目中使用了 POST 或 PUT 方法类型的接口,基本上都是以对象形式来接收参数的。

```
@ApiOperation(value = "登录接口", notes = "返回token")
@RequestMapping(value = "/users/admin/login", method = RequestMethod.POST)
public Result<String> login(@RequestBody @Valid AdminLoginParam
adminLoginParam) {
    省略部分代码
}
```

上面这段代码是管理员用户登录接口的方法定义,前端在请求体中放入 JSON 格式的请求参数,后端使用@RequestBody 注解进行接收,并将这些参数转换为对应的实体类。

为了传参形式的统一,对于 POST 或 PUT 方法类型的请求参数,前端传过来的格式要求为 JSON 形式,Content-Type 统一设置为 application/json。

4. 复杂对象接收

当然,有时会出现复杂对象传参的情况。比如,一个传参对象中包含另外一个实体对象或多个对象。这种方式与对象参数接收方式一样,前端开发人员需要进行简单的格式转换，在 JSON 串中加一层对象。后端在接收参数时，需要在原有多个对象的基础上再封装一个对象参数。

笔者以订单生成接口的传参来介绍，该方法的源代码定义在 ltd.order.cloud.newbee.controller.NewBeeMallOrderController 类中，代码如下：

```
@PostMapping("/saveOrder")
@ApiOperation(value = "生成订单接口", notes = "传参为地址id和待结算的购物项id数组")
public Result<String> saveOrder(@ApiParam(value = "订单参数") @RequestBody SaveOrderParam saveOrderParam, @TokenToMallUser MallUserToken loginMallUserToken) {
    省略部分代码
}
```

后端需要重新定义一个参数对象，并使用@RequestBody 注解进行接收和对象转换，SaveOrderParam 类的定义如下：

```
public class SaveOrderParam implements Serializable {

    @ApiModelProperty("订单项id数组")
    private Long[] cartItemIds;

    @ApiModelProperty("地址id")
    private Long addressId;
}
```

前端需要将用户所勾选的购物项 id 数组和收货地址的 id 传过来，这个接口的传参如下：

```
{
  "addressId": 0,
  "cartItemIds": [
    1,2,3
  ]
}
```

关于接口参数的处理，前端开发人员按照后端给出的接口文档进行参数的封装即可，后端开发人员则需要根据接口的实际情况进行灵活的设计，同时注意@RequestParam、@PathVariable、@RequestBody 三个注解的使用。希望读者可以根据本书的内容及项目源代码进行举一反三，能够灵活地设计和开发出适合自己项目的接口。

统一结果响应的设计和使用可以参考 8.3.4 节中的内容，这里不再赘述。

传参的规范和结果响应的统一，都会使控制层、业务层处理的数据格式一致化，保证了接口和编码规范的统一。这种做法不仅体现在本项目中，而且对开发人员在今后的企业级项目开发工作中也有着非常重要的意义，规范的参数定义和结果响应会极大程度地降低开发成本及沟通成本。

1.6　微服务架构项目打包和部署的注意事项

项目源代码拿到了，在 IDEA 中对项目进行修改和功能测试也完成了，终于到了打包和部署的环节。其实打包和部署的步骤并不复杂，只需要在命令行中进入当前项目的顶级目录，之后执行 Maven 的打包命令即可。

命令如下：

```
mvn clean package '-Dmaven.test.skip'
```

经过一段时间的等待后，项目中的所有模块都按照 pom.xml 文件中的配置打包成功了，如图 1-46 所示。

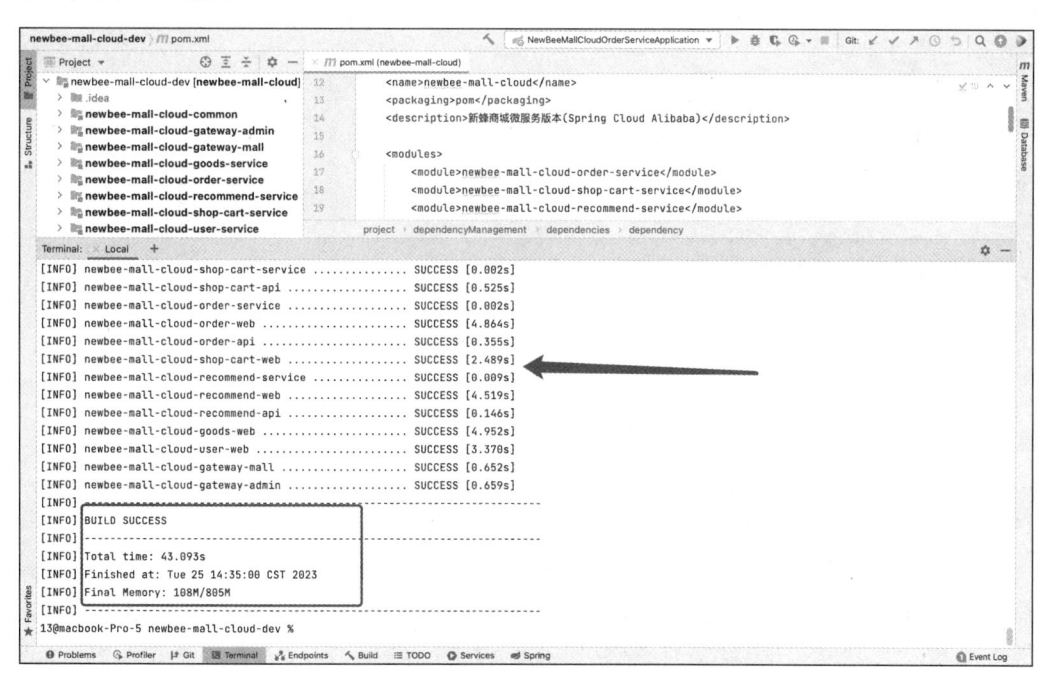

图 1-46　newbee-mall-cloud 项目的打包过程

用户微服务、购物车微服务、推荐微服务、商品微服务、订单微服务和两个网关服务实例都会被打包成一个可执行的 JAR 包，如 newbee-mall-cloud-user-web-0.0.1-SNAPSHOT.jar、newbee-mall-cloud-goods-web-0.0.1-SNAPSHOT.jar 等可执行文件。打包成功后，进入对应的 target 目录启动 JAR 包，即可启动不同的微服务实例。

执行的命令如下：

```
java -jar xxxx.jar
```

以启动商品微服务为例，进入 newbee-mall-cloud-goods-web/target 目录，执行 java -jar newbee-mall-cloud-goods-web-0.0.1-SNAPSHOT.jar 命令即可启动商品微服务实例，启动过程及报错信息如图 1-47 所示。

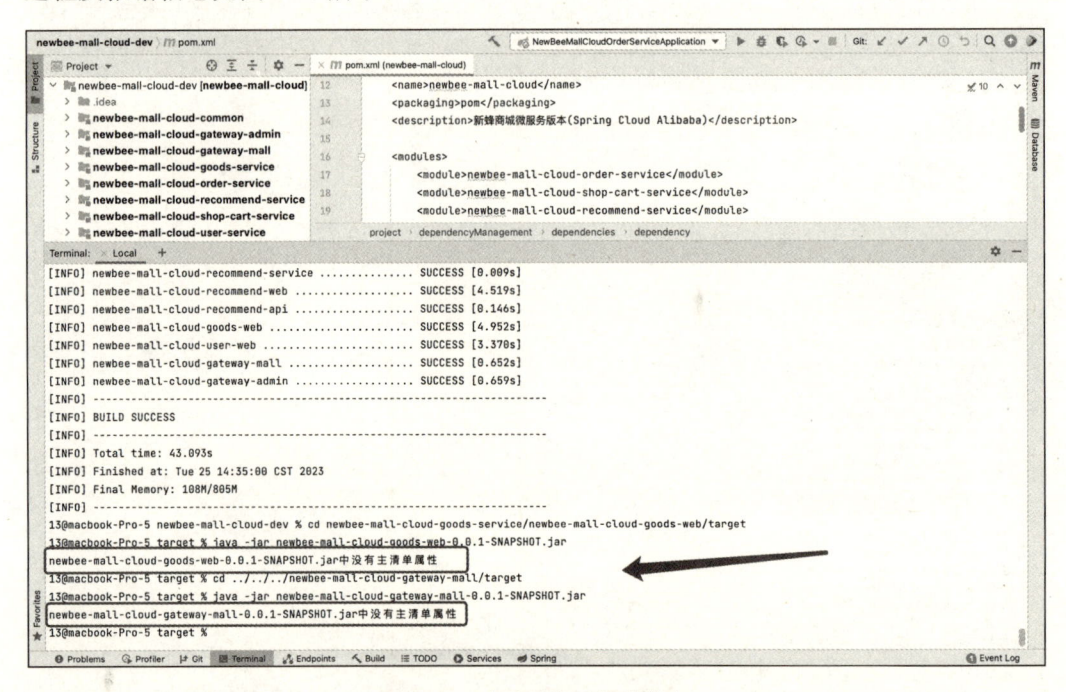

图 1-47　启动过程及报错信息

是的，没有成功，启动时报错：××.jar 中没有主清单属性。不止是商品微服务，网关服务实例也没有启动成功，报错内容相同。这就说明打出的包存在问题。

这是一个微服务架构的多模块项目，多模块项目中包含多个 Spring Boot 项目，用户微服务、购物车微服务、推荐微服务、商品微服务、订单微服务和两个网关服务实例其实都是 Spring Boot 项目。在这种多模块、多 Spring Boot 实例的项目中，需要在 pom.xml 文件中增加一些配置，否则打包后的 JAR 包就无法正常启动。

这里涉及一个知识点：**使用 Maven 构建多个 Spring Boot 实例**。这个知识点其实很少会碰到。我们平时开发的基本上是单个的 Spring Boot 项目，在打包时肯定不会遇到这个问题，而在做复杂架构下的项目时会遇到这个问题。解决办法不复杂，在对应的 pom.xml 文件中加入一些打包所需的 plugin 插件并指定对应实例的主启动类的**全类名**。

打开用户微服务、购物车微服务、推荐微服务、商品微服务、订单微服务和两个网关服务实例所对应的 newbee-mall-cloud-×××-web 工程目录，并在各自的 pom.xml 文件中增加一些配置。下面以商品微服务和商城端的网关服务为例进行讲解。

打开 newbee-mall-cloud-goods-web 目录下的 pom.xml 文件，在原配置的基础上增加如下代码：

```xml
<!-- 打包 -->
<build>
 <resources>
   <resource>
     <filtering>true</filtering>
     <directory>src/main/resources</directory>
   </resource>
 </resources>
 <plugins>
   <plugin>
     <groupId>org.apache.maven.plugins</groupId>
     <artifactId>maven-compiler-plugin</artifactId>
     <configuration>
       <source>${java.version}</source>
       <target>${java.version}</target>
       <encoding>${project.build.sourceEncoding}</encoding>
     </configuration>
   </plugin>
   <plugin>
     <groupId>org.apache.maven.plugins</groupId>
     <artifactId>maven-resources-plugin</artifactId>
     <configuration>
       <encoding>${project.build.sourceEncoding}</encoding>
     </configuration>
   </plugin>
   <plugin>
     <groupId>org.apache.maven.plugins</groupId>
     <artifactId>maven-jar-plugin</artifactId>
   </plugin>
   <plugin>
     <groupId>org.springframework.boot</groupId>
```

```xml
      <artifactId>spring-boot-maven-plugin</artifactId>
      <configuration>
        <mainClass>ltd.goods.cloud.newbee.NewBeeMallCloudGoodsServiceApplication</mainClass>
      </configuration>
      <executions>
        <execution>
          <goals>
            <goal>repackage</goal>
          </goals>
        </execution>
      </executions>
    </plugin>
  </plugins>
</build>
```

增加打包所需的 plugin 插件，并指定商品微服务主启动类 NewBeeMallCloudGoodsServiceApplication 类的全类名。

打开 newbee-mall-cloud-gateway-mall 目录下的 pom.xml 文件，在原配置的基础上增加如下代码：

```xml
<!-- 打包 -->
<build>
  <resources>
    <resource>
      <filtering>true</filtering>
      <directory>src/main/resources</directory>
    </resource>
  </resources>
  <plugins>
    <plugin>
      <groupId>org.apache.maven.plugins</groupId>
      <artifactId>maven-compiler-plugin</artifactId>
      <configuration>
        <source>${java.version}</source>
        <target>${java.version}</target>
        <encoding>${project.build.sourceEncoding}</encoding>
      </configuration>
    </plugin>
    <plugin>
      <groupId>org.apache.maven.plugins</groupId>
      <artifactId>maven-resources-plugin</artifactId>
```

```xml
        <configuration>
          <encoding>${project.build.sourceEncoding}</encoding>
        </configuration>
      </plugin>
      <plugin>
        <groupId>org.apache.maven.plugins</groupId>
        <artifactId>maven-jar-plugin</artifactId>
      </plugin>
      <plugin>
        <groupId>org.springframework.boot</groupId>
        <artifactId>spring-boot-maven-plugin</artifactId>
        <configuration>
          <mainClass>ltd.gateway.cloud.newbee.NewBeeMallCloudMallGatewayApplication</mainClass>
        </configuration>
        <executions>
          <execution>
            <goals>
              <goal>repackage</goal>
            </goals>
          </execution>
        </executions>
      </plugin>
    </plugins>
</build>
```

同样，增加打包所需的 plugin 插件，并指定网关服务层主启动类 NewBeeMall-CloudMallGatewayApplication 类的全类名。其他需要打包成 JAR 包模块下的 pom.xml 配置文件，也需要增加上述打包配置，这里就不再赘述了。

配置完成后，进入当前项目的顶级目录，执行 Maven 的打包命令。打包成功后，再次执行启动 JAR 包的命令，可以看到 JAR 包中没有主清单属性的问题已经不存在了，Jar 包可以顺利启动，一切正常，如图 1-48 所示。

项目打包和项目部署是两个步骤，得到各个微服务实例的可执行 JAR 包后，就可以进入部署环节了。不管是部署在服务器上，还是部署在本地，基本上都是在命令行执行 java-jar 命令。

下面是在 Linux 服务器上部署时的注意事项。

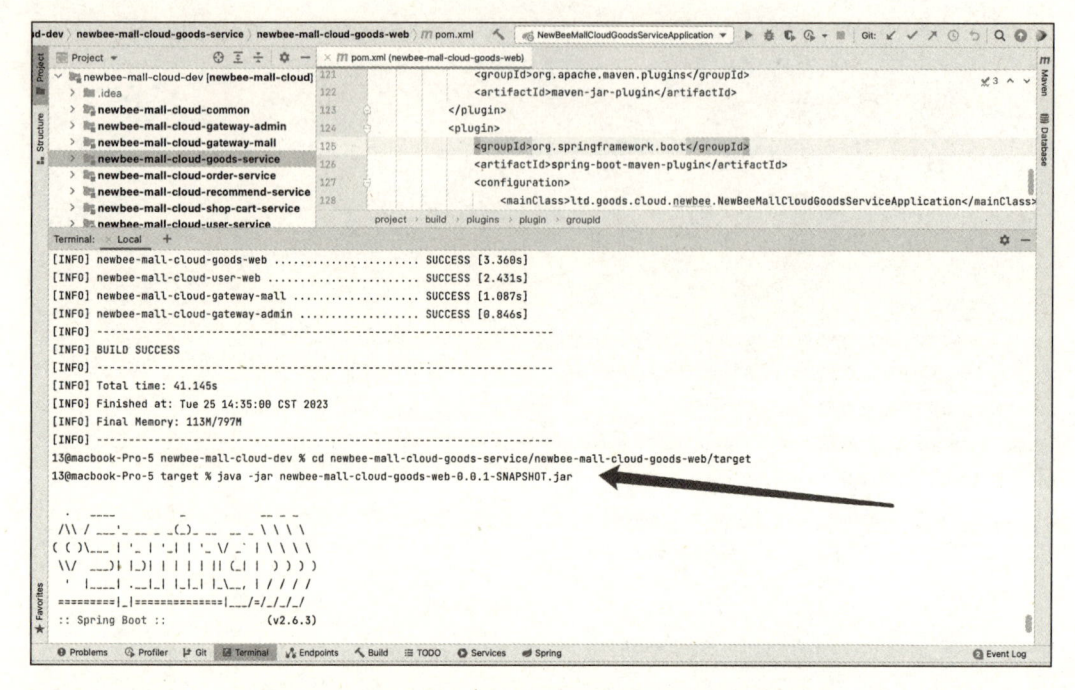

图 1-48　newbee-mall-cloud 项目打包和项目启动过程

启动命令：

```
java -jar newbee-mall-cloud-goods-web-0.0.1-SNAPSHOT.jar
```

这样启动的项目并没有在后台运行，一旦退出终端项目，基本上就跟着停掉了，可以在命令行前后分别添加 nohup 命令和 & 符号，这样就能够让项目一直在后台运行，并且项目的运行日志会输出到 nohup.out 文件中，此时的命令如下：

```
nohup java -jar newbee-mall-cloud-goods-web-0.0.1-SNAPSHOT.jar &
```

想要关闭这个项目，可以先查出它运行时的进程号，然后用 kill 命令关闭它。

想要部署服务集群，可以使用 --server.port 参数指定多个端口号。以部署商品微服务集群为例，执行的命令如下：

```
#启动时使用application.properties配置文件中配置的端口号29010
nohup java -jar newbee-mall-cloud-goods-web-0.0.1-SNAPSHOT.jar &

#启动时自定义端口号
nohup java -jar newbee-mall-cloud-goods-web-0.0.1-SNAPSHOT.jar --server.port=29011 &
```

```
nohup java -jar newbee-mall-cloud-goods-web-0.0.1-SNAPSHOT.jar --server.port=
29012 &
```

这样就可以部署一个微服务实例的集群了。启动后的 JAR 包实例会将自己注册到 Nacos Server 中，在被其他微服务调用时也会根据负载均衡算法提供服务。

另外，如果服务器中的内存并不宽裕，启动过多的微服务实例就可能有些吃力。此时，可以在启动命令中添加 JVM 参数限制项目消耗的内存资源，执行的命令如下：

```
nohup java -jar -server -Xms128m -Xmx384m newbee-mall-cloud-goods-web-
0.0.1-SNAPSHOT.jar &
```

至此，本章的内容介绍完毕。在后续章节中，笔者将结合编码介绍整个微服务架构项目从零到一的开发过程。

第 2 章 实战项目基础构建及公用模块引入

从本章开始就要动手搭建最终的微服务架构项目了，主要根据一个单体的商城 API 项目进行微服务化改造，开发过程中将 Spring Cloud Alibaba 的各个组件像添砖加瓦一样集成到项目里，让 newbee-mall-cloud 从一个空文件夹慢慢成为一个多模块、多组件、多功能的微服务架构项目。

2.1 编码前的准备

最终的微服务架构项目为 newbee-mall-cloud，该项目是由 newbee-mall-api 单体项目改造而来的，相关的业务和代码都已经开发完成。后续的实战过程是将单体项目改造为微服务架构项目的一个开发过程。读者在学习后续的实战章节前，一定要了解和体验 newbee-mall-api 项目，并仔细阅读第 1 章中的内容，其中讲解了实战项目的功能及改造前的微服务拆分思路。读者了解这部分内容对后续的实战操作更有帮助。

后续的实战过程就是笔者根据拆分思路一点一点去完成的，过程中修改了原单体项目中的一些实现方式和代码。笔者会介绍开发步骤中遇到的一些小问题，并且把每个步骤的源代码都分享给读者。虽然笔者对原来 newbee-mall-api 单体项目非常熟悉，但是在微服务改造的过程中还是遇到了不少问题，毕竟使用的是不同的架构和不同的知识点，出现了一些意料之外的情况。只有真正地去做、去编码才知道有哪些不足、哪些知识点没掌握、编码的时候遗漏了什么。

因此，如果有时间的话，读者一定要亲自动手改一下这个项目。如果没时间或觉得暂时还无法独立完成，那就跟着笔者的思路和编码过程来学习、体验。接下来的实战过程都是笔者的开发思路和开发过程，每个重要步骤都有一份单独的源代码，保证读者能够学习微服务架构项目开发的每个步骤，体验从零到一的微服务架构项目开发过程。读者也可以根据掌握的知识和相关的项目，自行改造和开发一个微服务架构项目。

前几章的讲解会相对详细一些，后几章则会重点讲解开发思路和过程。因为万事开头难，项目的主体部分一旦完成，后续就是一些添砖加瓦的重复工作，就水到渠成了。

2.2 搭建项目骨架

万丈高楼平地起。

接下来，笔者进行微服务架构项目第一个步骤的编码：新建一个文件夹，把项目的基础骨架搭建出来。

由于已经构建过 Spring Cloud Alibaba 整合服务治理后的多模块模板项目，因此这里直接使用 spring-cloud-alibaba-multi-service-demo 进行简单的改造即可。

2.2.1 构建项目并整理依赖关系

先把项目名称修改为 newbee-mall-cloud-dev-step01，再将 order-service-demo 修改为 newbee-mall-cloud-gateway-admin、将 shopcart-service-demo 修改为 newbee-mall-cloud-user-service，并删除 goods-service-demo 模块（本开发步骤中只需要两个 Maven 模块），然后删除 newbee-mall-cloud-gateway-admin 项目和 newbee-mall-cloud-user-service 项目中已存在的包和代码，只保留项目中基础的 Maven 目录结构。

下面整理项目中的依赖项和依赖关系。

修改 root 节点的 pom.xml 配置文件，代码如下：

```xml
<?xml version="1.0" encoding="UTF-8"?>
<project xmlns="http://maven.apa***.org/POM/4.0.0" xmlns:xsi=
"http://www.w*.org/2001/XMLSchema-instance"
         xsi:schemaLocation="http://maven.apa***.org/POM/4.0.0
https://maven.apa***.org/xsd/maven-4.0.0.xsd">
    <modelVersion>4.0.0</modelVersion>
    <groupId>ltd.newbee.cloud</groupId>
    <artifactId>newbee-mall-cloud</artifactId>
    <version>0.0.1-SNAPSHOT</version>
    <name>newbee-mall-cloud</name>
    <packaging>pom</packaging>
```

```xml
    <description>新蜂商城微服务版本(Spring Cloud Alibaba)</description>

    <modules>
        <module>newbee-mall-cloud-user-service</module>
        <module>newbee-mall-cloud-gateway-admin</module>
    </modules>

    <properties>
        <spring.boot.version>2.6.3</spring.boot.version>
        <spring.cloud.dependencies.version>2021.0.1</spring.cloud.dependencies.version>
        <spring.cloud.alibaba.version>2021.0.1.0</spring.cloud.alibaba.version>
        <java.version>1.8</java.version>
        <project.build.sourceEncoding>UTF-8</project.build.sourceEncoding>
        <project.reporting.outputEncoding>UTF-8</project.reporting.outputEncoding>
    </properties>

    <dependencyManagement>
        <dependencies>
            <dependency>
                <groupId>org.springframework.cloud</groupId>
                <artifactId>spring-cloud-dependencies</artifactId>
                <version>${spring.cloud.dependencies.version}</version>
                <type>pom</type>
                <scope>import</scope>
            </dependency>

            <dependency>
                <groupId>org.springframework.boot</groupId>
                <artifactId>spring-boot-dependencies</artifactId>
                <version>${spring.boot.version}</version>
                <type>pom</type>
                <scope>import</scope>
            </dependency>

            <dependency>
                <groupId>com.alibaba.cloud</groupId>
                <artifactId>spring-cloud-alibaba-dependencies</artifactId>
                <version>${spring.cloud.alibaba.version}</version>
                <type>pom</type>
                <scope>import</scope>
```

```xml
            </dependency>

            <dependency>
                <groupId>org.springframework.boot</groupId>
                <artifactId>spring-boot-starter-web</artifactId>
                <version>${spring.boot.version}</version>
            </dependency>

            <dependency>
                <groupId>org.springframework.boot</groupId>
                <artifactId>spring-boot-starter-test</artifactId>
                <version>${spring.boot.version}</version>
            </dependency>
        </dependencies>
</dependencyManagement>

<build>
    <plugins>
        <plugin>
            <groupId>org.apache.maven.plugins</groupId>
            <artifactId>maven-compiler-plugin</artifactId>
            <version>3.8.0</version>
            <configuration>
                <source>${java.version}</source>
                <target>${java.version}</target>
                <encoding>UTF-8</encoding>
            </configuration>
        </plugin>
        <plugin>
            <groupId>org.springframework.boot</groupId>
            <artifactId>spring-boot-maven-plugin</artifactId>
            <version>${spring.boot.version}</version>
        </plugin>
    </plugins>
</build>

<repositories>
    <repository>
        <id>central</id>
        <url>https://maven.aliy**.com/repository/central</url>
        <name>aliyun</name>
    </repository>
</repositories>
```

```
</project>
```

主要工作是配置项目的名称和模块的依赖关系，同时管理和配置 Spring Cloud Alibaba 基础依赖、微服务发现依赖和网关依赖。

子节点 newbee-mall-cloud-gateway-admin 的 pom.xml 配置代码如下：

```xml
<?xml version="1.0" encoding="UTF-8"?>
<project xmlns="http://maven.apa***.org/POM/4.0.0" xmlns:xsi=
"http://www.w*.org/2001/XMLSchema-instance"
     xsi:schemaLocation="http://maven.apa***.org/POM/4.0.0
https://maven.apa***.org/xsd/maven-4.0.0.xsd">
    <modelVersion>4.0.0</modelVersion>
    <groupId>ltd.newbee.cloud</groupId>
    <artifactId>newbee-mall-cloud-gateway-admin</artifactId>
    <version>0.0.1-SNAPSHOT</version>
    <name>newbee-mall-cloud-gateway-admin</name>
    <description>网关模块</description>

    <parent>
        <groupId>ltd.newbee.cloud</groupId>
        <artifactId>newbee-mall-cloud</artifactId>
        <version>0.0.1-SNAPSHOT</version>
    </parent>

    <properties>
        <java.version>1.8</java.version>
        <project.build.sourceEncoding>UTF-8</project.build.sourceEncoding>
        <project.reporting.outputEncoding>UTF-8</project.reporting.outputEncoding>
    </properties>

    <dependencies>
        <dependency>
            <groupId>org.springframework.cloud</groupId>
            <artifactId>spring-cloud-starter-gateway</artifactId>
        </dependency>

        <dependency>
            <groupId>com.alibaba.cloud</groupId>
            <artifactId>spring-cloud-starter-alibaba-nacos-discovery</artifactId>
        </dependency>

        <dependency>
            <groupId>org.springframework.cloud</groupId>
```

```xml
        <artifactId>spring-cloud-loadbalancer</artifactId>
    </dependency>

  </dependencies>
</project>
```

这是一个网关微服务，主要引入 Spring Cloud Gateway 依赖、微服务发现依赖和负载均衡依赖。

子节点 newbee-mall-cloud-user-service 的 pom.xml 配置代码如下：

```xml
<?xml version="1.0" encoding="UTF-8"?>
<project xmlns="http://maven.apa***.org/POM/4.0.0" xmlns:xsi=
"http://www.w*.org/2001/XMLSchema-instance"
         xsi:schemaLocation="http://maven.apa***.org/POM/4.0.0
https://maven.apa***.org/xsd/maven-4.0.0.xsd">
    <modelVersion>4.0.0</modelVersion>
    <groupId>ltd.newbee.cloud</groupId>
    <artifactId>newbee-mall-cloud-user-service</artifactId>
    <version>0.0.1-SNAPSHOT</version>
    <name>newbee-mall-cloud-user-service</name>
    <description>用户微服务</description>

    <parent>
        <groupId>ltd.newbee.cloud</groupId>
        <artifactId>newbee-mall-cloud</artifactId>
        <version>0.0.1-SNAPSHOT</version>
    </parent>

    <properties>
        <java.version>1.8</java.version>
        <project.build.sourceEncoding>UTF-8</project.build.sourceEncoding>
        <project.reporting.outputEncoding>UTF-8</project.reporting.
outputEncoding>
    </properties>

    <dependencies>
        <dependency>
            <groupId>org.springframework.boot</groupId>
            <artifactId>spring-boot-starter-web</artifactId>
        </dependency>

        <dependency>
            <groupId>org.springframework.boot</groupId>
            <artifactId>spring-boot-starter-test</artifactId>
```

```xml
            <scope>test</scope>
        </dependency>

        <dependency>
            <groupId>com.alibaba.cloud</groupId>
            <artifactId>spring-cloud-starter-alibaba-nacos-discovery</artifactId>
        </dependency>

    </dependencies>
</project>
```

这是整个实战项目中的用户微服务，主要引入微服务发现依赖和 Web 开发依赖。因为这是开发的第一个步骤，所以主要测试基础功能和代码，后续会引入 MyBatis、MySQL、Swagger 等实际开发时的依赖。

以上操作主要保证基础目录搭建成功、微服务注册成功、网关层能够正常与微服务实例通信，并且项目中没有代码报错。以上操作顺利完成后，再往项目中加入业务代码。不急，一步一步来。

2.2.2 编写测试代码

1. newbee-mall-cloud-user-service 项目编码

打开 newbee-mall-cloud-user-service 项目，新建 ltd.user.cloud.newbee 包。之后创建启动类 NewBeeMallCloudUserServiceApplication，代码如下：

```java
@SpringBootApplication
@EnableDiscoveryClient
public class NewBeeMallCloudUserServiceApplication {
    public static void main(String[] args) {
        SpringApplication.run(NewBeeMallCloudUserServiceApplication.class, args);
    }
}
```

新建 ltd.user.cloud.newbee.controller 包，在该包下新建 NewBeeMallCloudAdminUserController 类，并添加一个测试接口，代码如下：

```java
package ltd.user.cloud.newbee.controller;

import org.springframework.web.bind.annotation.GetMapping;
import org.springframework.web.bind.annotation.PathVariable;
```

```
import org.springframework.web.bind.annotation.RestController;

@RestController
public class NewBeeMallCloudAdminUserController {

    @GetMapping("/users/admin/test/{userId}")
    public String test(@PathVariable("userId") int userId) {
        String userName = "user:" + userId;
        // 返回信息给调用端
        return userName;
    }
}
```

修改该项目的配置文件，application.properties 文件代码如下：

```
server.port=29000

server.servlet.encoding.charset=UTF-8
server.servlet.encoding.force=true
server.servlet.encoding.enabled=true

# 微服务名称
spring.application.name=newbee-mall-cloud-user-service
# Nacos 地址
spring.cloud.nacos.discovery.server-addr=localhost:8848
# Nacos 登录用户名(默认为 nacos，在生产环境中一定要修改)
spring.cloud.nacos.username=nacos
# Nacos 登录密码(默认为 nacos，在生产环境中一定要修改)
spring.cloud.nacos.password=nacos
```

这里主要配置微服务名称、启动端口号，修改服务中心的相关配置项。

2. newbee-mall-cloud-gateway-admin 项目编码

打开 newbee-mall-cloud-gateway-admin 项目，并新建 ltd.gateway.cloud.newbee 包。之后创建启动类 NewBeeMallCloudAdminGatewayApplication，代码如下：

```
@SpringBootApplication
@EnableDiscoveryClient
public class NewBeeMallCloudAdminGatewayApplication {
    public static void main(String[] args) {
        SpringApplication.run(NewBeeMallCloudAdminGatewayApplication.class, args);
    }
}
```

修改该项目的配置文件，application.properties 文件中的配置项如下：

```
server.port=29100

server.servlet.encoding.charset=UTF-8
server.servlet.encoding.force=true
server.servlet.encoding.enabled=true

# 微服务名称
spring.application.name=newbee-mall-cloud-gateway-admin
# Nacos 地址
spring.cloud.nacos.discovery.server-addr=localhost:8848
# Nacos 登录用户名(默认为 nacos，在生产环境中一定要修改)
spring.cloud.nacos.username=nacos
# Nacos 登录密码(默认为 nacos，在生产环境中一定要修改)
spring.cloud.nacos.password=nacos
# 网关开启微服务注册与微服务发现
spring.cloud.gateway.discovery.locator.enabled=true
spring.cloud.gateway.discovery.locator.lower-case-service-id=true

# 用户微服务的路由配置
spring.cloud.gateway.routes[0].id=user-service-route
spring.cloud.gateway.routes[0].uri=lb://newbee-mall-cloud-user-service
spring.cloud.gateway.routes[0].order=1
spring.cloud.gateway.routes[0].predicates[0]=Path=/users/admin/**
```

这里主要配置微服务名称、启动端口号、路由，修改服务中心的相关配置项。

上述操作完成后，项目的目录结构如图 2-1 所示。

3. 效果演示

接下来需要启动 Nacos Server，之后依次启动这两个项目。如果未能成功启动，那么开发者需要查看控制台中的日志是否报错，并及时确认问题和修复。启动成功后进入 Nacos 控制台，单击"服务管理"中的服务列表，就可以看到列表中已经存在这两个微服务的信息了。

打开浏览器并在地址栏中输入如下网址：http://localhost:29000/users/admin/test/13。

响应结果如图 2-2 所示。

第 2 章 实战项目基础构建及公用模块引入

```
∨ newbee-mall-cloud-dev-step01
    > .idea
    ∨ newbee-mall-cloud-gateway-admin
        ∨ src
            ∨ main
                ∨ java
                    ∨ ltd.gateway.cloud.newbee
                        © NewBeeMallCloudAdminGatewayApplication
                    ∨ resources
                        application.properties
        m pom.xml
    ∨ newbee-mall-cloud-user-service
        ∨ src
            ∨ main
                ∨ java
                    ∨ ltd.user.cloud.newbee
                        ∨ controller
                            © NewBeeMallCloudAdminUserController
                        © NewBeeMallCloudUserServiceApplication
                    ∨ resources
                        application.properties
        m pom.xml
    m pom.xml
```

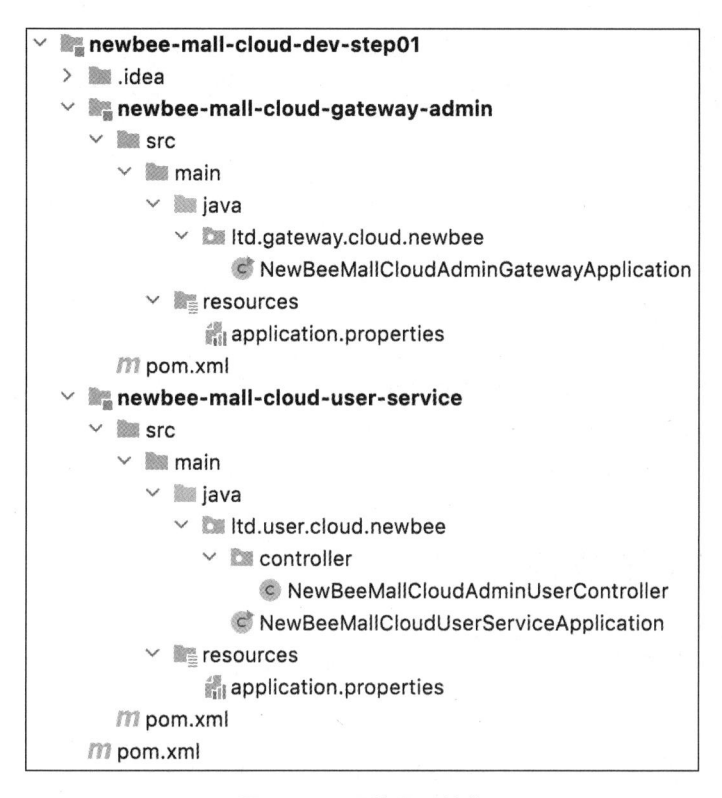

图 2-1　项目的目录结构

```
localhost:29000/users/admin/test/13
user:13
```

图 2-2　响应结果

验证网关层与用户微服务是否能够正常通信。打开浏览器并在地址栏中输入如下网址：http://localhost:29100/users/admin/test/13。

响应结果如图 2-3 所示。

```
localhost:29100/users/admin/test/13
user:13
```

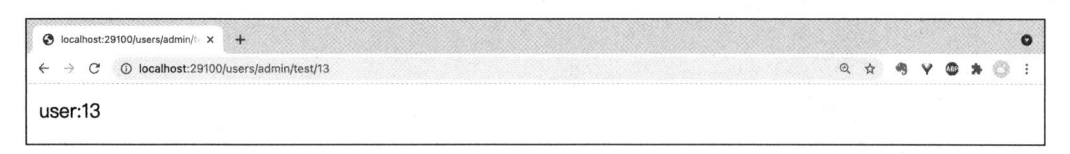

图 2-3　通过网关层访问的响应结果

希望读者能够根据笔者提供的开发步骤顺利地完成开发工作。

万事开头难。这句话对 newbee-mall-cloud 实战项目的开发同样适用。凭空开发一个微服务架构项目，对大部分开发人员来说都会有一定的难度，笔者刚开始做这个实战项目教程的时候同样有些束手束脚，会遇到如"先做哪一步呢？""先引入哪个组件呢？""先改造哪个微服务呢？""先配置什么东西呢？"之类的问题。

在开发前期肯定要保证项目结构、依赖关系都配置正确。同时，要保证微服务实例能够正常注册到服务中心，服务接口能够正常调用。当然，这里也加了网关模块，没有选择在开发后期引入网关服务，主要是为了在保证服务正常通信的同时，网关模块能够正常使用。虽然只是短短的一节内容，但是涉及的知识点是非常多的，如 Spring Cloud Alibaba 依赖配置、微服务实例与服务中心 Nacos 的通信、网关服务 Spring Cloud Gateway 的配置等。

2.3 用户微服务编码

下面改造用户微服务中的业务代码，把原单体项目 newbee-mall-api 中的功能模块一点一点地整合到这个工程里，开发出一个微服务架构项目。

2.3.1 引入业务依赖

原来的单体 API 项目中有 MyBatis、Swagger 等依赖项，在改造时需要一一引入微服务架构项目，在 pom.xml 主文件中增加依赖配置，代码如下：

```xml
<mybatis.starter.version>2.1.3</mybatis.starter.version>
<swagger.version>3.0.0</swagger.version>
<lombok.version>1.18.8</lombok.version>

<dependency>
    <groupId>org.mybatis.spring.boot</groupId>
    <artifactId>mybatis-spring-boot-starter</artifactId>
    <version>${mybatis.starter.version}</version>
</dependency>

<dependency>
    <groupId>io.springfox</groupId>
    <artifactId>springfox-boot-starter</artifactId>
```

```xml
    <version>${swagger.version}</version>
</dependency>

<dependency>
    <groupId>org.projectlombok</groupId>
    <artifactId>lombok</artifactId>
    <version>${lombok.version}</version>
    <scope>provided</scope>
</dependency>
```

打开用户微服务 newbee-mall-cloud-user-service 的工程目录,在 pom.xml 子文件中增加依赖项,代码如下:

```xml
<dependency>
    <groupId>org.springframework.boot</groupId>
    <artifactId>spring-boot-starter-validation</artifactId>
</dependency>

<dependency>
    <groupId>org.mybatis.spring.boot</groupId>
    <artifactId>mybatis-spring-boot-starter</artifactId>
</dependency>

<dependency>
    <groupId>org.projectlombok</groupId>
    <artifactId>lombok</artifactId>
    <scope>provided</scope>
</dependency>

<dependency>
    <groupId>io.springfox</groupId>
    <artifactId>springfox-boot-starter</artifactId>
</dependency>

<dependency>
    <groupId>mysql</groupId>
    <artifactId>mysql-connector-java</artifactId>
    <scope>runtime</scope>
</dependency>
```

项目所需的 Maven 依赖配置完成。

2.3.2 商城用户模块中的接口改造

打开用户微服务 newbee-mall-cloud-user-service 的工程目录，在 ltd.user.cloud.newbee 包下依次创建 config 包、dao 包、entity 包、service 包和 util 包，在 resources 目录下新增 mapper 文件夹用于存放 Mapper 文件，将原单体 API 项目中与管理员用户相关的业务代码和 Mapper 文件依次复制进来，完成后还需要在主类上添加 MyBatis 的扫描注解，让这些 Mapper 文件能够被正确地扫描和加载。

由于代码量较大，这里就不一一介绍和讲解了，读者按照对应的文件目录将代码从单体项目复制过来即可。

修改 newbee-mall-cloud-user-service 工程的 application.properties 配置文件，添加数据库连接及 MyBatis 扫描配置，代码如下：

```
# datasource config (MySQL)
spring.datasource.name=newbee-mall-cloud-user-datasource
spring.datasource.driverClassName=com.mysql.cj.jdbc.Driver
spring.datasource.url=jdbc:mysql://localhost:3306/newbee_mall_cloud_user_db?useUnicode=true&serverTimezone=Asia/Shanghai&characterEncoding=utf8&autoReconnect=true&useSSL=false&allowMultiQueries=true
spring.datasource.username=root
spring.datasource.password=123456
spring.datasource.hikari.minimum-idle=5
spring.datasource.hikari.maximum-pool-size=15
spring.datasource.hikari.auto-commit=true
spring.datasource.hikari.idle-timeout=60000
spring.datasource.hikari.pool-name=hikariCP
spring.datasource.hikari.max-lifetime=600000
spring.datasource.hikari.connection-timeout=30000
spring.datasource.hikari.connection-test-query=SELECT 1

# mybatis config
mybatis.mapper-locations=classpath:mapper/*Mapper.xml
```

上述工作完成后，目录结构如图 2-4 所示。

本步骤中的源代码，涉及的数据库为 newbee_mall_cloud_user_db，数据库表为 tb_newbee_mall_admin_user 和 tb_newbee_mall_admin_user_token，在开发过程中均会用到。

第 2 章 实战项目基础构建及公用模块引入

```
> newbee-mall-cloud-step02 [newbee-mall-cloud]
  > .idea
  > newbee-mall-cloud-gateway-admin
    > src
      > main
        > java
          > ltd.gateway.cloud.newbee
            NewBeeMallCloudAdminGatewayApplication
        > resources
    pom.xml
  > newbee-mall-cloud-user-service
    > src
      > main
        > java
          > ltd.user.cloud.newbee
            > config
              > annotation
              > handler
                AdminUserServiceExceptionHandler
                AdminUserSwagger3Config
                AdminUserWebMvcConfigurer
            > controller
              > param
              NewBeeMallCloudAdminUserController
            > dao
              AdminUserMapper
              NewBeeAdminUserTokenMapper
            > entity
            > service
              > impl
                AdminUserServiceImpl
              AdminUserService
            > util
          NewBeeMallCloudUserServiceApplication
        > resources
    pom.xml
  pom.xml
```

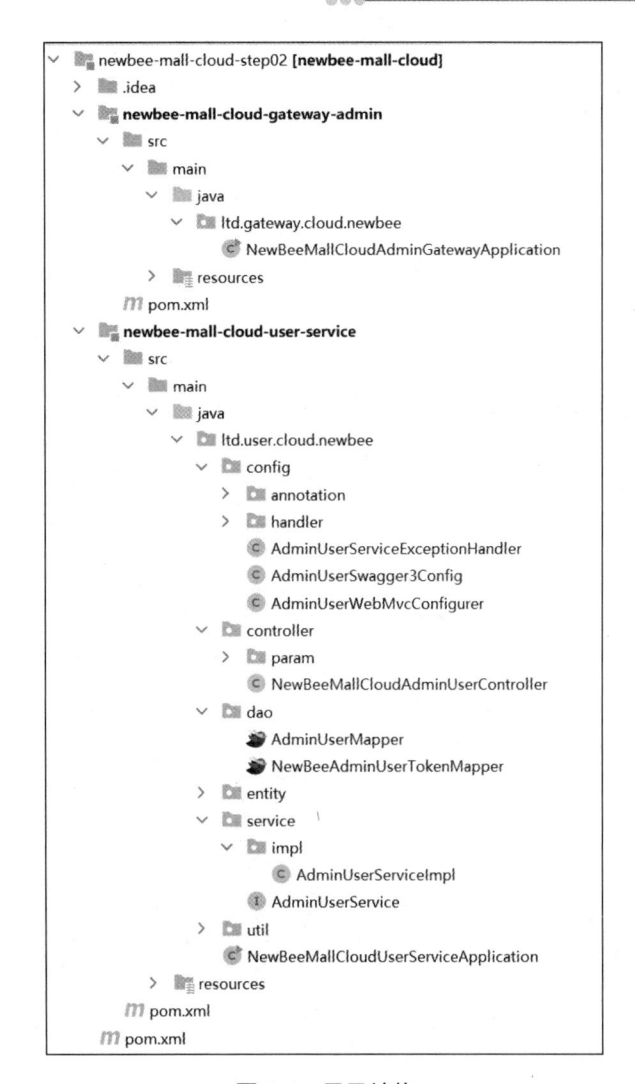

图 2-4　目录结构

建表语句如下：

```
CREATE DATABASE /*!32312 IF NOT EXISTS*/'newbee_mall_cloud_user_db' /*!40100
DEFAULT CHARACTER SET utf8 */;

USE 'newbee_mall_cloud_user_db';

DROP TABLE IF EXISTS 'tb_newbee_mall_admin_user';

# 创建管理员用户表
```

```sql
CREATE TABLE 'tb_newbee_mall_admin_user' (
    'admin_user_id' bigint(20) NOT NULL AUTO_INCREMENT COMMENT '管理员用户id',
    'login_user_name' varchar(50) NOT NULL COMMENT '管理员用户登录名称',
    'login_password' varchar(50) NOT NULL COMMENT '管理员用户登录密码',
    'nick_name' varchar(50) NOT NULL COMMENT '管理员用户显示昵称',
    'locked' tinyint(4) DEFAULT '0' COMMENT '是否锁定：0 未锁定，1 已锁定无法登录',
    PRIMARY KEY ('admin_user_id') USING BTREE
) ENGINE=InnoDB DEFAULT CHARSET=utf8 ROW_FORMAT=DYNAMIC;

INSERT INTO 'tb_newbee_mall_admin_user' ('admin_user_id', 'login_user_name', 'login_password', 'nick_name', 'locked')
VALUES
(1,'admin','e10adc3949ba59abbe56e057f20f883e','十三',0),
(2,'newbee-admin1','e10adc3949ba59abbe56e057f20f883e','新蜂01',0),
(3,'newbee-admin2','e10adc3949ba59abbe56e057f20f883e','新蜂02',0);

DROP TABLE IF EXISTS 'tb_newbee_mall_admin_user_token';

CREATE TABLE 'tb_newbee_mall_admin_user_token' (
  'admin_user_id' bigint(20) NOT NULL COMMENT '用户主键id',
  'token' varchar(32) NOT NULL COMMENT 'token值(32位字符串)',
  'update_time' datetime NOT NULL DEFAULT CURRENT_TIMESTAMP COMMENT '修改时间',
  'expire_time' datetime NOT NULL DEFAULT CURRENT_TIMESTAMP COMMENT 'token过期时间',
  PRIMARY KEY ('admin_user_id'),
  UNIQUE KEY 'uq_token' ('token')
) ENGINE=InnoDB DEFAULT CHARSET=utf8;
```

2.3.3 用户微服务改造过程中遇到的问题

笔者在用户微服务改造过程中遇到了一个小问题，在这里分享给各位读者。如果读者在实战时遇到这个问题，可以直接参考解决办法。

在把管理员模块的代码按照对应的目录复制到 newbee-mall-cloud-user-service 工程

后，无法正常启动项目，而是报告异常，报错日志如下：

```
org.springframework.context.ApplicationContextException: Failed to start
bean 'documentationPluginsBootstrapper'; nested exception is
java.lang.NullPointerException
```

这个问题是 Spring Boot 版本与 Swagger 版本不兼容导致的。因为当前工程中使用的是 Spring Boot 2.6.3，开发时引入的是 Swagger 2.8.0，所以报告异常。将 Swagger 2.8.0 改成 Swagger 3.0.0 之后就一切正常了。

当然，Swagger 3.0.0 与 Swagger 2.8.0 在整合时的区别还是挺大的，依赖项和配置类的写法、默认接口的访问地址都有变化，这里需要注意，在实现编码时直接参考或使用笔者提供的源代码即可。

编码完成后，准备好数据库和表就可以进行功能测试了。当然，在项目启动前需要启动 Nacos Server，之后依次启动这两个项目。启动成功后，打开浏览器并在地址栏中输入如下网址：http://localhost:29000/swagger-ui/index.html。

响应结果如图 2-5 所示。这样，读者就可以在 Swagger 提供的 UI 页面进行接口测试了。

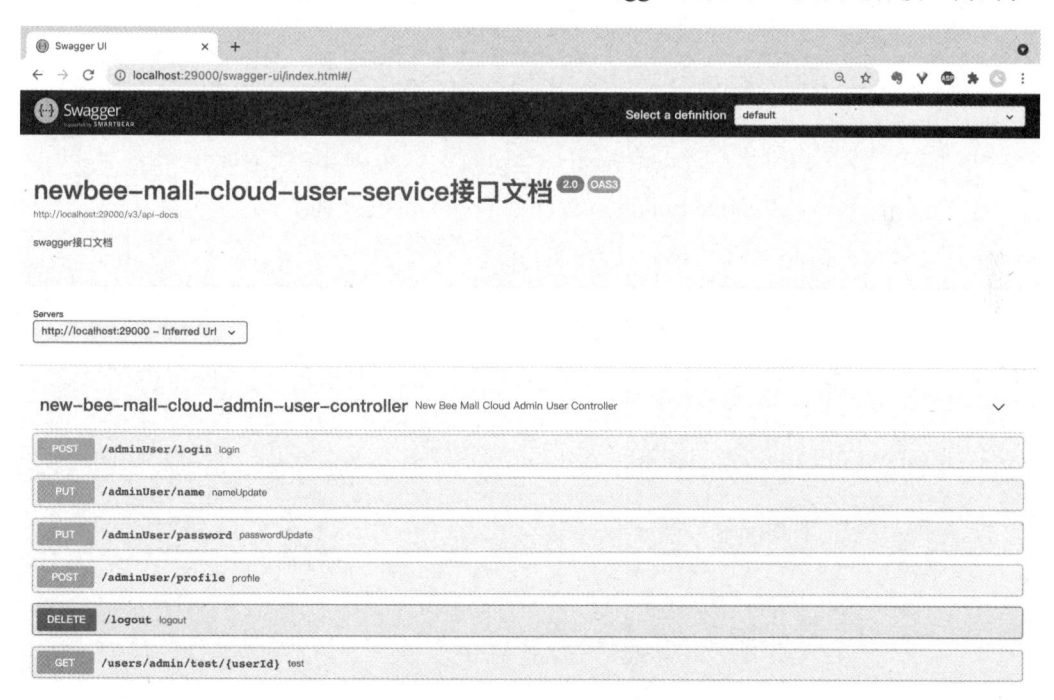

图 2-5　用户微服务接口文档的响应结果

2.4 引入公用模块

在前文中,笔者根据开发步骤整理了两份源代码文件,分别是 newbee-mall-cloud-dev-step01 和 newbee-mall-cloud-dev-step02,本节的源代码是在 newbee-mall-cloud-dev-step02 工程的基础上改造的,将工程命名为 newbee-mall-cloud-dev-step03。

随着工程中模块的增加,有些类重复出现在各个模块中。对于这些公用的类,大部分模块都会使用,但是在开发时不可能在每个模块中都保留这些重复的内容。因此,在实际开发时会创建一个公用模块用于存放这些公用的类,其他的业务模块将其作为依赖项引入。公用模块主要用于存放一些公用的类,如分页工具类、全局响应结果类、工具类等。

在工程中新增一个 newbee-mall-cloud-common 模块,并在 pom.xml 主文件中增加该模块的配置,代码如下:

```xml
<modules>
  <module>newbee-mall-cloud-user-service</module>
  <module>newbee-mall-cloud-gateway-admin</module>
  <!-- 新增common模块 -->
  <module>newbee-mall-cloud-common</module>
</modules>
```

子节点 newbee-mall-cloud-common 模块的 pom.xml 文件如下:

```xml
<?xml version="1.0" encoding="UTF-8"?>
<project xmlns="http://maven.apa***.org/POM/4.0.0" xmlns:xsi=
"http://www.w*.org/2001/XMLSchema-instance"
        xsi:schemaLocation="http://maven.apa***.org/POM/4.0.0
https://maven.apa***.org/xsd/maven-4.0.0.xsd">
    <modelVersion>4.0.0</modelVersion>
    <groupId>ltd.newbee.cloud</groupId>
    <artifactId>newbee-mall-cloud-common</artifactId>
    <version>0.0.1-SNAPSHOT</version>
    <packaging>jar</packaging>
    <name>newbee-mall-cloud-common</name>
    <description>公用模块</description>

    <parent>
        <groupId>ltd.newbee.cloud</groupId>
        <artifactId>newbee-mall-cloud</artifactId>
        <version>0.0.1-SNAPSHOT</version>
```

```xml
    </parent>

    <properties>
        <java.version>1.8</java.version>
    </properties>

    <dependencies>
        <dependency>
            <groupId>org.projectlombok</groupId>
            <artifactId>lombok</artifactId>
            <scope>provided</scope>
        </dependency>

        <dependency>
            <groupId>io.swagger</groupId>
            <artifactId>swagger-annotations</artifactId>
            <version>1.5.14</version>
            <scope>compile</scope>
        </dependency>
        <dependency>
            <groupId>org.springframework</groupId>
            <artifactId>spring-beans</artifactId>
        </dependency>
    </dependencies>
</project>
```

主要的配置项是模块名称、模块的打包方式。配置成功后，打开 newbee-mall-cloud-common 工程目录，新建 ltd.common.cloud.newbee 包，接着分别创建 dto 包、exception 包、util 包，并把 newbee-mall-cloud-step02 中用户微服务的一些公用类移动到公用模块中，编码完成后的目录结构如图 2-6 所示。

在后续实战过程中，还要创建其他微服务，使用公用模块就可以直接在 pom.xml 文件中引入该依赖配置。比如，当前工程中的用户微服务已经把公用的类都移动到公用模块中了，如果想要使用这些类，就可以在 pom.xml 文件中增加如下依赖配置：

```xml
<dependency>
  <groupId>ltd.newbee.cloud</groupId>
  <artifactId>newbee-mall-cloud-common</artifactId>
  <version>0.0.1-SNAPSHOT</version>
</dependency>
```

```
  v  newbee-mall-cloud-common
     v  src
        v  main
           v  java
              v  ltd.common.cloud.newbee
                 v  dto
                      © PageQueryUtil
                      © PageResult
                      © Result
                      © ResultGenerator
                 v  exception
                      ⚡ NewBeeMallException
                 v  util
                      © BeanUtil
                      © MD5Util
                      © NewBeeMallUtils
                      © NumberUtil
                      © SystemUtil
     newbee-mall-cloud-common.iml
     m pom.xml
  >  newbee-mall-cloud-gateway-admin
  >  newbee-mall-cloud-user-service
     newbee-mall-cloud.iml
     m pom.xml
```

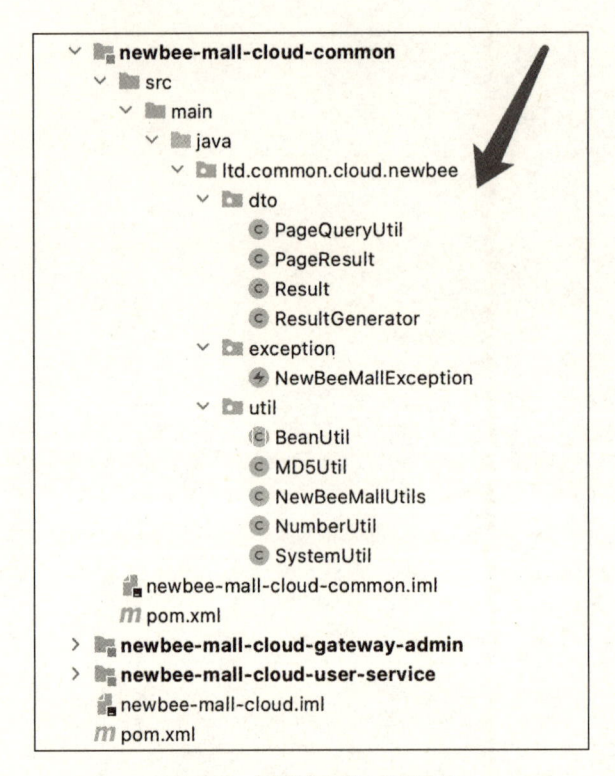

图 2-6 公用模块的目录结构

2.5 用户微服务模块改造

在 newbee-mall-cloud-step02 工程中，用户微服务是一个单独的模块。而在最终的微服务架构项目中，用户微服务模块的结构如下：

```
newbee-mall-cloud-user-service            // 用户微服务
    ├── newbee-mall-cloud-user-api        // 存放商城用户模块中暴露的用于远程调用的
                                              FeignClient 类
    └── newbee-mall-cloud-user-web        // 商城用户模块 API 的代码及逻辑
```

用户微服务所在的模块包含 user-web 和 user-api 两个模块，它们各自的作用都标注出来了。

这个做法其实很好理解，微服务架构项目中存在不同的微服务实例间的远程通信过程，在当前的实战项目中是通过 OpenFeign 来实现的，因此笔者就对用户微服务做了一次结构整理，单独创建了一个 newbee-mall-cloud-user-api 模块，用于存放商城用户模块中暴露的用于远程调用的 FeignClient 类。这样，如果有哪个微服务需要调用用户微服务中的一些接口，直接将 newbee-mall-cloud-user-api 添加到其依赖项中就可以使用 FeignClient 类中的方法来调用了。不止是用户微服务，其他业务微服务的模块结构也如此。

另外，newbee-mall-cloud 项目中存放 FeignClient 类的模块，其命名方式都是×××-api。这个命名方式参考了 Dubbo 项目中的命名方式，其实也可以定义为 newbee-mall-cloud-user-feign、newbee-mall-cloud-user-feignclient 等。

接下来，就是实际的改造过程，在 newbee-mall-cloud-user-service 模块下新增 newbee-mall-cloud-user-api 和 newbee-mall-cloud-user-web 两个模块。修改各模块下的 pom.xml 文件，梳理三个模块间的关联关系。

newbee-mall-cloud-user-service 模块的 pom.xml 文件代码修改如下：

```xml
<?xml version="1.0" encoding="UTF-8"?>
<project xmlns="http://maven.apa***.org/POM/4.0.0" xmlns:xsi=
"http://www.w*.org/2001/XMLSchema-instance"
         xsi:schemaLocation="http://maven.apa***.org/POM/4.0.0
https://maven.apa***.org/xsd/maven-4.0.0.xsd">
    <modelVersion>4.0.0</modelVersion>
    <groupId>ltd.newbee.cloud</groupId>
    <artifactId>newbee-mall-cloud-user-service</artifactId>
    <version>0.0.1-SNAPSHOT</version>
    <packaging>pom</packaging>
    <name>newbee-mall-cloud-user-service</name>
    <description>商城用户模块</description>

    <parent>
        <groupId>ltd.newbee.cloud</groupId>
        <artifactId>newbee-mall-cloud</artifactId>
        <version>0.0.1-SNAPSHOT</version>
    </parent>

    <properties>
        <java.version>1.8</java.version>
    </properties>

    <modules>
        <module>newbee-mall-cloud-user-web</module>
        <module>newbee-mall-cloud-user-api</module>
    </modules>

    <dependencies>
```

```
        </dependencies>
</project>
```

因为业务依赖和相关的业务代码都需要移到 newbee-mall-cloud-user-web 模块中，所以这里只定义模块间的关系即可，将该模块打包方式 packaging 配置项的值改为 pom，父模块是 newbee-mall-cloud，两个子模块分别是 newbee-mall-cloud-user-api 和 newbee-mall-cloud-user-web。

newbee-mall-cloud-user-api 模块的 pom.xml 文件代码修改如下：

```xml
<?xml version="1.0" encoding="UTF-8"?>
<project xmlns="http://maven.apa***.org/POM/4.0.0" xmlns:xsi=
"http://www.w*.org/2001/XMLSchema-instance"
         xsi:schemaLocation="http://maven.apa***.org/POM/4.0.0
https://maven.apa***.org/xsd/maven-4.0.0.xsd">
    <modelVersion>4.0.0</modelVersion>
    <groupId>ltd.user.newbee.cloud</groupId>
    <artifactId>newbee-mall-cloud-user-api</artifactId>
    <packaging>jar</packaging>
    <version>0.0.1-SNAPSHOT</version>
    <name>newbee-mall-cloud-user-api</name>
    <description>用户微服务 openfeign</description>

    <parent>
        <groupId>ltd.newbee.cloud</groupId>
        <artifactId>newbee-mall-cloud-user-service</artifactId>
        <version>0.0.1-SNAPSHOT</version>
    </parent>

    <properties>
        <java.version>1.8</java.version>
    </properties>

    <dependencies>

        <dependency>
            <groupId>org.springframework.cloud</groupId>
            <artifactId>spring-cloud-starter-openfeign</artifactId>
        </dependency>

    </dependencies>
</project>
```

这里定义了与父模块 newbee-mall-cloud-user-service 的关系，打包方式 packaging 配置项的值为 jar。同时，由于后面要添加 FeignClient，因此这里引入了 OpenFeign 的依赖项。

newbee-mall-cloud-user-web 模块的 pom.xml 文件代码修改如下：

```xml
<?xml version="1.0" encoding="UTF-8"?>
<project xmlns="http://maven.apa***.org/POM/4.0.0" xmlns:xsi=
"http://www.w*.org/2001/XMLSchema-instance"
         xsi:schemaLocation="http://maven.apa***.org/POM/4.0.0
https://maven.apa***.org/xsd/maven-4.0.0.xsd">
    <modelVersion>4.0.0</modelVersion>
    <groupId>ltd.user.newbee.cloud</groupId>
    <artifactId>newbee-mall-cloud-user-web</artifactId>
    <version>0.0.1-SNAPSHOT</version>
    <name>newbee-mall-cloud-user-web</name>
    <description>用户微服务</description>

    <parent>
        <groupId>ltd.newbee.cloud</groupId>
        <artifactId>newbee-mall-cloud-user-service</artifactId>
        <version>0.0.1-SNAPSHOT</version>
    </parent>

    <properties>
        <java.version>1.8</java.version>
    </properties>

    <dependencies>
        <dependency>
            <groupId>org.springframework.boot</groupId>
            <artifactId>spring-boot-starter-web</artifactId>
        </dependency>

        <dependency>
            <groupId>org.springframework.boot</groupId>
            <artifactId>spring-boot-starter-test</artifactId>
            <scope>test</scope>
        </dependency>

        <dependency>
            <groupId>com.alibaba.cloud</groupId>
            <artifactId>spring-cloud-starter-alibaba-nacos-discovery</artifactId>
        </dependency>
```

```xml
<dependency>
    <groupId>org.springframework.boot</groupId>
    <artifactId>spring-boot-starter-validation</artifactId>
</dependency>

<dependency>
    <groupId>org.mybatis.spring.boot</groupId>
    <artifactId>mybatis-spring-boot-starter</artifactId>
</dependency>

<dependency>
    <groupId>org.projectlombok</groupId>
    <artifactId>lombok</artifactId>
    <scope>provided</scope>
</dependency>

<dependency>
    <groupId>io.springfox</groupId>
    <artifactId>springfox-boot-starter</artifactId>
</dependency>

<dependency>
    <groupId>mysql</groupId>
    <artifactId>mysql-connector-java</artifactId>
    <scope>runtime</scope>
</dependency>

<dependency>
    <groupId>ltd.newbee.cloud</groupId>
    <artifactId>newbee-mall-cloud-common</artifactId>
    <version>0.0.1-SNAPSHOT</version>
</dependency>

    </dependencies>
</project>
```

这里定义了与父模块 newbee-mall-cloud-user-service 的关系，同时将相关的业务依赖项也移到配置文件中。

接下来就是把 newbee-mall-cloud-step02 工程中用户微服务的业务代码和配置项移到 newbee-mall-cloud-user-web 模块中，这个过程比较简单，复制并粘贴即可。改造后的商城用户模块的目录结构如图 2-7 所示。

```
> newbee-mall-cloud-common
> newbee-mall-cloud-gateway-admin
v    newbee-mall-cloud-user-service
  v    newbee-mall-cloud-user-api
    v    src
      v    main
        >    java
         newbee-mall-cloud-user-api.iml
       m pom.xml
  v    newbee-mall-cloud-user-web
    v    src
      v    main
        >    java
        >    resources
         newbee-mall-cloud-user-web.iml
       m pom.xml
     newbee-mall-cloud-user-service.iml
   m pom.xml
 newbee-mall-cloud.iml
m pom.xml
```

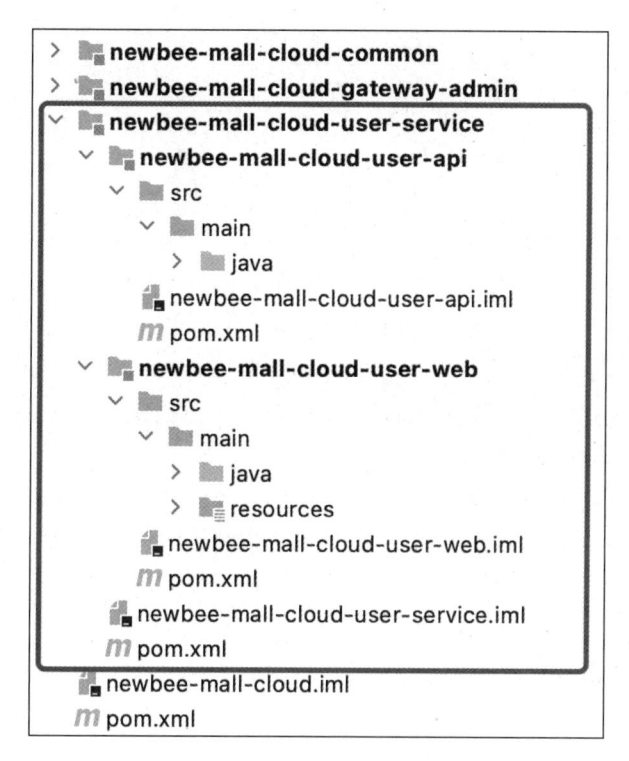

图 2-7　改造后的商城用户模块的目录结构

商城用户模块改造完成后，测试步骤是不能漏掉的。一定要验证项目是否能正常启动、接口是否能正常调用，防止在代码移动过程中出现的一些小问题导致项目无法启动或代码报错。

2.6　OpenFeign编码暴露远程接口

下面在 newbee-mall-cloud-user-api 模块中新增需要暴露的接口 FeignClient。

新建 ltd.user.cloud.newbee.openfeign 包，之后在该包下新增 NewBeeCloudAdminUserServiceFeign 类，用于创建对管理员用户相关接口的 Feign 调用。

NewBeeCloudAdminUserServiceFeign 类代码如下：

```
package ltd.user.cloud.newbee.openfeign;

import org.springframework.cloud.openfeign.FeignClient;
import org.springframework.web.bind.annotation.GetMapping;
import org.springframework.web.bind.annotation.PathVariable;
```

```java
@FeignClient(value = "newbee-mall-cloud-user-service", path = "/users")
public interface NewBeeCloudAdminUserServiceFeign {

    @GetMapping(value = "/admin/{token}")
    String getAdminUserByToken(@PathVariable(value = "token") String token);
}
```

这个方法的作用就是根据 token 值来获取管理员用户数据，主要用作鉴权。调用的接口是 newbee-mall-cloud-user-web 模块下 NewBeeMallCloudAdminUserController 类中的 getAdminUserByToken() 方法。如果其他微服务需要与用户微服务进行远程通信获取管理员用户数据，就可以引入 newbee-mall-cloud-user-api 模块作为依赖，并且直接调用 getAdminUserByToken() 方法。

另外，上述代码只是测试写法，后续会根据实际的接口设计进行修改。这里主要为了讲解 newbee-mall-cloud-user-service 模块的目录结构设计，以及该模块下的两个子模块 newbee-mall-cloud-user-api 和 newbee-mall-cloud-user-web 的作用与目录规范。

2.7 远程调用OpenFeign应该如何设置

这是一个拓展性的问题：微服务架构项目中的远程调用 FeignClent 该写在哪里？远程调用 Feign 包该建在哪个模块里？

当然，这里只是笔者根据过往的开发经验来总结的，并不是标准答案，只是作为一个引申的知识点介绍给各位读者。

比如，订单微服务需要调用用户微服务下的/getUserDetail/{userId}接口来获取用户信息。这个接口的功能逻辑已经在用户微服务中实现且功能正常，如果想要通过 OpenFeign 远程调用，就需要实现一个 FeignClient 类，在订单微服务里引入 FeignClient 类并开启 OpenFeign 调用就完成了。这个 FeignClient 类可以在订单微服务工程中编写，也可以在用户微服务工程中编写，甚至有些团队会单独创建一个 Feign 模块用于存放所有的 FeignClient 类。毕竟，只要引入 OpenFeign 的依赖 JAR 包就可以编写 FeignClient 类，不需要与业务代码耦合。那么，也就引出了一个问题：FeignClient 是写在调用端（如案例中的订单微服务）、被调用端（如案例中的用户微服务），还是写在单独创建的 Feign 模块里呢？

newbee-mall-cloud 项目中的目录设计是写在被调用端的，各个微服务下都单独创建了对应的×××-api 模块用于存放 FeignClient 类，如果有其他微服务需要远程调用，直接引入对应的×××-api 依赖，之后就能够发起远程调用了。如此设计的原因如下。

在正常的企业开发模式下，微服务的细节是不对外公开的，如果 A 业务组想要 B 业务组提供一个远程调用接口，那么直接让 B 业务组编码即可，因为 A 业务组的开发人员不熟悉 B 业务组的代码，在这种情况下让 B 业务组来提供更合理。不仅如此，这种方式更加灵活和安全，B 业务组想要暴露哪些业务接口，自行实现并暴露即可，不想暴露的就不用写在 Feign 包里了。比如，用户微服务下有 100 个接口，需要暴露的接口有 3 个，开发团队只需要在 FeignClient 类中暴露这 3 个接口即可。

如果写在 A 业务组的微服务里，那么 A 业务组的开发人员可能还需要查看由 B 业务组负责的微服务里有哪些接口，之后再去写 FeignClient 类，这种做法是不合理的，增加了上游微服务开发团队的工作量，并且如果每个开发团队都自行实现 FeignClient 类，代码就重复了。所以，由被调用方来做这个工作更好一点。

当然，所有的 FeignClient 类都被统一放在一个单独的模块下，也是一种实现方式。比如，在 newbee-mall-cloud 项目中，可以新建一个 newbee-mall-cloud-feign 模块，当前的所有×××-api 模块下的 FeignClient 类都可以统一放在 newbee-mall-cloud-feign 模块中，在需要远程调用时直接引入 newbee-mall-cloud-feign 模块即可。

项目骨架搭建完成后又引入了公用模块，并介绍了项目的目录规范，之后引入服务通信 OpenFeign 进行简单的编码，后续实战章节中的目录结构基本都是按照这个规范来做的，希望读者能够根据笔者提供的开发步骤顺利地完成本章的项目改造。

第 3 章 用户微服务编码实践及功能讲解

除极少数功能不需要用户登录才能使用外,实战项目中的大部分接口都需要用户在登录状态下才能正常使用。前面章节中已经把管理员用户的源代码整理了进来,不过对于源代码的细节并未介绍。用户登录和身份鉴权是必不可少的内容,因此本章会结合 newbee-mall-cloud-dev-step03 这份源代码重点讲解这两个功能的源代码实现,当读者明白了其中的实现细节后,再进行微服务架构下的改造及网关层鉴权的编码就更加得心应手。

3.1 登录流程介绍

3.1.1 什么是登录

在互联网上,供多人使用的网站或应用系统会为每位用户都配置一套独特的用户名和密码,用户可以使用自己的用户名和密码进入系统,以便系统识别该用户的身份,从而保存该用户的使用习惯、使用数据。用户使用这套用户名和密码进入系统,以及系统验证进入成功或失败的过程,被称为登录。

在登录成功后,用户就可以合法地使用该账号具有的各项功能。例如,淘宝网用户可以正常浏览商品和购买商品等;论坛用户可以查看或更改资料、发表和回复帖子等;OA(办公自动化)等系统管理员用户可以正常地处理各种数据和信息。从最简单的角度来说,登录就是输入用户名和密码进入一个系统进行访问和操作。

3.1.2 用户登录状态

客户端（通常是浏览器）在连上 Web 服务器后，要想获得 Web 服务器中的各种资源，就需要遵守一定的通信规则。Web 项目通常使用 HTTP，它用于定义客户端与 Web 服务器通信的方式。HTTP 是无状态的协议，也就是说，这个协议是无法记录用户访问状态的，其每次请求都是独立的、没有任何关联的。

以新蜂商城（本书实战项目）的后台管理系统为例，它拥有多个页面。在页面跳转过程中和通过接口进行数据交互时，系统需要知道用户的状态，尤其是用户登录的状态，以便服务器验证用户状态是否正常。这样系统才能判断是否可以让当前用户使用某些功能或获取某些数据。

这时，就需要在每个页面上都对用户的身份进行验证和确认，但现实情况却不是如此。一个网站不可能让用户在每个页面上都输入用户名和密码，这是一个违反操作逻辑的设计，也没有用户愿意使用这样的系统。

因此，在设计登录流程时，只让用户进行一次登录操作。为了实现这个功能，需要一些辅助技术，用得最多的技术就是浏览器的 Cookie。而在 Java Web 开发中，比较常见的是使用 Session 技术来实现。将用户登录的信息存放在 Cookie 或 Session 中，这样就可以通过读取在 Cookie 或 Session 中的用户登录信息，达到记录用户状态、验证用户状态的目的。

3.1.3 登录流程设计

登录的本质是身份验证和登录状态的保持，在实际编码中是如何实现的呢？

首先，在数据库中查询这条用户记录，伪代码如下：

```
select * from xxx_user where account_number = 'xxxx';
```

如果不存在这条记录，则表示身份验证失败，登录流程终止；如果存在这条记录，则表示身份验证成功。

其次，进行登录状态的存储和验证，存储的伪代码如下：

```
//通过 Cookie 存储
Cookie cookie = new Cookie("userName",xxxxx);

//通过 Session 存储
session.setAttribute("userName",xxxxx);
```

验证逻辑的伪代码如下：

```
//通过 Cookie 获取需要验证的数据并进行校验
Cookie cookies[] = request.getCookies();
if (cookies != null){
    for (int i = 0; i < cookies.length; i++)
        {
            Cookie cookie = cookies[i];
            if (name.equals(cookie.getName()))
            {
                return cookie;
            }
        }
}

//通过 Session 获取需要验证的数据并进行校验
session.getAttribute("userName");
```

登录的本质是身份验证和用户状态的保持，还有一点不能忽略，就是登录功能的安全验证设计。一般的做法是将密码加密存储，不过千万不要在 Cookie 中存放用户密码，加密的密码也不行。因为这个密码可以被人获取并尝试离线穷举。同样，有些网站会在 Cookie 中存储一些用户的敏感信息，这些都是不安全的行为。

在本书的实战项目中，通过生成用户令牌 token 的形式进行用户状态的保持和身份验证。简单理解，这里所说的 token 就是后端生成的一个字符串，该字符串与用户信息关联，token 字符串通过一些无状态的数据生成，并不包含用户的敏感信息。

简易的登录流程如图 3-1 所示。

当然，还有一些验证操作是必需的。比如，前端在发送数据时需要验证数据格式及有效性，后端接口在访问之前也需要验证用户信息是否有效，因此完整的登录流程如图 3-2 所示。

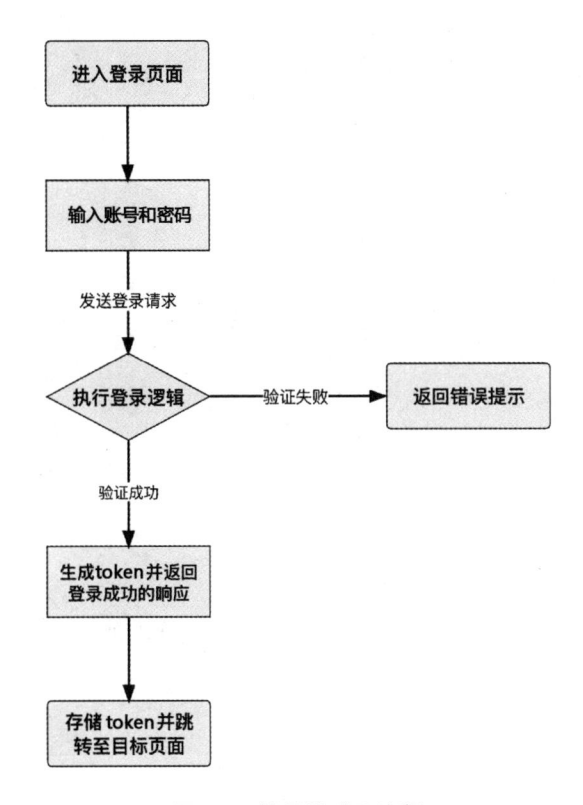

图 3-1　简易的登录流程

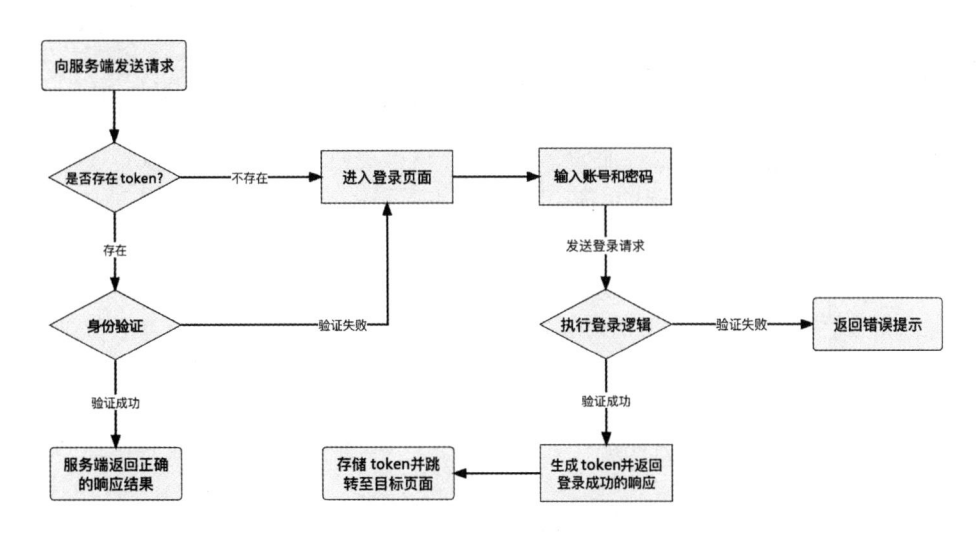

图 3-2　完整的登录流程

3.2 登录功能的源代码介绍

由于当前已经完成了商城后台管理系统管理员用户的相关功能,因此这里以管理员用户功能模块下的登录功能进行源代码讲解。

登录接口的代码如下:

```
@RequestMapping(value = "/users/admin/login", method = RequestMethod.POST)
public Result<String> login(@RequestBody @Valid AdminLoginParam
adminLoginParam) {
  String loginResult = adminUserService.login(adminLoginParam.getUserName(),
adminLoginParam.getPasswordMd5());
  logger.info("manage login api,adminName={},loginResult={}",
adminLoginParam.getUserName(), loginResult);

  //登录成功
  if (StringUtils.hasText(loginResult) && loginResult.length() == 32) {
    Result result = ResultGenerator.genSuccessResult();
    result.setData(loginResult);
    return result;
  }
  //登录失败
  return ResultGenerator.genFailResult(loginResult);
}
```

源代码在 newbee-mall-cloud-user-web 模块下的 ltd.user.cloud.newbee.controller.NewBeeMallCloudAdminUserController 类中。

在实现用户登录功能时,前端需要向后端传输两个参数:登录名和密码。在设计登录接口时通常会使用 POST 方法来处理,这两个参数会被封装成一个对象传递给后端接口。因此,项目中定义了 AdminLoginParam 类来接收登录参数。

在登录接口定义中,@RequestMapping(value = "/users/admin/login", method = RequestMethod.POST)表示登录请求为 POST 方式,请求路径为/users/admin/login。使用@RequestBody 注解对登录参数进行接收并封装成 AdminLoginParam 对象用于业务层的逻辑处理。@Valid 注解的作用为参数验证,在定义登录参数对象时使用了@NotEmpty 注解,表示该参数不能为空,如果在这里不添加@Valid 注解,则非空验证不会执行。之后就是调用业务层的 login()方法进行登录逻辑的处理,根据业务层返回的内容封装请求结果并响应给前端。

登录功能的业务层代码如下：

```java
public String login(String userName, String password) {
  AdminUser loginAdminUser = adminUserMapper.login(userName, password);
  if (loginAdminUser != null) {
    //登录后即执行修改token值的操作
    String token = getNewToken(System.currentTimeMillis() + "",
loginAdminUser.getAdminUserId());
    AdminUserToken adminUserToken = newBeeAdminUserTokenMapper.
selectByPrimaryKey(loginAdminUser.getAdminUserId());
    //当前时间
    Date now = new Date();
    //过期时间
    Date expireTime = new Date(now.getTime() + 2 * 24 * 3600 * 1000);//过期
时间为48小时
    if (adminUserToken == null) {
      adminUserToken = new AdminUserToken();
      adminUserToken.setAdminUserId(loginAdminUser.getAdminUserId());
      adminUserToken.setToken(token);
      adminUserToken.setUpdateTime(now);
      adminUserToken.setExpireTime(expireTime);
      //新增一个token值
      if (newBeeAdminUserTokenMapper.insertSelective(adminUserToken) > 0) {
        //新增成功后返回
        return token;
      }
    } else {
      adminUserToken.setToken(token);
      adminUserToken.setUpdateTime(now);
      adminUserToken.setExpireTime(expireTime);
      //更新
      if (newBeeAdminUserTokenMapper.updateByPrimaryKeySelective
(adminUserToken) > 0) {
        //修改成功后返回
        return token;
      }
    }
  }
  return "登录失败";
}
```

源代码在 newbee-mall-cloud-user-web 模块下的 ltd.user.cloud.newbee.service.impl.AdminUserServiceImpl 类中。

管理员用户登录的方法中共有 30 行左右代码，总结一下就是先查询并验证管理员用户身份，然后进行 token 值的生成和过期时间的设置，最后将管理员用户的 token 值保存到数据库中。

结合前文中的登录流程来理解，管理员用户登录的详细过程如下。

① 根据用户名和密码查询管理员用户数据，如果存在，则继续后续流程。

② 生成 token 值，这里可以简单理解为生成一个随机字符串，在这一步其实已经完成了登录逻辑，只是后续需要对 token 值进行查询，所以还需要将管理员用户的 token 值入库。

③ 根据管理员用户 id 查询商城管理员 token 表，决定是进行更新操作还是进行新增操作。

④ 根据当前时间获取过期时间。

⑤ 封装管理员用户的 token 值并进行入库操作（新增或修改）。

⑥ 返回 token 值。

管理员用户登录功能的实现逻辑就讲解完了，具体实现主要用到了商城管理员 token 表。不过，这是单体版本的实现方式，后续在实现网关层鉴权改造时，会对这部分逻辑进行修改。

3.3　token值处理及鉴权源代码介绍

登录功能代码主要用于验证登录信息并生成 token 值，当然这只是第一步——生成身份验证信息，接下来笔者介绍登录模块中的用户身份保持和身份验证。用户身份保持和身份验证这两个概念并不复杂，结合前文中的登录流程理解即可。前文中处理的流程分支是 token 值不存在，进入登录页面。那么，如果 token 值存在，即已经登录成功，又该怎样进行身份验证？登录成功后会获得一个 token 值，该怎样使用这个 token 值？本节就来讲解这些知识。

后端处理 token 值的步骤总结如下。

① 生成 token 值，这在 3.2 节已经介绍过。

② 获取请求中的 token 值。

③ 验证 token 值，查看是否存在、是否过期等。

实现登录功能后，需要对用户的登录状态进行验证，这里所说的登录状态即"token

值是否存在及 token 值是否有效",而 token 值是否有效,则通过后端代码验证。由于大部分接口都需要进行登录验证,如果每个方法都添加查询用户数据的语句,则有些多余,因此对方法做了抽取,通过注解切面的形式返回用户信息。

对于管理员用户的身份鉴权,在功能实现时定义了一个@TokenToAdminUser注解。之后自定义了一个方法参数解析器,在需要用户身份信息的方法定义上添加@TokenToAdminUser注解,通过方法参数解析器来获得当前登录的对象信息,自定义的 TokenToAdminUserMethodArgumentResolver 代码如下:

```java
package ltd.user.cloud.newbee.config.handler;

import ltd.user.cloud.newbee.config.annotation.TokenToAdminUser;
import ltd.user.cloud.newbee.dao.NewBeeAdminUserTokenMapper;
import ltd.user.cloud.newbee.entity.AdminUserToken;
import ltd.common.cloud.newbee.exception.NewBeeMallException;
import org.springframework.beans.factory.annotation.Autowired;
import org.springframework.core.MethodParameter;
import org.springframework.stereotype.Component;
import org.springframework.web.bind.support.WebDataBinderFactory;
import org.springframework.web.context.request.NativeWebRequest;
import org.springframework.web.method.support.HandlerMethodArgumentResolver;
import org.springframework.web.method.support.ModelAndViewContainer;

@Component
public class TokenToAdminUserMethodArgumentResolver implements HandlerMethodArgumentResolver {

    @Autowired
    private NewBeeAdminUserTokenMapper newBeeAdminUserTokenMapper;

    public TokenToAdminUserMethodArgumentResolver() {
    }

    public boolean supportsParameter(MethodParameter parameter) {
        if (parameter.hasParameterAnnotation(TokenToAdminUser.class)) {
            return true;
        }
        return false;
    }

    public Object resolveArgument(MethodParameter parameter, ModelAndViewContainer mavContainer, NativeWebRequest webRequest,
```

```java
WebDataBinderFactory binderFactory) {
        if (parameter.getParameterAnnotation(TokenToAdminUser.class)
instanceof TokenToAdminUser) {
            String token = webRequest.getHeader("token");
            if (null != token && !"".equals(token) && token.length() == 32) {
                AdminUserToken adminUserToken = newBeeAdminUserTokenMapper.
selectByToken(token);
                if (adminUserToken == null) {
                    NewBeeMallException.fail("ADMIN_NOT_LOGIN_ERROR");
                } else if (adminUserToken.getExpireTime().getTime() <=
System.currentTimeMillis()) {
                    NewBeeMallException.fail("ADMIN_TOKEN_EXPIRE_ERROR");
                }
                return adminUserToken;
            } else {
                NewBeeMallException.fail("ADMIN_NOT_LOGIN_ERROR");
            }
        }
        return null;
    }
}
```

源代码在 newbee-mall-cloud-user-web 模块下的 ltd.user.cloud.newbee.service.config.handler.TokenToAdminUserMethodArgumentResolver 类中。

执行逻辑如下。

① 获取请求头中的 token 值，若不存在，则返回错误信息给前端；若存在，则继续后续流程。

② 通过 token 值来查询 AdminUserToken 对象，查看是否存在或是否过期，若不存在或已过期，则返回错误信息给前端；若正常，则继续后续流程。

③ 在 WebMvcConfigurer 中配置 TokenToAdminUserMethodArgumentResolver，使其生效，代码如下：

```java
@Configuration
public class AdminUserWebMvcConfigurer extends WebMvcConfigurationSupport {

    @Autowired
    private TokenToAdminUserMethodArgumentResolver
tokenToAdminUserMethodArgumentResolver;

    /**
```

```
     * @param argumentResolvers
     * @tip @TokenToAdminUser 注解处理方法
     */
    public void addArgumentResolvers(List<HandlerMethodArgumentResolver>
argumentResolvers) {
        argumentResolvers.add(tokenToAdminUserMethodArgumentResolver);
    }
}
```

源代码在 newbee-mall-cloud-user-web 模块下的 ltd.user.cloud.newbee.config.AdminUserWebMvcConfigurer 类中。

如此，在需要进行用户鉴权的 API 定义上添加@TokenToAdminUser 注解，之后进行相应的代码逻辑处理。在本实战项目中，有两种用户身份，分别是管理员用户和商城用户，两种身份的登录功能及鉴权功能的实现方式基本一致，后续就不再占用篇幅单独介绍商城用户的登录功能和鉴权功能了。

3.4 用户微服务代码改造

改造成微服务架构后，一般会在网关层进行鉴权操作，所以这部分代码会进行一些调整。不过，改造的只是代码层面和实现细节，其底层的原理并没有太多改动，依然可以根据笔者提供的登录流程来理解和学习。

本节的源代码是在 newbee-mall-cloud-dev-step03 工程的基础上改造的，将工程命名为 newbee-mall-cloud-dev-step04。改造思路是将原来存储在商城管理员 token 表中的数据改为存储到 Redis 中，这样在网关层也可以通过读取 Redis 中的数据实现用户鉴权操作。

3.4.1 引入 Redis 进行鉴权改造

引入与 Redis 相关的依赖并配置好相关的连接参数。

第一步，引入依赖。

打开 newbee-mall-cloud-user-web 模块下的 pom.xml 文件，在 dependencies 标签下引入与 Redis 相关的依赖，新增配置代码如下：

```
<dependency>
  <groupId>org.springframework.boot</groupId>
```

```xml
    <artifactId>spring-boot-starter-data-redis</artifactId>
</dependency>
```

第二步，新增连接 Reids 数据库的配置项。

在 newbee-mall-cloud-user-web 模块下的 application.properties 配置文件中增加连接 Redis 的配置项，代码如下：

```properties
##Redis 配置
# Redis 数据库索引（默认为 0）
spring.redis.database=13
# Redis 服务器地址
spring.redis.host=127.0.0.1
# Redis 服务器连接端口
spring.redis.port=6379
# Redis 服务器连接密码
spring.redis.password=123456
# 连接池最大连接数（使用负值表示没有限制）
spring.redis.jedis.pool.max-active=8
# 连接池最大阻塞等待时间（使用负值表示没有限制）
spring.redis.jedis.pool.max-wait=-1
# 连接池中的最大空闲连接
spring.redis.jedis.pool.max-idle=8
# 连接池中的最小空闲连接
spring.redis.jedis.pool.min-idle=0
# 连接超时时间（毫秒）
spring.redis.timeout=5000
```

连接 Redis 的配置项主要包括 Redis 服务器地址、数据库索引、账号密码、连接超时时间等。

第三步，新增 Reids 配置类。

在 newbee-mall-cloud-user-web 模块下的 ltd.user.cloud.newbee.config 包中新增自定义的 Redis 配置类，代码如下：

```java
package ltd.user.cloud.newbee.config;

import org.springframework.boot.autoconfigure.AutoConfigureAfter;
import org.springframework.boot.autoconfigure.data.redis.RedisAutoConfiguration;
import org.springframework.cache.CacheManager;
import org.springframework.cache.annotation.CachingConfigurerSupport;
import org.springframework.cache.annotation.EnableCaching;
import org.springframework.cache.interceptor.KeyGenerator;
```

```java
import org.springframework.context.annotation.Bean;
import org.springframework.context.annotation.Configuration;
import org.springframework.data.redis.cache.RedisCacheManager;
import org.springframework.data.redis.connection.lettuce.
LettuceConnectionFactory;
import org.springframework.data.redis.core.RedisTemplate;
import org.springframework.data.redis.serializer.
GenericJackson2JsonRedisSerializer;
import org.springframework.data.redis.serializer.StringRedisSerializer;

import javax.annotation.Resource;
import java.io.Serializable;
import java.lang.reflect.Method;
import java.util.HashSet;
import java.util.Set;

@Configuration
@EnableCaching
@AutoConfigureAfter(RedisAutoConfiguration.class)
public class RedisConfig extends CachingConfigurerSupport {

    @Resource
    private LettuceConnectionFactory lettuceConnectionFactory;

    public RedisConfig(){}

    @Bean
    public RedisTemplate<String, Serializable> redisCacheTemplate
(LettuceConnectionFactory redisConnectionFactory){
        RedisTemplate<String,Serializable> template = new RedisTemplate
<String,Serializable>();
        template.setKeySerializer(new StringRedisSerializer());
        template.setValueSerializer(new GenericJackson2JsonRedisSerializer());
        template.setHashKeySerializer(new StringRedisSerializer());
        template.setHashValueSerializer(new GenericJackson2JsonRedisSerializer());
        template.setConnectionFactory(redisConnectionFactory);
        return template;
    }

    @Bean
    public CacheManager cacheManager() {
        RedisCacheManager.RedisCacheManagerBuilder builder =
RedisCacheManager.RedisCacheManagerBuilder
```

```java
            .fromConnectionFactory(lettuceConnectionFactory);
        @SuppressWarnings("serial")
        Set<String> cacheNames = new HashSet<String>() {
            {
                add("codeNameCache");
            }
        };
        builder.initialCacheNames(cacheNames);
        return builder.build();
    }

    @Bean
    public KeyGenerator keyGenerator() {
        return new KeyGenerator() {
            @Override
            public Object generate(Object target, Method method, Object... params) {
                StringBuffer stringBuffer = new StringBuffer();
                stringBuffer.append(target.getClass().getName());
                stringBuffer.append(method.getName());
                for (Object obj : params) {
                    stringBuffer.append(obj.toString());
                }
                return stringBuffer.toString();
            }
        };
    }
}
```

以上代码主要配置 RedisTemplate 对象，并自定义一些序列化方式。以上代码是 Redis 配置类常见的配置方式，可以直接在代码中使用。

另外，在项目开发中使用 Redis 时，一般会自定义一个静态工具类，把一些常见的缓存操作方法抽取出来方便调用。在本项目中对缓存的操作并不复杂，因此并未单独抽取 Redis 工具类，而是直接使用 RedisTemplate 类中提供的底层方法。

3.4.2 用户微服务中登录代码及鉴权代码修改

原来的登录逻辑中会生成 AdminUserToken 的相关数据并保存到 MySQL 数据库中，现在引入了 Redis 数据库，把这部分数据保存到 Redis 数据库中即可，修改后的登录方法代码如下：

```java
@Autowired
private RedisTemplate redisTemplate;

@Override
public String login(String userName, String password) {
  AdminUser loginAdminUser = adminUserMapper.login(userName, password);
  if (loginAdminUser != null) {
    //登录后即执行修改 token 值的操作
    String token = getNewToken(System.currentTimeMillis() + "",
loginAdminUser.getAdminUserId());
    AdminUserToken adminUserToken = new AdminUserToken();
    adminUserToken.setAdminUserId(loginAdminUser.getAdminUserId());
    adminUserToken.setToken(token);
    ValueOperations<String, AdminUserToken> setToken =
redisTemplate.opsForValue();
    setToken.set(token, adminUserToken, 2 * 24 * 60 * 60, TimeUnit.SECONDS);
//过期时间为 48 小时
    return token;
  }
  return "登录失败";
}
```

修改的类是 newbee-mall-cloud-user-web 模块下的 ltd.user.cloud.newbee.service.impl.AdminUserServiceImpl 类。

删除原有保存和修改 MySQL 数据库中 AdminUserToken 数据的逻辑代码,之后引入 RedisTemplate 对象,并且添加将 AdminUserToken 数据存储到 Redis 数据库中的代码。简单理解就是改变数据的存储方式,不过登录方法中生成 token 值并保存等待后续鉴权的底层思路没有太多变化。

登录方法的逻辑修改完成后,相应地修改鉴权逻辑。原来的逻辑是获取请求头中的 token 变量,并根据这个变量查询 MySQL 数据库是否存在和是否过期。现在引入了 Redis 数据库,就不需要查询 MySQL 数据库了,直接读取 Redis 数据库中是否有对应的 token 值即可。修改后的鉴权代码如下:

```java
@Autowired
private RedisTemplate redisTemplate;

public Object resolveArgument(MethodParameter parameter,
ModelAndViewContainer mavContainer, NativeWebRequest webRequest,
WebDataBinderFactory binderFactory) {
  if (parameter.getParameterAnnotation(TokenToAdminUser.class) instanceof
TokenToAdminUser) {
    String token = webRequest.getHeader("token");
    if (null != token && !"".equals(token) && token.length() == 32) {
```

```
    ValueOperations<String, AdminUserToken> opsForAdminUserToken =
redisTemplate.opsForValue();
    AdminUserToken adminUserToken = opsForAdminUserToken.get(token);
    if (adminUserToken == null) {
      NewBeeMallException.fail("ADMIN_NOT_LOGIN_ERROR");
    }
    return adminUserToken;
  } else {
    NewBeeMallException.fail("ADMIN_NOT_LOGIN_ERROR");
  }
}
  return null;
}
```

修改的类是 newbee-mall-cloud-user-web 模块下的 ltd.user.cloud.newbee.service.config.handler.TokenToAdminUserMethodArgumentResolver 类。

删除原有查询 MySQL 数据库中 AdminUserToken 数据的逻辑代码，之后引入 RedisTemplate 对象，并且添加从 Redis 数据库中读取 AdminUserToken 数据的代码，后续逻辑并没有更改。

至此，对 tb_newbee_mall_admin_user_token 表的相关操作就可以全部删除了，直接从数据库中删除这张表及工程中对这张表操作的相关代码。这一步就比较简单了，删除 Mapper 文件及 dao 包中的接口即可。

3.5 网关层鉴权

接下来，就要在网关层增加与 Redis 相关的代码，实现鉴权功能，在网关层对请求进行前置的身份验证操作。

3.5.1 在网关层引入 Redis

这个步骤与在用户微服务中引入 Redis 的步骤一致，主要是在网关模块引入 Redis 连接依赖、新增 Redis 连接配置项，之后在 newbee-mall-cloud-gateway-admin 模块下新建 ltd.gateway.cloud.newbee.config 包，并新增自定义的 Redis 配置类。

代码和配置项与前文类似，这里就不继续粘贴代码了。网关层的 Redis 连接配置需要与用户微服务中的 Redis 连接配置一致，这样才能够读取登录时的 token 值，读者可以根据本章提供的源代码学习和理解。

3.5.2 鉴权的全局过滤器编码实现

网关服务组件是客户端到微服务架构内部的一座桥梁，通过网关服务为请求提供了统一的访问入口，也能够对请求做一些定制化的前置处理。比如，在当前场景下，需要在网关层进行统一鉴权，这样就能够避免无正确身份 id 的请求直接进入微服务实例。如果请求头中有正确的身份 id，则放行，让后方的微服务实例进行请求处理；如果没有正确的身份 id，则直接在网关层响应一个错误提示即可。

在具体编码时，会涉及网关服务的过滤器知识点。本实战项目使用 Spring Cloud Gateway 的全局过滤器实现鉴权功能。

打开 newbee-mall-cloud-gateway-admin 工程，新建 ltd.gateway.cloud.newbee.filter 包，并新建全局过滤器 ValidTokenGlobalFilter，代码如下：

```java
package ltd.gateway.cloud.newbee.filter;

import com.fasterxml.jackson.databind.ObjectMapper;
import com.fasterxml.jackson.databind.node.ObjectNode;
import ltd.common.cloud.newbee.dto.Result;
import ltd.common.cloud.newbee.dto.ResultGenerator;
import ltd.common.cloud.newbee.pojo.AdminUserToken;
import org.springframework.beans.factory.annotation.Autowired;
import org.springframework.cloud.gateway.filter.GatewayFilterChain;
import org.springframework.cloud.gateway.filter.GlobalFilter;
import org.springframework.core.Ordered;
import org.springframework.core.io.buffer.DataBuffer;
import org.springframework.data.redis.core.RedisTemplate;
import org.springframework.data.redis.core.ValueOperations;
import org.springframework.http.HttpHeaders;
import org.springframework.http.HttpStatus;
import org.springframework.stereotype.Component;
import org.springframework.util.StringUtils;
import org.springframework.web.server.ServerWebExchange;
import reactor.core.publisher.Flux;
import reactor.core.publisher.Mono;

import java.nio.charset.StandardCharsets;

@Component
public class ValidTokenGlobalFilter implements GlobalFilter, Ordered {
```

```java
    @Autowired
    private RedisTemplate redisTemplate;

    @Override
    public Mono<Void> filter(ServerWebExchange exchange, GatewayFilterChain chain) {

        // 登录接口，直接放行
        if (exchange.getRequest().getURI().getPath().equals("/users/admin/login")){
            return chain.filter(exchange);
        }

        HttpHeaders headers = exchange.getRequest().getHeaders();

        if (headers == null || headers.isEmpty()) {
            // 返回错误提示
            return wrapErrorResponse(exchange,chain);
        }

        String token = headers.getFirst("token");

        if (!StringUtils.hasText(token)) {
            // 返回错误提示
            return wrapErrorResponse(exchange,chain);
        }
        ValueOperations<String, AdminUserToken> opsForAdminUserToken = redisTemplate.opsForValue();
        AdminUserToken tokenObject = opsForAdminUserToken.get(token);
        if (tokenObject == null) {
            // 返回错误提示
            return wrapErrorResponse(exchange,chain);
        }

        return chain.filter(exchange);
    }

    @Override
    public int getOrder() {
        return Ordered.HIGHEST_PRECEDENCE;
    }

    Mono<Void> wrapErrorResponse(ServerWebExchange exchange, GatewayFilterChain chain) {
```

```
        Result result = ResultGenerator.genErrorResult(419, "无权限访问");
        ObjectMapper mapper = new ObjectMapper();
        ObjectNode resultNode = mapper.valueToTree(result);
        byte[] bytes = resultNode.toString().getBytes(StandardCharsets.UTF_8);
        DataBuffer dataBuffer = exchange.getResponse().bufferFactory().wrap(bytes);
        exchange.getResponse().setStatusCode(HttpStatus.OK);
        return exchange.getResponse().writeWith(Flux.just(dataBuffer));
    }
}
```

主要的代码逻辑就在 filter()方法中，处理步骤如下。

① 判断请求路径，如果是登录接口，则直接放行；如果不是登录接口，则进行后续处理。

② 获取请求头对象，如果为空，则直接返回错误提示，不会将请求转发到后方的微服务实例中；如果请求头不为空，则取出其中名称为 token 的值，如果该值不存在，则直接返回错误提示，不会将请求转发到后方的微服务实例中。

③ 根据 token 值查询 Redis 数据库中是否存在对应的数据。如果 Redis 数据库中不存在对应的数据，则直接返回错误提示，不会将请求转发到后方的微服务实例中；如果 Redis 数据库中存在对应的数据，则表示鉴权成功，将请求转发到后方的微服务实例中去处理请求。

至此，网关层鉴权功能所需的基本配置与功能代码已经完备，下一步进行功能测试。

3.5.3 功能测试

编码完成后，准备好数据库和表就可以启动项目了。当然，在项目启动前需要启动 Nacos Server 和 Redis Server，之后依次启动这两个项目。启动成功后，就可以进行用户登录功能与网关层鉴权功能的测试。

由于在网关层已经配置了路由转发，因此可以直接通过请求网关层的地址获得与请求后方微服务实例相同的效果。本来网关层只有请求转发的功能，经过代码整改和添加鉴权代码后，在网关层也能进行鉴权操作了。

这里，笔者使用 Postman 工具进行接口请求和功能测试，在地址栏中输入如下网址：http://localhost:29100/users/admin/profile。

设置请求方法为 POST，测试效果如图 3-3 所示。

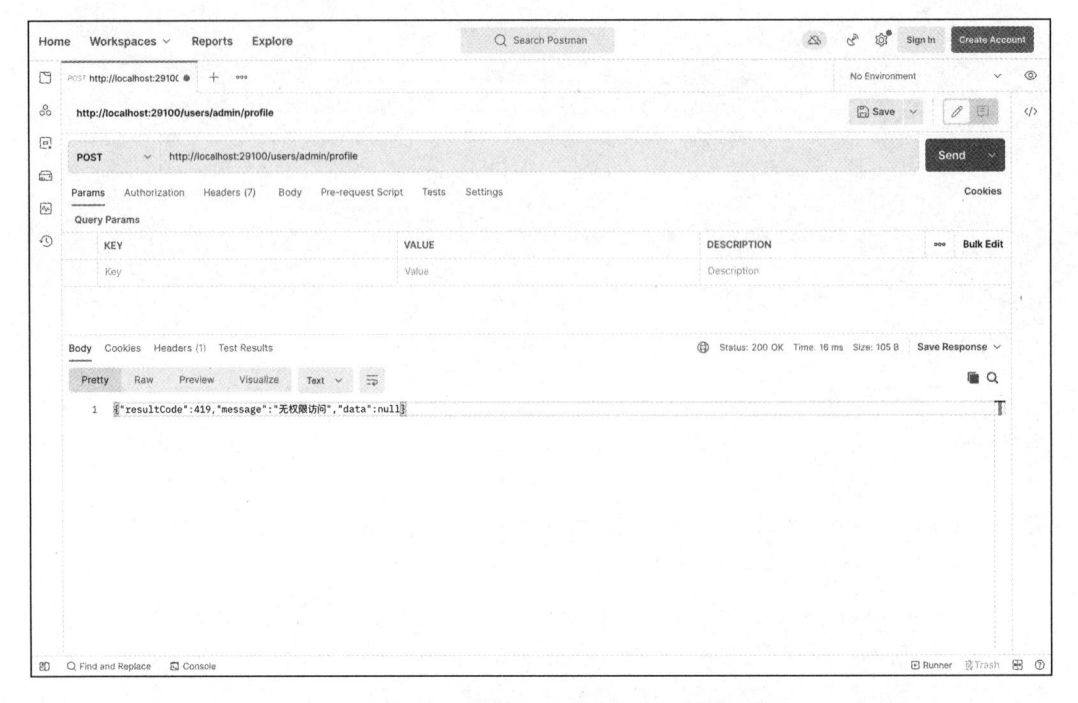

图 3-3　不传 token 值时管理员用户信息请求测试效果

29100 是网关服务的端口号，这里通过直接访问网关层地址来获取管理员用户信息。因为没有在请求头中添加 token 字段，所以网关层的 token 过滤器直接拦截了这个请求，并返回错误提示。这个请求根本没有进入用户微服务实例，网关层鉴权生效了。

上述测试过程演示的是鉴权失败的情况，接下来通过请求登录接口获取正确的 token 值演示鉴权成功的情况。

首先获取一个 token 值，打开 Postman 工具，在地址栏中输入如下网址：http://localhost:29100/users/admin/login。

设置请求方法为 POST，同时输入登录时所需要的用户名和密码，相关请求测试如图 3-4 所示。

这里通过直接访问网关层地址来请求用户微服务的登录接口，因为登录接口会被直接放行，所以由用户微服务来处理这个请求，最终得到了登录成功的结果及 token 值。

此时，可以查看 Redis 数据库中索引为 13 的数据，如图 3-5 所示。token 值被正确地存入 Redis 数据库，并且设置了过期时间。

第 3 章　用户微服务编码实践及功能讲解

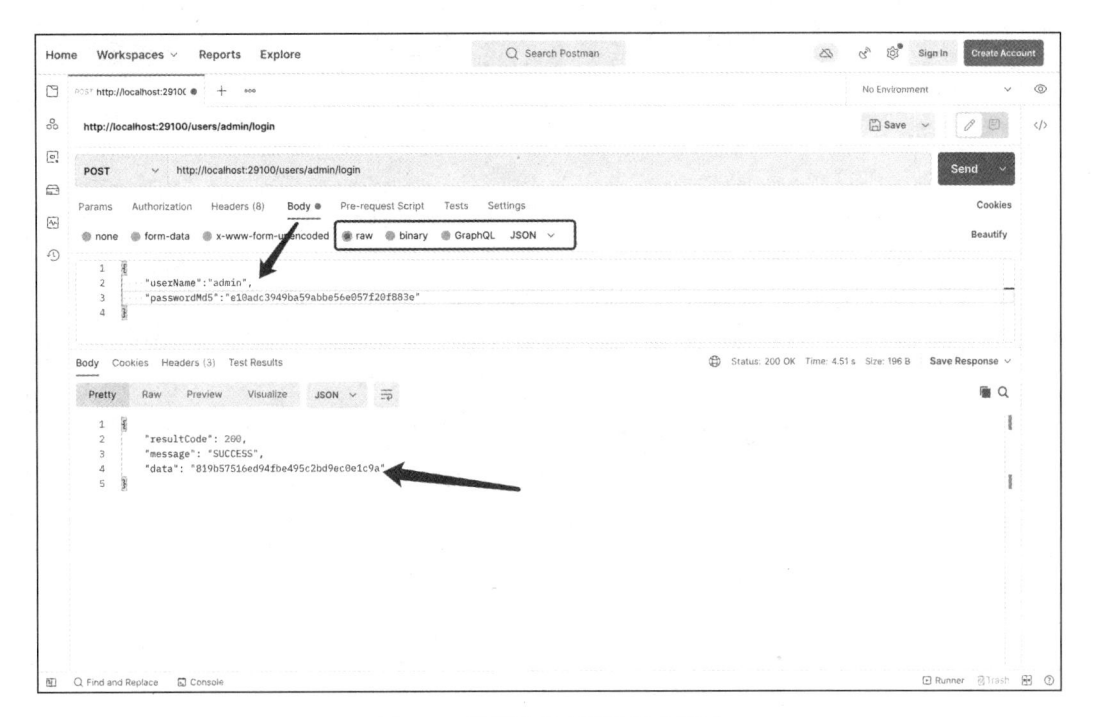

图 3-4　管理员用户登录请求测试

图 3-5　Redis 数据库中的数据

获取 token 值后再次打开 Postman 工具，请求管理员用户信息接口，在地址栏中输入如下网址：http://localhost:29100/users/admin/profile。

设置请求方法为 POST，同时添加一个请求头参数，Key 为 "token"，Value 为上一个步骤获取的 token 值，管理员用户信息请求测试结果如图 3-6 所示。

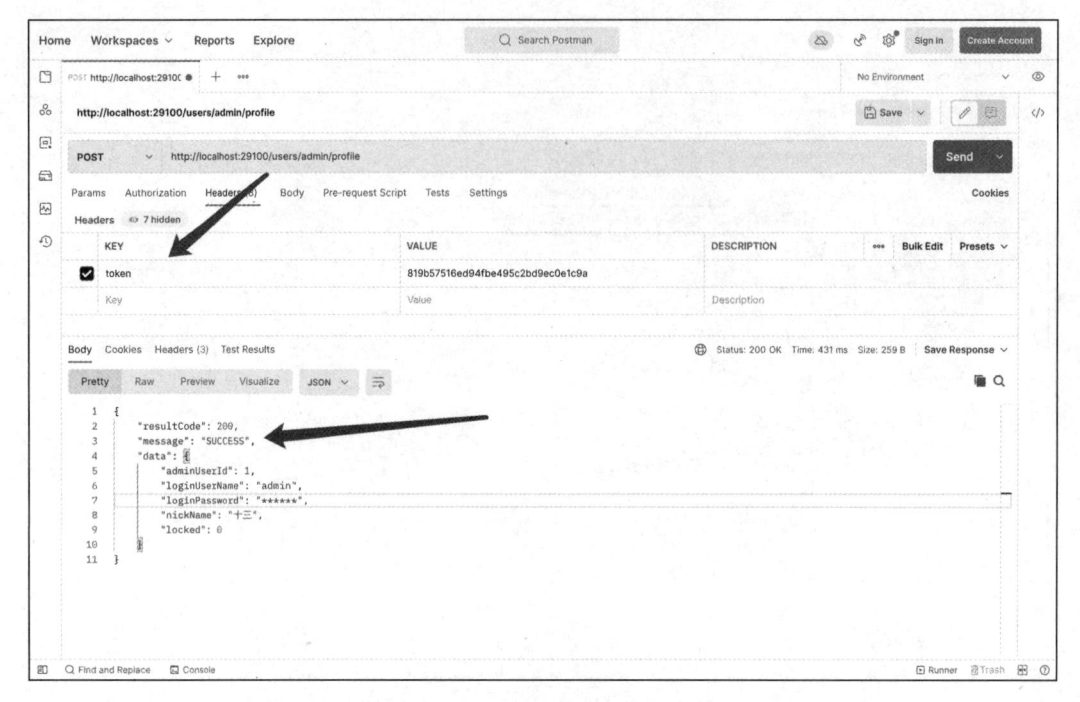

图 3-6　管理员用户信息请求测试结果

　　最终，成功获取了正确的数据，功能测试完成。当然，读者在测试时也可以选择 debug 模式启动两个项目，并打上几个断点，发起请求后看一下代码的执行步骤，会理解得更透彻一些。

　　本章涉及的业务编码并不复杂，主要是登录功能的改造和网关层鉴权功能的添加，涉及的知识点主要有对 Redis 数据库的操作、网关服务 Spring Cloud Gateway、全局过滤器的引入。在前面章节中学习的知识点和微服务组件开发经验正在一点一点地渗透到微服务架构项目中，这种实战过程的体验还是非常令人兴奋的，希望读者能够根据笔者提供的开发步骤顺利地完成本章的项目改造。

第 4 章

商品微服务编码实践及功能讲解

本章继续讲解微服务架构项目的改造过程,主要是商品微服务的功能介绍及开发步骤。商品微服务主要包括两个模块,分别是商品分类管理模块和商品管理模块。

4.1 商品微服务介绍

4.1.1 商品分类管理模块介绍

1. 商品分类简介

分类是通过比较事物之间的相似性,把具有某些共同点或相似特征的事物归于一个不确定集合的逻辑方法。对事物进行分类的作用是使一个大集合中的内容条理清楚、层次分明。分类在电商中也叫类目。要设计一个商品管理系统,首先需要把分类系统做好,因为它是商品管理系统非常基础和重要的一个环节。

商品分类就是对商品分门别类。有的商品是衣服,有的商品是数码产品,还有的商品是美妆/护理产品等。这样处理的好处是方便用户筛选和辨别。以天猫商城和京东商城为例,在商城首页中很大一部分版面都可以进行分类的选择。用户在这里可以通过分类设置快速进入对应的商品列表页面并进行商品选择。

天猫商城分类页面如图 4-1 所示。

京东商城分类页面如图 4-2 所示。

图 4-1　天猫商城分类页面

图 4-2　京东商城分类页面

2. 分类层级设计

通过观察天猫商城和京东商城的分类页面，能够看出二者分类层级的设计方法。在不同的层级下，商城系统需要对商品做进一步的归类操作。因为商品规模和业务不同，所以不同层级的展现效果也不同。天猫商城和京东商城在分类层级的设计思路上是相同的，即三级分层。

如果不设置一定的分类层级，过多的商品类目会给用户筛选带来困难。在设置分类层级后，用户在查找商品时可以遵循"先大类后小类"的原则。比如，用户想买一部手机，可以先在一级分类中筛选并定位到"手机/数码"，然后在二级分类下进行筛选。

当然，也有人提出设计更多层级的分类，如四级分类、五级分类等。层级太多，一是对用户不太友好，不利于搜索；二是对后台管理人员不友好，不方便管理。目前，大部分商城系统选择的分类层级是三级，新蜂商城的商品分类也被设置成三级。

3. 商品分类管理模块的主要功能

关于商品分类管理模块的项目改造，主要是针对后台管理系统进行接口开发，具体功能可以整理为以下两点。

① 分类数据的设置。

② 商品与分类的挂靠和关联。

分类数据的设置包括分类信息的添加、修改等操作，这些数据的存在使得用户可以在商城端进行筛选。商品与分类的挂靠和关联是指将商品信息与分类信息建立联系。比如，在商品表中设置一个分类 id 的关联字段，商品与分类二者之间就产生了关联关系，这样就能够通过对应的分类搜索到对应的商品列表。

在设置分类数据时，对应的页面效果如图 4-3 所示。可以进行添加、修改和删除操作，页面中的操作会调用后端所开发的接口。

4. 商品分类表结构设计

商品分类被设置成三个层级，不过在设计表时并没有做成三张表，因为大部分字段都是一样的，所以就选择增加一个 category_level 字段来区分分类的层级，同时使用 parent_id 字段进行上下级类目之间的关联，分类表的字段设计如下：

```
# 创建商品微服务所需数据
CREATE DATABASE /*!32312 IF NOT EXISTS*/'newbee_mall_cloud_goods_db'
/*!40100 DEFAULT CHARACTER SET utf8 */;
```

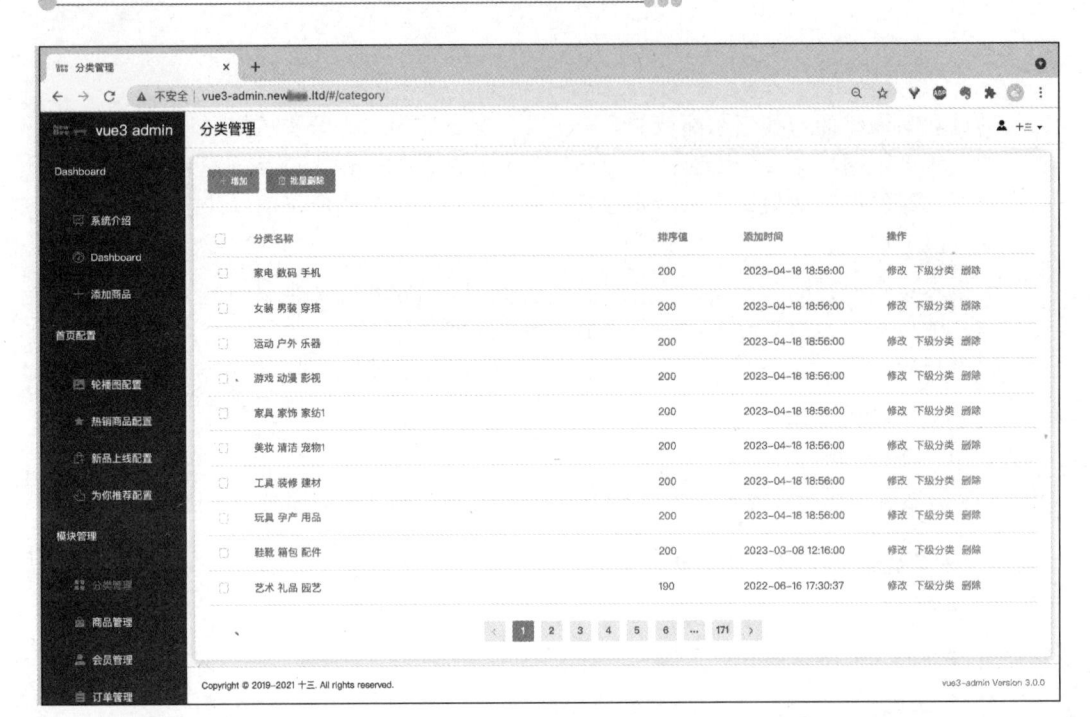

图 4-3　分类管理页面效果

```
USE 'newbee_mall_cloud_goods_db';

# 创建商品分类表

DROP TABLE IF EXISTS 'tb_newbee_mall_goods_category';

CREATE TABLE 'tb_newbee_mall_goods_category' (
  'category_id' bigint(20) NOT NULL AUTO_INCREMENT COMMENT '分类id',
  'category_level' tinyint(4) NOT NULL DEFAULT '0' COMMENT '分类级别(1-一级分类 2-二级分类 3-三级分类)',
  'parent_id' bigint(20) NOT NULL DEFAULT '0' COMMENT '父分类id',
  'category_name' varchar(50) NOT NULL DEFAULT '' COMMENT '分类名称',
  'category_rank' int(11) NOT NULL DEFAULT '0' COMMENT '排序值(字段越大越靠前)',
  'is_deleted' tinyint(4) NOT NULL DEFAULT '0' COMMENT '删除标识字段(0-未删除 1-已删除)',
  'create_time' timestamp NOT NULL DEFAULT CURRENT_TIMESTAMP COMMENT '创建时间',
  'create_user' int(11) NOT NULL DEFAULT '0' COMMENT '创建者id',
  'update_time' timestamp NOT NULL DEFAULT CURRENT_TIMESTAMP COMMENT '修改时间',
```

```
'update_user' int(11) DEFAULT '0' COMMENT '修改者id',
PRIMARY KEY ('category_id') USING BTREE
) ENGINE=InnoDB DEFAULT CHARSET=utf8 ROW_FORMAT=DYNAMIC;
```

商品分类表的字段及每个字段对应的含义都在上面的代码中做了介绍，读者可以对照理解，并正确地把建表 SQL 语句导入数据库。这里只给出了建表语句，如果需要一些测试分类数据，可以到本实战项目的开源仓库下载商品微服务初始化的 SQL 语句，笔者准备了 100 多条分类数据用于功能测试。另外，如果想对功能进行改造，读者也可以自行根据该 SQL 语句进行扩展。

4.1.2 商品管理模块介绍

在购物网站中，购物行为必然建立在商品的基础上。作为交易的基础，商品管理模块可以说是电商系统最重要的部分，是整个电商系统的数据基础，用于记录与商品有关的数据，虽然系统逻辑不复杂，但是操作的数据比较多，需要掌控细节，订单、营销、支付、物流等环节都需要从商品管理模块中获取和操作数据。

1. 商品管理模块的主要功能

本节的项目改造主要针对后台管理系统中商品管理模块进行接口开发，主要功能如下：

① 商品添加接口。

② 商品修改接口。

③ 商品列表接口。

④ 商品上架接口与商品下架接口。

在设置商品数据时，对应的页面效果如图 4-4 所示。可以进行添加、修改等操作，页面中的操作会调用后端所开发的这些接口。

2. 商品信息表结构设计

为了设计商品信息表的字段，笔者参考了一些知名线上商城的设计，如天猫商城和京东商城的商品详情页，通过观察其中的内容可以大致得出一些商品信息的必要字段。天猫商城和京东商城的商品详情页显示内容分别如图 4-5 和图 4-6 所示。

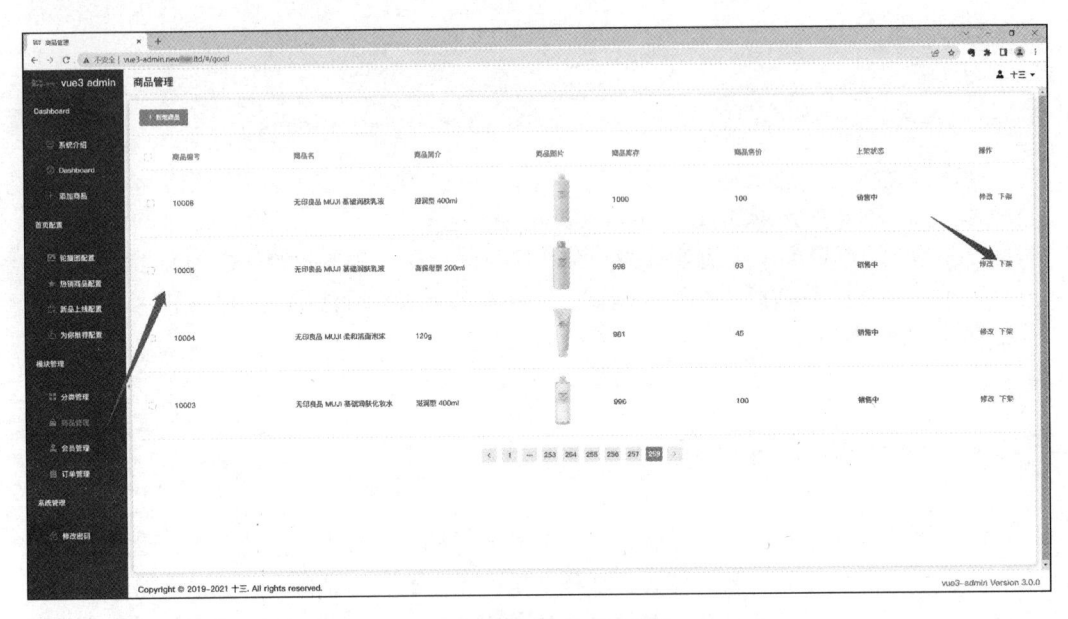

图 4-4 商品管理页面效果

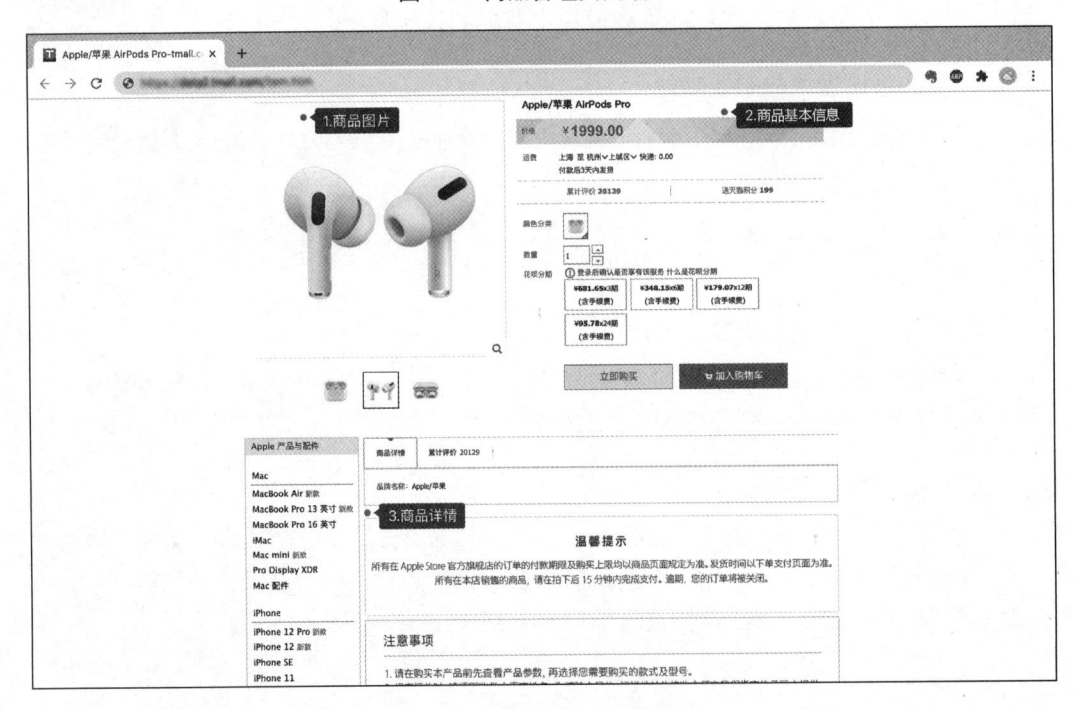

图 4-5 天猫商城的商品详情页

图 4-6　京东商城的商品详情页

简单来说，用户在阿里系、京东系等商品信息长期引导下形成了使用习惯，商品信息的设计和模块划分比较一致，这种模式已经渐渐地沉淀下来并形成一种无形的规则，主要分为：

① 商品图片；

② 商品基本信息；

③ 商品详情。

通过图 4-5 和图 4-6 中的一些信息可以得出商品信息表的必要字段：

① 商品名称；

② 商品简介；

③ 商品原价；

④ 商品实际售价；

⑤ 商品详情内容；

⑥ 商品主图/商品封面图。

以上字段是商品信息实体应该具有的基本字段，新蜂商城在此基础上增加了几个字段：

① 商品库存（订单出库及仓库数据统计）；

② 分类 id（与商品类目的关联关系）；

③ 销售状态（可以控制商品是否能够正常销售）。

最终，商品信息表的设计如下，在数据库中直接执行如下 SQL 语句即可。

```sql
USE 'newbee_mall_cloud_goods_db';

DROP TABLE IF EXISTS 'tb_newbee_mall_goods_info';

# 创建商品信息表

CREATE TABLE 'tb_newbee_mall_goods_info' (
  'goods_id' bigint(20) unsigned NOT NULL AUTO_INCREMENT COMMENT '商品信息表主键id',
  'goods_name' varchar(200) NOT NULL DEFAULT '' COMMENT '商品名',
  'goods_intro' varchar(200) NOT NULL DEFAULT '' COMMENT '商品简介',
  'goods_category_id' bigint(20) NOT NULL DEFAULT '0' COMMENT '关联分类id',
  'goods_cover_img' varchar(200) NOT NULL DEFAULT '/admin/dist/img/no-img.png' COMMENT '商品主图',
  'goods_carousel' varchar(500) NOT NULL DEFAULT '/admin/dist/img/no-img.png' COMMENT '商品轮播图',
  'goods_detail_content' text NOT NULL COMMENT '商品详情',
  'original_price' int(11) NOT NULL DEFAULT '1' COMMENT '商品价格',
  'selling_price' int(11) NOT NULL DEFAULT '1' COMMENT '商品实际售价',
  'stock_num' int(11) unsigned NOT NULL DEFAULT '0' COMMENT '商品库存数量',
  'tag' varchar(20) NOT NULL DEFAULT '' COMMENT '商品标签',
  'goods_sell_status' tinyint(4) NOT NULL DEFAULT '0' COMMENT '商品上架状态 1-下架 0-上架',
  'create_user' int(11) NOT NULL DEFAULT '0' COMMENT '添加者主键id',
  'create_time' datetime NOT NULL DEFAULT CURRENT_TIMESTAMP COMMENT '商品添加时间',
  'update_user' int(11) NOT NULL DEFAULT '0' COMMENT '修改者主键id',
  'update_time' datetime NOT NULL DEFAULT CURRENT_TIMESTAMP COMMENT '商品修改时间',
  PRIMARY KEY ('goods_id') USING BTREE
) ENGINE=InnoDB DEFAULT CHARSET=utf8 ROW_FORMAT=DYNAMIC;
```

商品信息表的字段及每个字段对应的含义都在上面的 SQL 语句中做了介绍，读者可以对照理解，并正确地把建表 SQL 语句导入数据库。这里只给出了建表语句，如果需要测试商品数据，可以到本实战项目的开源仓库下载商品微服务初始化的 SQL 语句，笔者在其中准备了近千条商品数据用于功能测试。另外，想要对功能进行改造，读者也可以自行根据该 SQL 语句进行扩展。

4.2 创建商品微服务编码

前文主要介绍商品信息表的结构字段及主要的功能，接下来讲解具体的代码改造过程。本节的源代码是在 newbee-mall-cloud-dev-step04 工程的基础上改造的，将工程命名为 newbee-mall-cloud-dev-step05。

在工程中新增一个 newbee-mall-cloud-goods-service 模块，并在 pom.xml 主文件中增加该模块的配置，代码如下：

```xml
<modules>
    <!-- 新增商品微服务 -->
    <module>newbee-mall-cloud-goods-service</module>
    <module>newbee-mall-cloud-user-service</module>
    <module>newbee-mall-cloud-gateway-admin</module>
    <module>newbee-mall-cloud-common</module>
</modules>
```

该模块的目录结构设置与 newbee-mall-cloud-user-service 模块的目录结构设置类似，如下所示：

```
newbee-mall-cloud-goods-service          // 商品微服务
    ├── newbee-mall-cloud-goods-api      // 存放商品模块中暴露的用于远程调用
    │                                       的 FeignClient 类
    └── newbee-mall-cloud-goods-web      // 商品模块 API 的代码及逻辑
```

在开发时，笔者直接将 newbee-mall-cloud-user-service 模块中的代码复制了过来，之后对模块名称和目录名称进行了修改。

子节点 newbee-mall-cloud-goods-service 模块的 pom.xml 文件代码如下：

```xml
<?xml version="1.0" encoding="UTF-8"?>
<project xmlns="http://maven.apa***.org/POM/4.0.0"
xmlns:xsi="http://www.w*.org/2001/XMLSchema-instance"
         xsi:schemaLocation="http://maven.apa***.org/POM/4.0.0
https://maven.apa***.org/xsd/maven-4.0.0.xsd">
    <modelVersion>4.0.0</modelVersion>
```

```xml
    <groupId>ltd.newbee.cloud</groupId>
    <artifactId>newbee-mall-cloud-goods-service</artifactId>
    <version>0.0.1-SNAPSHOT</version>
    <packaging>pom</packaging>
    <name>newbee-mall-cloud-goods-service</name>
    <description>商品模块</description>

    <parent>
        <groupId>ltd.newbee.cloud</groupId>
        <artifactId>newbee-mall-cloud</artifactId>
        <version>0.0.1-SNAPSHOT</version>
    </parent>

    <properties>
        <java.version>1.8</java.version>
    </properties>

    <modules>
        <module>newbee-mall-cloud-goods-web</module>
        <module>newbee-mall-cloud-goods-api</module>
    </modules>

    <dependencies>
    </dependencies>
</project>
```

主要的配置项是模块名称、模块的打包方式及模块间的父子关系。其中，重点配置了父模块 newbee-mall-cloud，两个子模块分别是 newbee-mall-cloud-goods-api 和 newbee-mall-cloud-goods-web。

newbee-mall-cloud-goods-api 模块的 pom.xml 文件代码修改如下：

```xml
<?xml version="1.0" encoding="UTF-8"?>
<project xmlns="http://maven.apa***.org/POM/4.0.0" xmlns:xsi=
"http://www.w*.org/2001/XMLSchema-instance"
         xsi:schemaLocation="http://maven.apa***.org/POM/4.0.0
https://maven.apa***.org/xsd/maven-4.0.0.xsd">
    <modelVersion>4.0.0</modelVersion>
    <groupId>ltd.user.newbee.cloud</groupId>
    <artifactId>newbee-mall-cloud-goods-api</artifactId>
    <packaging>jar</packaging>
    <version>0.0.1-SNAPSHOT</version>
    <name>newbee-mall-cloud-goods-api</name>
    <description>商品微服务 openfeign</description>
```

```xml
<parent>
    <groupId>ltd.newbee.cloud</groupId>
    <artifactId>newbee-mall-cloud-goods-service</artifactId>
    <version>0.0.1-SNAPSHOT</version>
</parent>

<properties>
    <java.version>1.8</java.version>
</properties>

<dependencies>

    <dependency>
        <groupId>org.springframework.cloud</groupId>
        <artifactId>spring-cloud-starter-openfeign</artifactId>
    </dependency>

</dependencies>
</project>
```

这里定义了与父模块 newbee-mall-cloud-goods-service 的关系，打包方式 packaging 配置项的值为 jar。同时，因为后面要添加 FeignClient 类，所以这里引入了 OpenFeign。

newbee-mall-cloud-goods-web 模块的 pom.xml 文件代码修改如下：

```xml
<?xml version="1.0" encoding="UTF-8"?>
<project xmlns="http://maven.apa***.org/POM/4.0.0" xmlns:xsi=
"http://www.w*.org/2001/XMLSchema-instance"
    xsi:schemaLocation="http://maven.apa***.org/POM/4.0.0
https://maven.apa***.org/xsd/maven-4.0.0.xsd">
    <modelVersion>4.0.0</modelVersion>
    <groupId>ltd.user.newbee.cloud</groupId>
    <artifactId>newbee-mall-cloud-goods-web</artifactId>
    <version>0.0.1-SNAPSHOT</version>
    <name>newbee-mall-cloud-goods-web</name>
    <description>商品微服务</description>

    <parent>
        <groupId>ltd.newbee.cloud</groupId>
        <artifactId>newbee-mall-cloud-goods-service</artifactId>
        <version>0.0.1-SNAPSHOT</version>
    </parent>
```

```xml
<properties>
    <java.version>1.8</java.version>
</properties>

<dependencies>
    <dependency>
        <groupId>org.springframework.boot</groupId>
        <artifactId>spring-boot-starter-web</artifactId>
    </dependency>

    <dependency>
        <groupId>org.springframework.boot</groupId>
        <artifactId>spring-boot-starter-test</artifactId>
        <scope>test</scope>
    </dependency>

    <dependency>
        <groupId>com.alibaba.cloud</groupId>
        <artifactId>spring-cloud-starter-alibaba-nacos-discovery</artifactId>
    </dependency>

    <dependency>
        <groupId>org.springframework.boot</groupId>
        <artifactId>spring-boot-starter-validation</artifactId>
    </dependency>

    <dependency>
        <groupId>org.mybatis.spring.boot</groupId>
        <artifactId>mybatis-spring-boot-starter</artifactId>
    </dependency>

    <dependency>
        <groupId>org.projectlombok</groupId>
        <artifactId>lombok</artifactId>
        <scope>provided</scope>
    </dependency>

    <dependency>
        <groupId>io.springfox</groupId>
```

```xml
            <artifactId>springfox-boot-starter</artifactId>
        </dependency>

        <dependency>
            <groupId>mysql</groupId>
            <artifactId>mysql-connector-java</artifactId>
            <scope>runtime</scope>
        </dependency>

        <dependency>
            <groupId>ltd.newbee.cloud</groupId>
            <artifactId>newbee-mall-cloud-common</artifactId>
            <version>0.0.1-SNAPSHOT</version>
        </dependency>

    </dependencies>
</project>
```

这里定义了与父模块 newbee-mall-cloud-goods-service 的关系，同时将相关的业务依赖项移到该配置文件中，因为商品功能模块的主要业务代码都在这个模块中。

由于改造过程中直接复制了 newbee-mall-cloud-user-service 模块的代码，因此在修改完依赖配置后，要修改包名，把 ltd.user.cloud.×××的名称改为 ltd.goods.cloud.×××，之后修改和删除复制过来的 Java 文件和 Mapper 文件，因为这些代码与商品微服务并无关系。唯一没有删除的是 config 包中的代码，因为全局异常处理配置类、Swagger 配置类、自定义 MVC 配置类在商品微服务中也是必要的，所以只修改了这些类的类名。最后，修改 application.properties 配置文件，删除与 Redis 连接的相关配置。

这样，就完成了一个新模块的初始构建工作，得到了一份较为纯净的模块代码，此时的 newbee-mall-cloud-goods-service 模块中并没有业务代码，目录结构如图 4-7 所示。在后续开发步骤中，要添加其他微服务到项目中，可以按照以上步骤来完成，或者直接复制 newbee-mall-cloud-dev-step05 源代码下的 newbee-mall-cloud-goods-service 模块到工程中，修改一下名称即可。

本节的内容并未涉及业务编码，主要介绍商品微服务的功能和表结构设计，以及在项目中完成商品微服务的初始化构建。读者可以根据这些代码自己动手完成微服务编码，如拿到这份代码后，以此为基础，自行完成商品微服务所有代码的功能。如果自己实现耗费时间，也可以按照笔者给出的步骤来完成微服务拆分。

```
v 📁 newbee-mall-cloud-dev-step05 [newbee-mall-cloud]
  > 📁 .idea
  > 📁 newbee-mall-cloud-common
  > 📁 newbee-mall-cloud-gateway-admin
  v 📁 newbee-mall-cloud-goods-service
    v 📁 newbee-mall-cloud-goods-api
      v 📁 src
        v 📁 main
          v 📁 java
            > 📁 ltd.goods.cloud.newbee.openfeign
      🅜 pom.xml
    v 📁 newbee-mall-cloud-goods-web
      v 📁 src
        v 📁 main
          v 📁 java
            > 📁 ltd.goods.cloud.newbee
          v 📁 resources
            🅜 application.properties
      🅜 pom.xml
    🅜 pom.xml
  > 📁 newbee-mall-cloud-user-service
  📄 .gitignore
  📄 LICENSE
  🅜 pom.xml
```

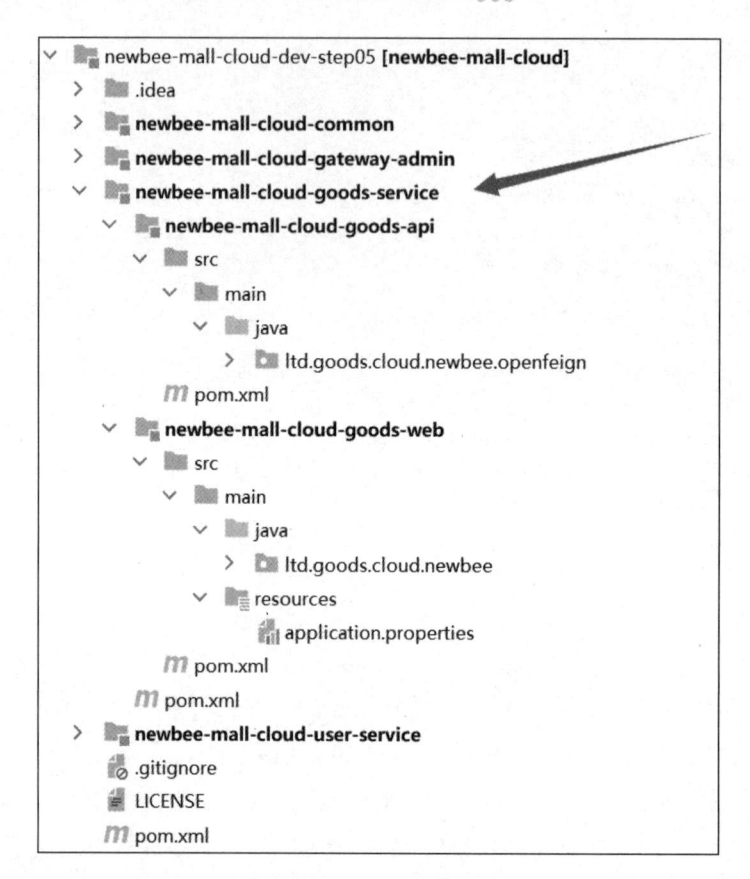

图 4-7　目录结构

4.3　商品微服务与用户微服务通信

  本节继续讲解微服务架构下商品微服务的编码改造。由于微服务化的原因，各个功能模块都被拆分为独立的微服务，如已经开发了一部分的用户微服务和当前正在开发的商品微服务。独立的微服务有独立的数据库，想要获取不是本服务中的数据，肯定要与另外一个微服务实例进行通信。比如，本节将要讲解的就是在其他微服务实例中，通过与用户微服务通信获取当前登录用户的信息及鉴权的编码。

  本节的源代码是在 newbee-mall-cloud-dev-step05 工程的基础上改造的，将工程命名为 newbee-mall-cloud-dev-step06。

4.3.1　为什么需要调用用户微服务

以单体模式 newbee-mall-api 项目中后台管理系统商品模块中接口的定义为例，在商品列表接口方法定义中，有一个标注了 @TokenToAdminUser 注解的参数 adminUser，如图 4-8 所示。读者对这个参数的含义应该不陌生，它就是通过处理请求头中的 token 字段，来鉴权和获取当前登录管理员用户的身份信息。在单体模式下，所有功能模块的表都在一个数据库中，所有功能模块的编码也都在一个工程中，因此完成对 token 字段的处理并不复杂。

图 4-8　newbee-mall-api 项目商品列表接口源代码

而到了微服务架构下，对 token 字段的处理过程就变了。商品微服务是无法直接连接用户微服务数据库的，因此想要完成对请求头中 token 字段的处理就必须与用户微服务进行通信。实现方式是通过调用用户微服务暴露的接口来验证 token 字段是否正确，同时也能够通过 token 字段来获取当前登录的用户信息。当然，本小节讲解的是后台管理系统中管理员用户的处理步骤，后续改造中想要获取商城用户的信息，也需要与用户微服务进行通信，与这里所讲解的处理步骤类似。

不过，肯定会有读者有如下疑问：网关层不是已经做了鉴权吗？微服务实例中为什么又要做一次？

网关层确实做了一次鉴权的操作，实现方式为验证 token 字段。而微服务实例中对 token 字段的处理不仅仅是鉴权，大部分接口都需要用户的身份信息，即当前登录用户的数据 login_user_data。根据这个数据可以完成更细粒度的身份验证，如对于某些数据，A 用户可以修改，但是 B 用户不可以修改；某个表中需要记录操作者信息，能够直接使用 login_user_data 中的数据；需要在日志中输出当前用户的身份信息，也可以直接使用 login_user_data 中的数据。因此，其他微服务实例调用用户微服务，主要是为了完成更细致的业务操作，以及获取当前登录用户的数据并用于后续的业务逻辑。

当然，也有读者会想另一个问题：如果某个接口中不需要获取当前登录用户的数据，网关层已经做了一次鉴权，是不是可以不处理 token 字段了呢？

在这种情况下是可以不处理 token 字段的，毕竟网关层做的鉴权操作已经满足基本要求了。不过，在真实的企业开发中，日志在输出时带上当前登录用户的信息是必不可少的步骤，有些开发团队的开发规范中也规定要获取当前登录用户的数据。因此，笔者建议最好不要省略对用户身份信息的处理。

4.3.2 商品微服务调用用户微服务编码实践

接下来进行实际的编码，在商品微服务中完成对用户微服务的调用及处理 token 字段。

第一步，修改用户微服务中暴露接口的 FeignClient 代码。

打开 newbee-mall-cloud-user-api 工程下的 pom.xml 文件,增加公用模块的依赖配置，代码如下：

```
<dependency>
  <groupId>ltd.newbee.cloud</groupId>
  <artifactId>newbee-mall-cloud-common</artifactId>
  <version>0.0.1-SNAPSHOT</version>
</dependency>
```

修改 NewBeeCloudAdminUserServiceFeign，把 getAdminUserByToken()方法的返回类型修改为 Result，代码修改如下：

```
package ltd.user.cloud.newbee.openfeign;

import org.springframework.cloud.openfeign.FeignClient;
```

第 4 章　商品微服务编码实践及功能讲解

```
import org.springframework.web.bind.annotation.GetMapping;
import org.springframework.web.bind.annotation.PathVariable;
import ltd.common.cloud.newbee.dto.Result;

@FeignClient(value = "newbee-mall-cloud-user-service", path = "/users")
public interface NewBeeCloudAdminUserServiceFeign {

    @GetMapping(value = "/admin/{token}")
    Result getAdminUserByToken(@PathVariable(value = "token") String token);
}
```

第二步，修改商品微服务，引入 OpenFeign、user-api 依赖。

打开 newbee-mall-cloud-goods-web 工程下的 pom.xml 文件，增加远程通信所需的依赖配置和 newbee-mall-cloud-user-api 模块，新增配置代码如下：

```
<dependency>
  <groupId>org.springframework.cloud</groupId>
  <artifactId>spring-cloud-starter-loadbalancer</artifactId>
</dependency>

<dependency>
  <groupId>ltd.user.newbee.cloud</groupId>
  <artifactId>newbee-mall-cloud-user-api</artifactId>
  <version>0.0.1-SNAPSHOT</version>
</dependency>
```

这里主要引入 LoadBalancer 依赖和 user-api 依赖，由于 user-api 中已经含有 OpenFeign 依赖配置，因此这里不需要额外增加。

第三步，增加配置，启用 OpenFeign 并使 FeignClient 类生效。

打开 newbee-mall-cloud-goods-web 工程，在项目的启动类 NewBeeMallCloudGoodsServiceApplication 上添加 @EnableFeignClients 注解，并配置相关的 FeignClient 类，代码如下：

```
@EnableFeignClients(basePackageClasses = {ltd.user.cloud.newbee.openfeign.NewBeeCloudAdminUserServiceFeign.class})
```

这里使用 basePackageClasses 配置了需要使用的 FeignClient 类，即 NewBeeCloudAdminUserServiceFeign 类。接下来就可以直接使用 NewBeeCloudAdminUserServiceFeign 与用户微服务进行远程通信了。

第四步，修改 token 字段处理类中的逻辑代码。

打开 newbee-mall-cloud-goods-web 工程，修改 TokenToAdminUserMethodArgument

Resolver 类中对 token 字段处理的逻辑代码，主要引入 NewBeeCloudAdminUser ServiceFeign 类，通过调用用户微服务来获取管理员用户的数据。

修改后的代码如下：

```
@Component
public class TokenToAdminUserMethodArgumentResolver implements
HandlerMethodArgumentResolver {

    @Autowired
    private NewBeeCloudAdminUserServiceFeign newBeeCloudAdminUserService;

    public TokenToAdminUserMethodArgumentResolver() {
    }

    public boolean supportsParameter(MethodParameter parameter) {
        if (parameter.hasParameterAnnotation(TokenToAdminUser.class)) {
            return true;
        }
        return false;
    }

    public Object resolveArgument(MethodParameter parameter,
ModelAndViewContainer mavContainer, NativeWebRequest webRequest,
WebDataBinderFactory binderFactory) {
        if (parameter.getParameterAnnotation(TokenToAdminUser.class)
instanceof TokenToAdminUser) {
            String token = webRequest.getHeader("token");
            if (null != token && !"".equals(token) && token.length() == 32) {
                // 通过用户微服务获取用户信息
                Result result = newBeeCloudAdminUserService.getAdminUserByToken(token);

                if (result == null || result.getResultCode() != 200 ||
result.getData() == null) {
                    NewBeeMallException.fail("ADMIN_NOT_LOGIN_ERROR");
                }

                LinkedHashMap resultData = (LinkedHashMap) result.getData();

                // 将返回的字段封装到 LoginAdminUser 对象中
                LoginAdminUser loginAdminUser = new LoginAdminUser();
                loginAdminUser.setAdminUserId(Long.valueOf(resultData.get("adminUserId").toString()));
                loginAdminUser.setLoginUserName((String) resultData.get("loginUserName"));
```

```
            loginAdminUser.setNickName((String) resultData.get("nickName"));
            loginAdminUser.setLocked(Byte.valueOf(resultData.get
("locked").toString()));
            return loginAdminUser;
        } else {
            NewBeeMallException.fail("ADMIN_NOT_LOGIN_ERROR");
        }
    }
    return null;
}
```

执行逻辑如下。

① 获取请求头中的 token 字段，若不存在，则返回错误信息给前端；若存在，则继续后续流程。

② 通过 token 字段的值远程调用用户微服务，查询管理员用户的数据。如果远程调用时返回的结果类为空或未成功调用，则返回错误信息给前端；如果成功，则继续后续流程。

③ 如果正确获取用户微服务响应的结果类，则将其转换为 LoginAdminUser 对象并配置到方法参数中供后续流程使用。

当然，这部分代码可以进行优化，即在 NewBeeCloudAdminUserServiceFeign 中通过注解的形式添加返回结果类中的类型，代码如下：

```
Result<LoginAdminUser> getAdminUserByToken(@PathVariable(value = "token")
String token);
```

如此，就不需要在远程通信的调用后使用 LinkedHashMap 接收返回结果，之后再一个字段接一个字段地转换和设置了。管理员用户 token 字段处理代码如此写的原因是为章节编排考虑，让读者更加清晰地了解这个步骤做了哪些事情，后续处理商城用户 token 字段时就会使用注解结果类的编码方式了。

至此，便完成了在商品微服务中通过远程通信获取当前登录用户的功能。

4.3.3 功能测试

为了让读者理解得更深刻，笔者新增了一个测试接口。在 newbee-mall-cloud-goods-web 工程中新建 ltd.goods.cloud.newbee.controller 包，并新建 NewBeeAdminGoodsCategoryController 测试类，新增代码如下：

```java
package ltd.goods.cloud.newbee.controller;

import io.swagger.annotations.Api;
import io.swagger.annotations.ApiOperation;
import ltd.common.cloud.newbee.dto.Result;
import ltd.common.cloud.newbee.dto.ResultGenerator;
import ltd.goods.cloud.newbee.config.annotation.TokenToAdminUser;
import ltd.goods.cloud.newbee.entity.LoginAdminUser;
import org.slf4j.Logger;
import org.slf4j.LoggerFactory;
import org.springframework.web.bind.annotation.RequestMapping;
import org.springframework.web.bind.annotation.RequestMethod;
import org.springframework.web.bind.annotation.RestController;

@RestController
@Api(value = "v1", tags = "后台管理系统分类模块接口")
@RequestMapping("/goods/admin")
public class NewBeeAdminGoodsCategoryController {

    private static final Logger logger = LoggerFactory.getLogger(NewBeeAdminGoodsCategoryController.class);

    @RequestMapping(value = "/testLoginAdminUser", method = RequestMethod.GET)
    @ApiOperation(value = "测试", notes = "测试")
    public Result testLoginAdminUser(@TokenToAdminUser LoginAdminUser adminUser) {
        logger.info("adminUser:{}", adminUser.toString());
        return ResultGenerator.genSuccessResult();
    }
}
```

因为测试接口的定义中标注了@TokenToAdminUser 注解，所以 TokenToAdminUserMethodArgumentResolver 类会对 testLoginAdminUser()方法中的 adminUser 参数进行前置处理。获取当前请求头中的 token 字段，并通过 OpenFeign 调用用户微服务中暴露的/users/admin/{token}接口获取用户数据，之后组装到当前方法的 adminUser 参数中。

编码完成后，准备好数据库和表就可以启动项目了。当然，在项目启动前需要启动 Nacos Server 和 Redis Server，之后依次启动 newbee-mall-cloud-goods-web 工程和 newbee-mall-cloud-goods-web 工程下的主类。启动成功后，就可以进行本章的功能测试了。如果能正常通过用户微服务获取用户信息并用于身份验证，则功能测试完成。

先打开用户微服务的 Swagger 页面，在浏览器中输入如下网址：http://localhost:29000/

swagger-ui/index.html。

然后在该页面使用登录接口获取一个 token 值用于后续的功能测试，如笔者在测试时获取了一个值为"6b9c0062841e2c9fd118002176b45cb7"的 token 字段。

接下来打开商品微服务的 Swagger 页面，在浏览器中输入如下网址：http://localhost:29010/swagger-ui/index.html。

最后就可以在 Swagger 提供的 UI 页面进行接口测试了。

如果没有在请求头中设置 token 值或设置了错误的 token 值，则会收到一个错误的响应结果，如图 4-9 所示。

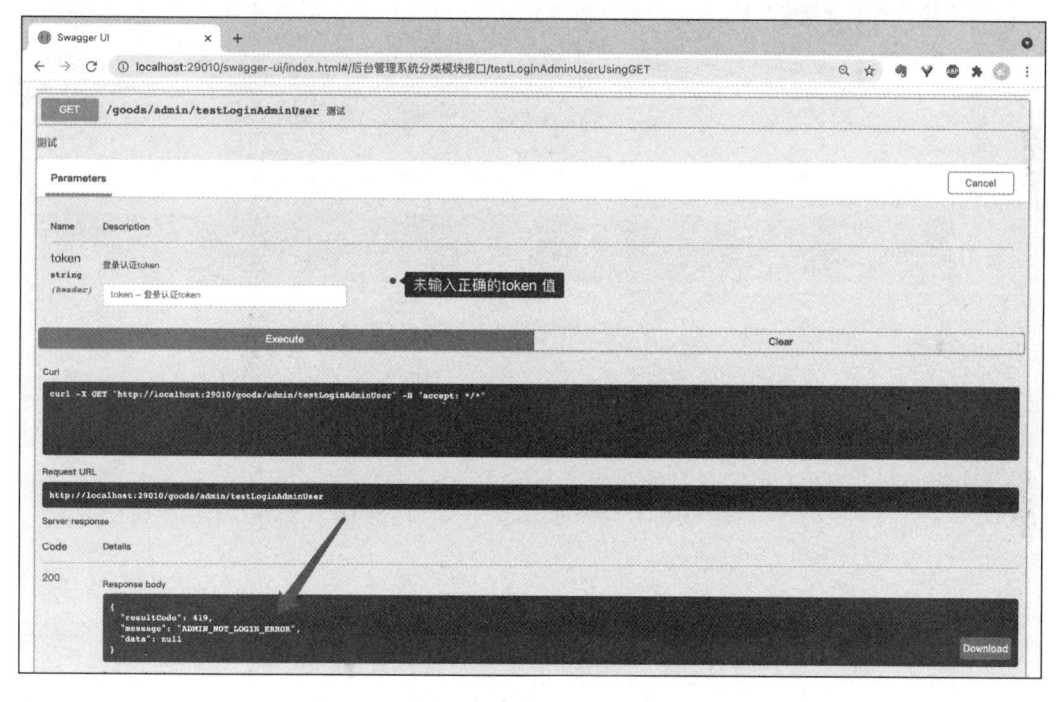

图 4-9 未登录状态访问接口时的响应结果

请求结果中的 resultCode 为 419，message 为 ADMIN_NOT_LOGIN_ERROR，表示当前请求被禁止访问了，因为当前用户的登录状态不正确。

如果在请求头中设置了正确的 token 值，则会收到一个正确的响应结果，如图 4-10 所示。

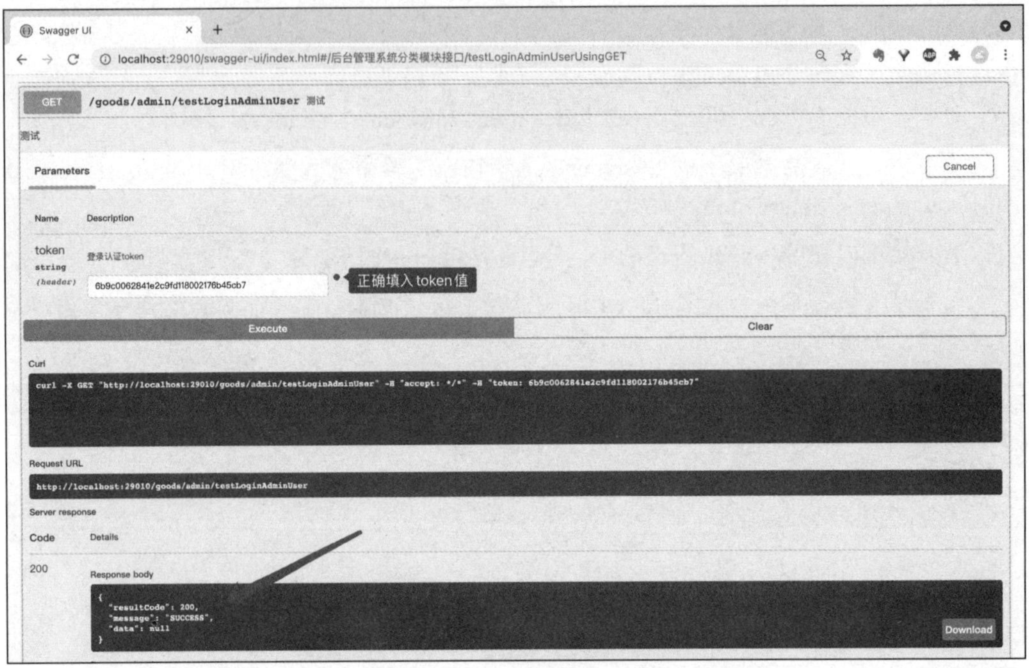

图 4-10　正常登录状态访问接口时的响应结果

请求结果中的 resultCode 为 200，message 为 SUCCESS，表示当前用户的登录状态正常，并且获得了正常的结果响应。同时，在控制台上会看到一条打印当前登录用户信息的日志，如下所示：

```
2023-07-05 15:53:17.246  INFO 79019 --- [io-29010-exec-5] c.n.c.NewBee
AdminGoodsCategoryController  : adminUser:LoginAdminUser(adminUserId=1,
loginUserName=admin, loginPassword=null, nickName=十三, locked=0)
```

功能测试通过，商品微服务可以正常与用户微服务通信，并实现微服务中用户的鉴权等功能。当然，读者在测试时，也可以用 debug 模式分别启动用户微服务和商品微服务，在涉及的代码中打上几个断点，在测试接口时就可以更好地观察整个流程。

本节主要介绍了调用用户微服务的原因，以及其他微服务该如何借助与用户微服务的远程通信来完成获取登录用户信息、鉴权等操作。不只是商品微服务，本实战项目中的其他微服务亦是如此，并且实现方式和编码思路基本一致。希望读者能够根据笔者提供的开发步骤顺利地完成本节的项目改造，后续章节中会继续完善商品微服务的功能。

第4章 商品微服务编码实践及功能讲解

4.4 商品微服务编码

承接前文，本节继续讲解微服务架构下商品微服务的编码改造，把原来单体项目中的功能模块一点一点整合到这个工程里。本节的源代码是在 newbee-mall-cloud-dev-step06 工程的基础上改造的，将工程命名为 newbee-mall-cloud-dev-step07。

4.4.1 商品微服务代码改造

打开商品微服务 newbee-mall-cloud-goods-web 的工程目录，在 ltd.goods.cloud.newbee 包下依次创建 config 包、dao 包、entity 包、service 包。在 resources 目录下新增 mapper 文件夹用于存放 Mapper 文件，直接将原单体 API 项目中与商品管理和商品分类管理相关的业务代码及 Mapper 文件（如图 4-11 所示）依次复制过来。

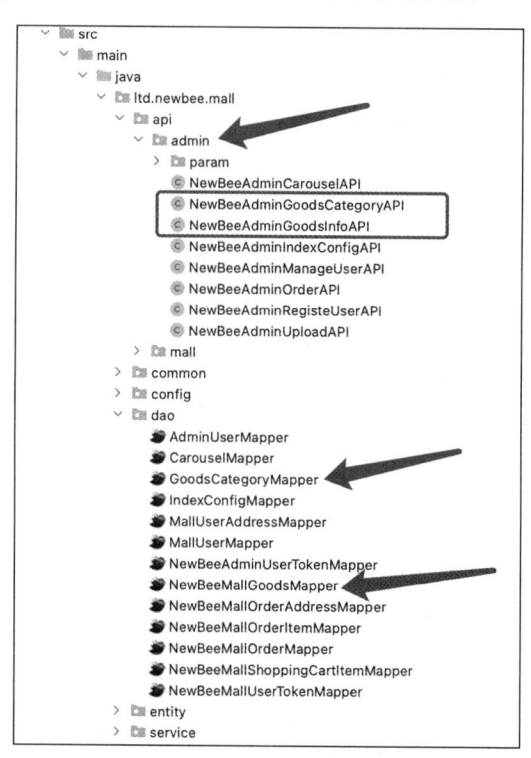

图 4-11 单体 API 项目中与商品管理和商品分类管理相关的业务代码及 Mapper 文件

由于代码量较大，这里就不一一讲解了，按照对应的文件目录复制过来即可。上述步骤完成后，最终的代码目录如图 4-12 所示。

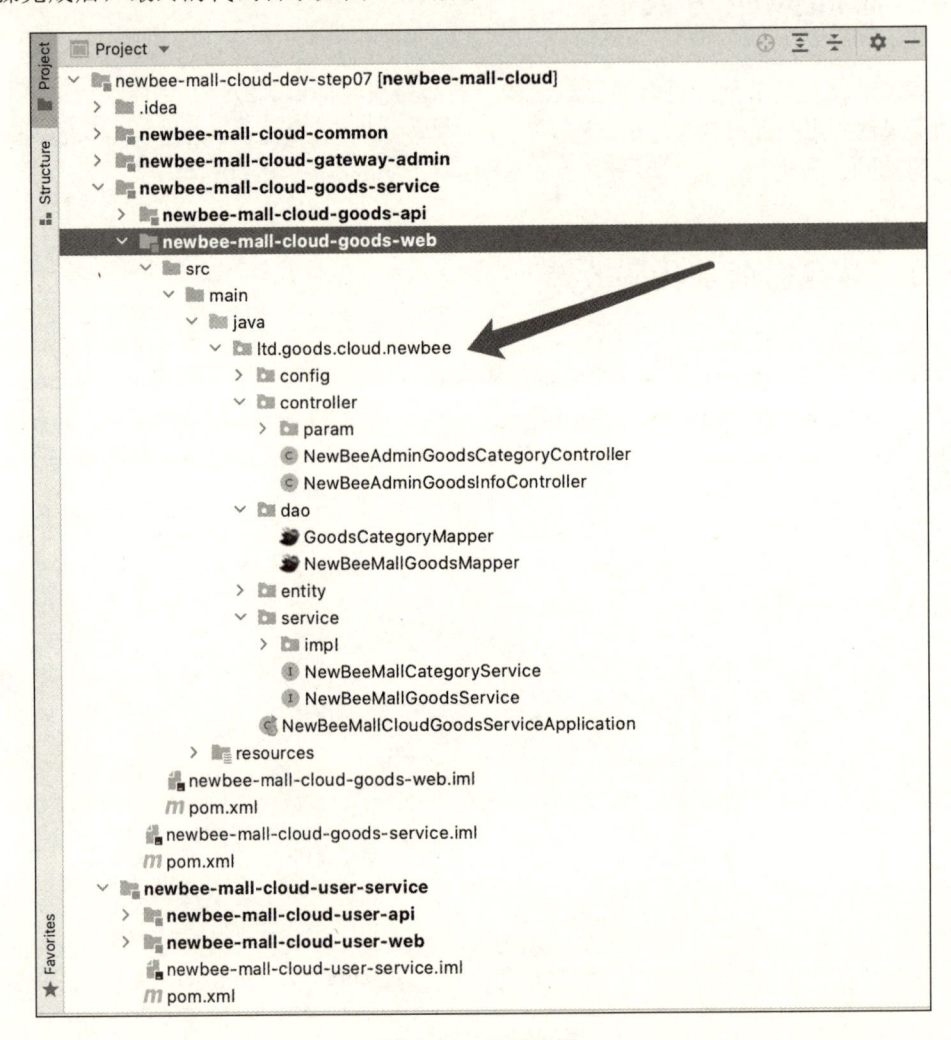

图 4-12 代码目录

修改 newbee-mall-cloud-goods-web 工程的 application.properties 配置文件，主要是数据库连接参数及 MyBatis 扫描配置，代码如下：

```
# datasource config (MySQL)
spring.datasource.name=newbee-mall-cloud-goods-datasource
spring.datasource.driverClassName=com.mysql.cj.jdbc.Driver
```

```
spring.datasource.url=jdbc:mysql://localhost:3306/newbee_mall_cloud_good
s_db?useUnicode=true&serverTimezone=Asia/Shanghai&characterEncoding=utf8
&autoReconnect=true&useSSL=false&allowMultiQueries=true
spring.datasource.username=root
spring.datasource.password=123456
spring.datasource.hikari.minimum-idle=5
spring.datasource.hikari.maximum-pool-size=15
spring.datasource.hikari.auto-commit=true
spring.datasource.hikari.idle-timeout=60000
spring.datasource.hikari.pool-name=hikariCP
spring.datasource.hikari.max-lifetime=600000
spring.datasource.hikari.connection-timeout=30000
spring.datasource.hikari.connection-test-query=SELECT 1

# mybatis config
mybatis.mapper-locations=classpath:mapper/*Mapper.xml
```

这部分源代码涉及的数据库为 newbee_mall_cloud_goods_db，数据库表为 tb_newbee_mall_goods_category 和 tb_newbee_mall_goods_info。

在商品微服务的代码调整前，有些工具类已经被移到公用模块 newbee-mall-cloud-common 中，所以在 pom.xml 文件中需要引入公用模块。同时，代码中使用这些工具类的地方也需要处理一下引用路径。

除此之外，商品微服务中还有一些公用类，都放到了公用模块 newbee-mall-cloud-common 中。

Controller 类中的接口地址都做了微调，与之前单体项目中定义的 URL 不同。调整的原因主要是在网关配置时方便一些。

打开 newbee-mall-cloud-gateway-admin 项目中的 application.properties 文件，新增关于商品微服务的路由信息，配置项为 spring.cloud.gateway.routes.*，新增代码如下：

```
# 商品微服务的路由配置-1
spring.cloud.gateway.routes[1].id=categories-admin-service-route
spring.cloud.gateway.routes[1].uri=lb://newbee-mall-cloud-goods-service
spring.cloud.gateway.routes[1].order=1
spring.cloud.gateway.routes[1].predicates[0]=Path=/categories/admin/**

# 商品微服务的路由配置-2
```

```
spring.cloud.gateway.routes[2].id=goods-admin-service-route
spring.cloud.gateway.routes[2].uri=lb://newbee-mall-cloud-goods-service
spring.cloud.gateway.routes[2].order=1
spring.cloud.gateway.routes[2].predicates[0]=Path=/goods/admin/**
```

这里主要配置 newbee-mall-cloud-gateway-admin 到商品微服务的路由信息。如果访问网关项目的路径是以/goods/admin 或/categories/admin 开头的，就路由到商品微服务实例。

4.4.2 OpenFeign 编码暴露远程接口

在 newbee-mall-cloud-goods-api 模块中新增需要暴露的接口 FeignClient。如果其他微服务实例需要获取商品微服务中的资源，就可以通过调用商品微服务下暴露的接口来实现。

笔者在 NewBeeAdminGoodsInfoController 类中新建了一个商品详情接口，并将其暴露，代码如下：

```
@GetMapping("/goodsDetail")
@ApiOperation(value = "获取单条商品信息", notes = "根据id查询")
public Result goodsDetail(@RequestParam("goodsId") Long goodsId) {
    NewBeeMallGoods goods = newBeeMallGoodsService.getNewBeeMallGoodsById(goodsId);
    return ResultGenerator.genSuccessResult(goods);
}
```

在 newbee-mall-cloud-goods-api 目录下新建 ltd.goods.cloud.newbee.openfeign 包，之后在该包下新增 NewBeeCloudGoodsServiceFeign 类，用于创建对商品模块中相关接口的 OpenFeign 调用。

NewBeeCloudGoodsServiceFeign 类的代码如下：

```
package ltd.goods.cloud.newbee.openfeign;

import ltd.common.cloud.newbee.dto.Result;
import ltd.goods.cloud.newbee.dto.NewBeeMallGoodsDTO;
import org.springframework.cloud.openfeign.FeignClient;
import org.springframework.web.bind.annotation.GetMapping;
import org.springframework.web.bind.annotation.RequestParam;

@FeignClient(value = "newbee-mall-cloud-goods-service", path = "/goods")
public interface NewBeeCloudGoodsServiceFeign {
```

```
@GetMapping(value = "/admin/goodsDetail")
    Result<NewBeeMallGoodsDTO> getGoodsDetail(@RequestParam(value =
"goodsId") Long goodsId);
}
```

这个方法的作用是根据商品 id 获取一条商品数据。调用的接口是 newbee-mall-cloud-goods-web 模块下 NewBeeAdminGoodsInfoController 类中的 goodsDetail() 方法。如果其他服务需要与商品微服务进行远程通信获取相关数据，就可以引入 newbee-mall-cloud-goods-api 模块作为依赖，并且直接调用 NewBeeCloudGoodsServiceFeign 类中的 getGoodsDetail() 方法。

在后续的开发过程中，如果商品微服务中有其他接口需要暴露以供其他微服务调用，就可以继续在 NewBeeCloudGoodsServiceFeign 类中编码。

4.4.3 功能测试

代码修改完成后，测试步骤是不能漏掉的，一定要验证项目是否能正常启动，接口是否能正常调用，防止在代码移动过程中出现一些小问题，导致项目无法启动或代码报错。在项目启动前需要分别启动 Nacos Server 和 Redis Server，之后依次启动 newbee-mall-cloud-user-web 工程、newbee-mall-cloud-goods-web 工程和 newbee-mall-cloud-gateway-admin 工程下的主类。启动成功后，就可以进行本节的功能测试了。

先打开用户微服务的 Swagger 页面，在浏览器中输入如下网址：http://localhost:29000/swagger-ui/index.html。

然后在该页面中使用登录接口获取一个 token 值，用于后续的功能测试。比如，笔者在测试时获取了一个值为 "95eb476a4ac9677488278b8a7e3b2834" 的 token 字段。

接着打开商品微服务的 Swagger 页面，在浏览器中输入如下网址：http://localhost:29010/swagger-ui/index.html。

最后就可以在 Swagger 页面提供的 UI 页面中进行商品微服务的接口测试了，商品微服务接口文档显示的内容如图 4-13 所示。

这里演示一下商品分类相关接口。单击 "新增分类" → "Try it out" 按钮，在参数栏中输入分类名称、分类等级等字段，在登录认证 token 的输入框中输入管理员用户登录接口返回的 token 值，如图 4-14 所示。

图 4-13　商品微服务接口文档显示的内容

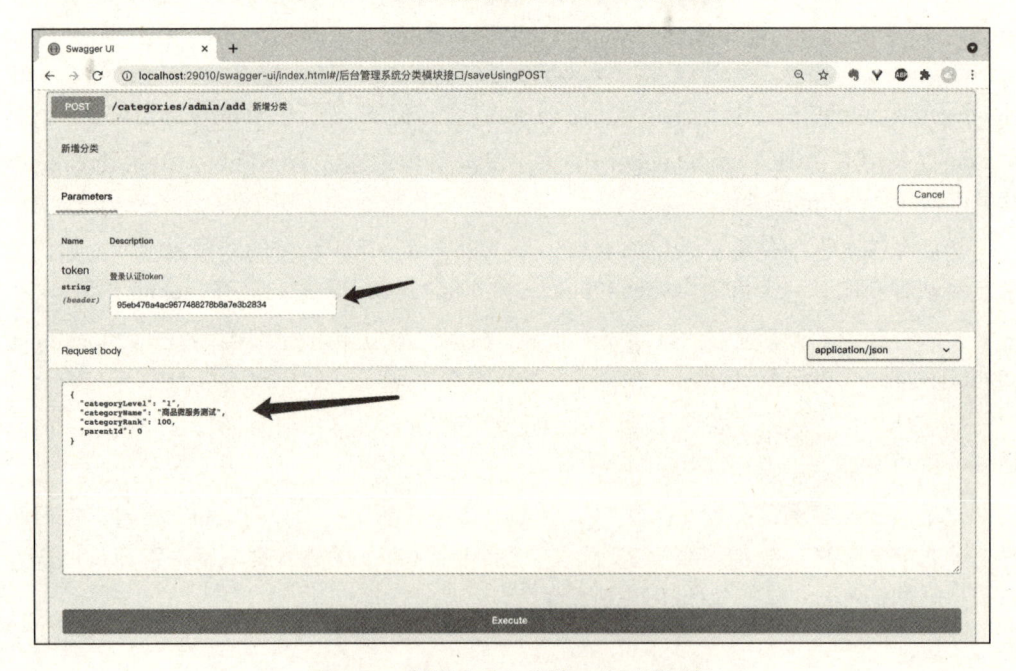

图 4-14　新增商品分类接口的测试过程

单击"Execute"按钮，接口的响应结果如图 4-15 所示。

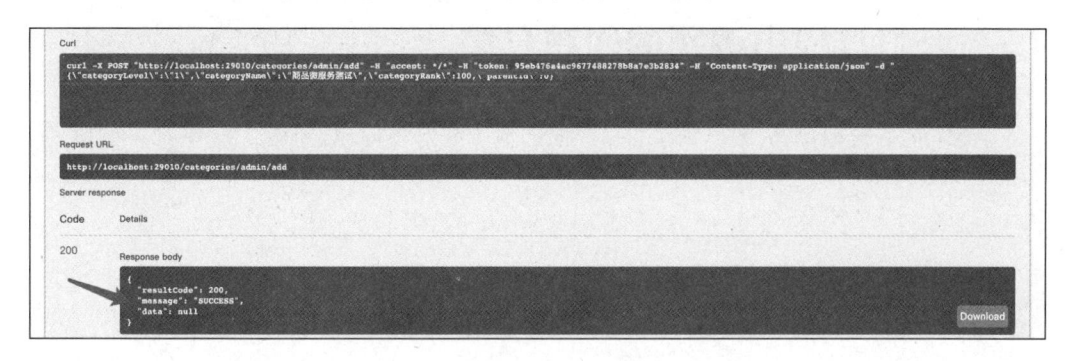

图 4-15 新增商品分类接口的响应结果

若后端接口的测试结果中有"SUCCESS"，则表示数据添加成功，查看商品微服务的数据库，分类表中已经新增了一条数据。笔者在测试时，输入的参数都是符合规范的。如果输入的参数没有通过基本的验证判断，就会报出对应的错误提示，读者在测试时需要注意这一点。

该接口对应到实际的项目页面中，是新蜂商城后台管理系统中的商品分类管理页面，添加商品分类的示例页面如图 4-16 所示。

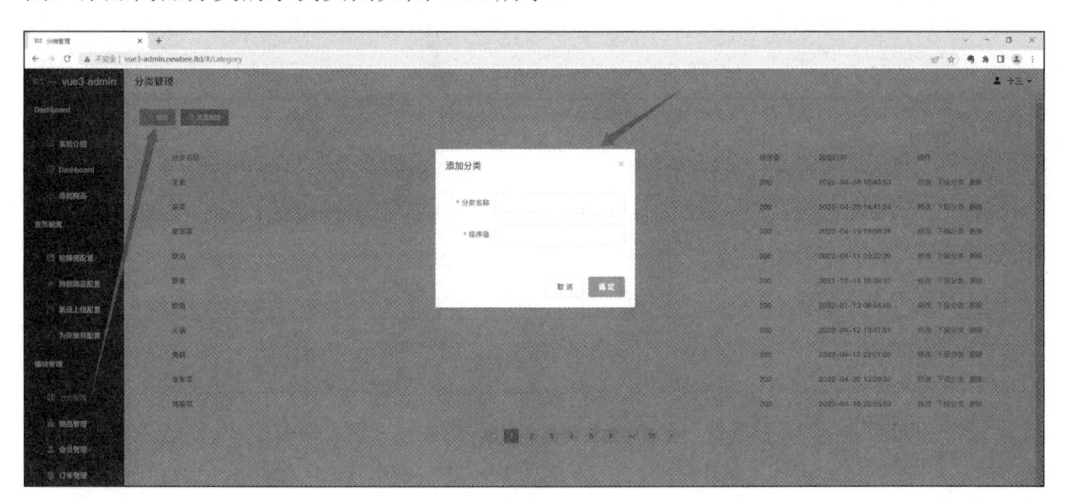

图 4-16 添加商品分类的示例页面

这里只演示了一个接口的测试过程，读者在测试时可以看看其他接口。商品管理模块和商品分类管理模块下的接口对应到实际的页面中，分别是新蜂商城后台管理系统中的商品管理页面和商品分类管理页面。

4.5 改造过程中遇到的问题总结

当然,笔者在商品微服务改造过程中也遇到了几个小问题,在这里分享给各位读者,读者在实战时可以参考。

4.5.1 问题 1:循环依赖

编码完成后,尝试启动项目。商品微服务无法启动,控制台报出了如下异常信息:

```
The dependencies of some of the beans in the application context form a cycle:

┌─────┐
|  goodsServiceWebMvcConfigurer (field private
ltd.goods.cloud.newbee.config.handler.TokenToAdminUserMethodArgumentResolver
ltd.goods.cloud.newbee.config.GoodsServiceWebMvcConfigurer.tokenToAdminUserMethodArgumentResolver)
↑     ↓
|  tokenToAdminUserMethodArgumentResolver (field private
ltd.user.cloud.newbee.openfeign.NewBeeCloudAdminUserServiceFeign
ltd.goods.cloud.newbee.config.handler.TokenToAdminUserMethodArgumentResolver.newBeeCloudAdminUserService)
└─────┘
```

日志中提示的信息足够清晰。在启动过程中发现了循环依赖的问题,从而导致项目无法启动。

以上就是问题的描述:项目无法启动,原因为循环依赖。

笔者想:明明是同样的代码,在单体项目中是正常的,到了微服务架构下怎么就出现问题了呢?

接下来尝试解决这个问题。其实,这个错误与"单体/微服务"无关,主要是 Spring Boot 自 2.6 版本开始就禁用了循环依赖,下面是 Spring Boot 2.6.0 版本发布时的说明:

```
Circular References Prohibited by Default
Circular references between beans are now prohibited by default. If your
application fails to start due to a BeanCurrentlyInCreationException you are
strongly encouraged to update your configuration to break the dependency cycle.
If you are unable to do so, circular references can be allowed again by setting
spring.main.allow-circular-references to true, or using the new setter
```

```
methods on SpringApplication and SpringApplicationBuilder This will restore
2.5's behaviour and automatically attempt to break the dependency cycle.
```

也就是说，如果在单体项目 newbee-mall-api 中升级 Spring Boot 到 2.6 或以上版本，也可能报这个错误。因此，要想解决该问题，只需要解决循环依赖或通过配置让 Spring Boot 不强制禁用循环依赖。

具体到编码层面，可以使用 @Lazy 注解延迟加载某个陷入循环依赖的类，或者通过修改 allow-circular-references 配置项让 Spring Boot 允许循环依赖。以上两种方式都可以解决这个问题，笔者选择的是暂时允许循环依赖。打开 application.properties 配置文件，添加如下配置项即可。

```
spring.main.allow-circular-references=true
```

4.5.2　问题 2：缺少 LoadBalancer 依赖

缺少 LoadBalancer 依赖依然会导致商品微服务无法正常启动，调试过程中控制台报出了如下异常信息：

```
org.springframework.beans.factory.BeanCreationException: Error creating bean
with name 'ltd.user.cloud.newbee.openfeign.NewBeeCloudAdminUserServiceFeign':
Unexpected exception during bean creation; nested exception is
java.lang.IllegalStateException: No Feign Client for loadBalancing defined.
Did you forget to include spring-cloud-starter-loadbalancer?
    at org.springframework.beans.factory.annotation.AutowiredAnnotation
BeanPostProcessor$AutowiredFieldElement.resolveFieldValue(AutowiredAnnot
ationBeanPostProcessor.java:659) ~[spring-beans-5.3.15.jar:5.3.15]
    at org.springframework.beans.factory.annotation.AutowiredAnnotation
BeanPostProcessor$AutowiredFieldElement.inject(AutowiredAnnotationBeanPo
stProcessor.java:639) ~[spring-beans-5.3.15.jar:5.3.15]
```

出现这个问题的原因非常简单：未引入负载均衡器。虽然简单，但是在实战过程中遇到了几次，主要是笔者在开发时粗心导致的。

在对应模块的 pom.xml 文件中加入 LoadBalancer 依赖即可。

```xml
<dependency>
  <groupId>org.springframework.cloud</groupId>
  <artifactId>spring-cloud-starter-loadbalancer</artifactId>
</dependency>
```

好的，后台管理系统中的商品管理模块接口和商品分类管理模块接口就改造完成了。不过，商品微服务中的接口不止这些，后续在改造商城端功能的时候还会继续在该模块下编码。希望读者能够根据笔者提供的开发步骤顺利地完成本节的项目改造。

第 5 章

推荐微服务编码实践及功能讲解

本章继续讲解微服务架构项目的改造内容,主要介绍推荐微服务的功能及开发步骤。推荐微服务主要包括两个功能模块,分别是轮播图管理模块和商品推荐管理模块。其中,涉及的接口依然是后台管理系统所需要的接口,即商品管理模块的接口,商城端所需要的接口会在后续章节中讲解。

5.1 推荐微服务主要功能模块介绍

5.1.1 轮播图管理模块介绍

横跨屏幕的轮播图是时下比较流行的网页设计。网站设计师通过这种覆盖用户视线的图片,给用户营造一种身临其境的视觉感受,非常符合人类视觉优先的信息获取方式。大部分网站会在首页使用这种设计方式,优质的首图能够让用户了解网站的推荐内容。

购物网站在首页轮播图中会展示各种推荐商品、优惠活动等。在这个区域,网站管理者可以放置吸引人的商品图片、不久后即将上线的主力产品,还可以放置用户最关心的促销通知等。淘宝网、京东商城、华为商城等都采取这种首页轮播图的网页设计方式。

华为商城首页的轮播图效果如图 5-1 所示。

图 5-1　华为商城首页的轮播图效果

京东商城首页的轮播图效果如图 5-2 所示。

图 5-2　京东商城首页的轮播图效果

笔者在新蜂商城首页顶部也添加了轮播图，显示效果如图 5-3 所示。

在首页接口中，只读取轮播图数据即可。配置轮播图可以在后台管理系统中操作，如添加、修改和删除轮播图，如图 5-4 所示。

图 5-3　新蜂商城首页顶部的轮播图效果

图 5-4　轮播图管理页面

接下来的两个小节主要讲解在推荐微服务中后台管理模块的接口开发与实现，获取首页数据的接口实现比较简单，后续章节中再进行介绍。

5.1.2　商品推荐管理模块介绍

除轮播图外，新蜂商城首页还有三个板块需要进行数据渲染，分别是热销商品板块、新品上线板块和推荐商品板块。

在商城首页中设计这三个板块，主要是为了丰富版面布局，使页面不单调。

当然，这部分的设计也参考了当前主流线上商城的商品推荐设计，不过这些线上商城都有大量的正式数据做支撑，做的肯定要比新蜂商城复杂得多。比如，热销商品一定是在大量实际订单的统计下做出来的数据渲染；又如，商品推荐也一定是在用户的浏览痕迹和下单习惯的基础上计算出来的。目前，新蜂商城的开发人员只有笔者一个人，订单也只有模拟数据，要做出淘宝网、京东商城那种效果是不现实的。因此，新蜂商城中的热销商品、新品上线、推荐商品这三个板块中的数据是在后台中配置的，首页渲染前直接读取数据就可以了，这些数据并没有进行实时的数据统计。

由于新蜂商城只是技术实战项目，因此其订单、浏览痕迹和用户习惯等数据都是模拟的，并非实时统计的。这三个板块中的内容是在后台进行配置的，首页配置管理页面如图 5-5 所示。

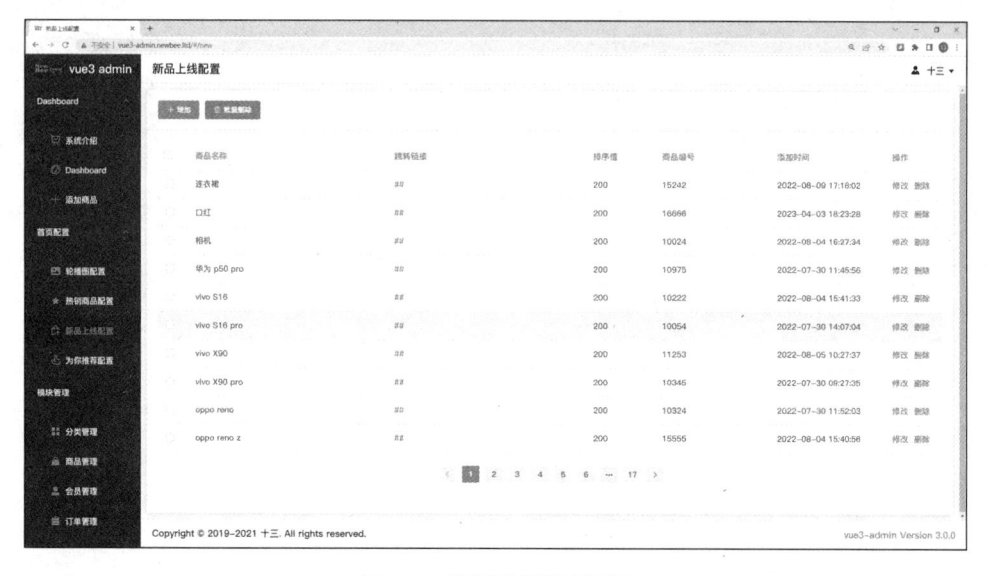

图 5-5　首页配置管理页面

5.1.3 表结构设计

首页的轮播图数据和推荐商品数据,主要是通过读取 tb_newbee_mall_carousel 表和 tb_newbee_mall_index_config 表获得的。在接口实现时,还需要查询商品表 tb_newbee_mall_goods_info 中的数据,但是该表在另一个微服务的数据库中。因此,笔者采用远程调用商品微服务的方式来获取这部分数据,完成响应功能的实现。推荐微服务涉及的两张表的表结构如下:

```sql
# 创建推荐微服务所需数据
CREATE DATABASE /*!32312 IF NOT EXISTS*/'newbee_mall_cloud_recommend_db' /*!40100 DEFAULT CHARACTER SET utf8 */;

USE 'newbee_mall_cloud_recommend_db';

# 创建首页推荐表

DROP TABLE IF EXISTS 'tb_newbee_mall_index_config';

CREATE TABLE 'tb_newbee_mall_index_config' (
  'config_id' bigint(20) NOT NULL AUTO_INCREMENT COMMENT '首页配置项主键id',
  'config_name' varchar(50) NOT NULL DEFAULT '' COMMENT '显示字符(配置搜索时不可为空,其他可为空)',
  'config_type' tinyint(4) NOT NULL DEFAULT '0' COMMENT '1-搜索框热搜 2-搜索下拉框热搜 3-(首页)热销商品 4-(首页)新品上线 5-(首页)为你推荐',
  'goods_id' bigint(20) NOT NULL DEFAULT '0' COMMENT '商品id 默认为0',
  'redirect_url' varchar(100) NOT NULL DEFAULT '##' COMMENT '单击后的跳转地址(默认不跳转)',
  'config_rank' int(11) NOT NULL DEFAULT '0' COMMENT '排序值(字段越大越靠前)',
  'is_deleted' tinyint(4) NOT NULL DEFAULT '0' COMMENT '删除标识字段(0-未删除 1-已删除)',
  'create_time' timestamp NOT NULL DEFAULT CURRENT_TIMESTAMP COMMENT '创建时间',
  'create_user' int(11) NOT NULL DEFAULT '0' COMMENT '创建者id',
  'update_time' timestamp NOT NULL DEFAULT CURRENT_TIMESTAMP COMMENT '最新修改时间',
  'update_user' int(11) DEFAULT '0' COMMENT '修改者id',
  PRIMARY KEY ('config_id')
) ENGINE=InnoDB DEFAULT CHARSET=utf8;
```

新增首页推荐表数据

```sql
INSERT INTO tb_newbee_mall_index_config
(config_name, config_type, goods_id, redirect_url, config_rank, is_deleted,
create_time, create_user, update_time, update_user)
VALUES
('热销商品 iPhone 12', 3, 10906, '##', 201, 0, '2021-03-08 18:55:49', 0,
'2021-03-08 18:55:49', 0),
('热销商品 华为Mate 40 Pro', 3, 10908, '##', 300, 0, '2021-03-08 18:55:49',
0, '2021-03-08 18:55:49', 0),
('新品上线 MackBook2021', 4, 10920, '##', 180, 0, '2021-03-08 18:55:49', 0,
'2021-03-08 18:55:49', 0),
('新品上线 华为 P50 Pro', 4, 10921, '##', 160, 0, '2021-03-08 18:55:49', 0,
'2021-03-08 18:55:49', 0),
('新品上线 Apple Watch', 4, 10919, '##', 101, 0, '2021-03-08 18:55:49', 0,
'2021-03-08 18:55:49', 0),
('纪梵希高定香榭天鹅绒唇膏', 5, 10233, '##', 80, 0, '2021-03-08 18:55:49', 0,
'2021-03-08 18:55:49', 0),
('P50 白色', 5, 10922, '##', 102, 0, '2021-03-08 18:55:49', 0, '2021-03-08
18:55:49', 0),
('free buds pro', 5, 10930, '##', 102, 0, '2021-03-08 18:55:49', 0, '2021-03-08
18:55:49', 0),
('iPhone 13', 5, 10916, '##', 101, 0, '2021-03-08 18:55:49', 0, '2021-03-08
18:55:49', 0),
('华为Mate 40 Pro', 5, 10907, '##', 80, 0, '2021-03-08 18:55:49', 0,
'2021-03-08 18:55:49', 0),
('MacBook Pro 2021', 5, 10920, '##', 100, 0, '2021-03-08 18:55:49', 0,
'2021-03-08 18:55:49', 0),
('WATCH 3 Pro', 5, 10928, '##', 99, 0, '2021-03-08 18:55:49', 0, '2021-03-08
18:55:49', 0),
('塑料浴室座椅', 5, 10154, '##', 80, 0, '2021-03-08 18:55:49', 0, '2021-03-08
18:55:49', 0),
('华为 soundx', 5, 10929, '##', 100, 0, '2021-03-08 18:55:49', 0, '2021-03-08
18:55:49', 0),
('matepad pro', 5, 10906, '##', 0, 0, '2021-03-08 18:55:49', 0, '2021-03-08
18:55:49', 0),
('热销商品 P40', 3, 10902, '##', 200, 0, '2021-03-08 18:55:49', 0, '2021-03-08
18:55:49', 0),
('新品上线 华为 P50 Pocket', 4, 10925, '##', 200, 0, '2021-03-08 18:55:49', 0,
'2021-03-08 18:55:49', 0),
('新品上线 华为Mate X Pro', 4, 10926, '##', 200, 0, '2021-03-08 18:55:49', 0,
'2021-03-08 18:55:49', 0),
```

```sql
('华为 Mate 30 Pro', 5, 10927, '##', 101, 0, '2021-03-08 18:55:49', 0,
'2021-03-08 18:55:49', 0),
('新品上线 iPhone13', 4, 10915, '##', 190, 0, '2021-03-08 18:55:49', 0,
'2021-03-08 18:55:49', 0),
('Air Pods 第三代', 3, 10918, '##', 301, 0, '2021-03-08 18:55:49', 0,
'2021-03-08 18:55:49', 0);

DROP TABLE IF EXISTS 'tb_newbee_mall_carousel';

# 创建轮播图表

CREATE TABLE 'tb_newbee_mall_carousel' (
  'carousel_id' int(11) NOT NULL AUTO_INCREMENT COMMENT '首页轮播图主键id',
  'carousel_url' varchar(100) NOT NULL DEFAULT '' COMMENT '轮播图',
  'redirect_url' varchar(100) NOT NULL DEFAULT '''##''' COMMENT '点击后的跳转地址(默认不跳转)',
  'carousel_rank' int(11) NOT NULL DEFAULT '0' COMMENT '排序值(字段越大越靠前)',
  'is_deleted' tinyint(4) NOT NULL DEFAULT '0' COMMENT '删除标识字段(0-未删除 1-已删除)',
  'create_time' timestamp NOT NULL DEFAULT CURRENT_TIMESTAMP COMMENT '创建时间',
  'create_user' int(11) NOT NULL DEFAULT '0' COMMENT '创建者id',
  'update_time' timestamp NOT NULL DEFAULT CURRENT_TIMESTAMP COMMENT '修改时间',
  'update_user' int(11) NOT NULL DEFAULT '0' COMMENT '修改者id',
  PRIMARY KEY ('carousel_id') USING BTREE
) ENGINE=InnoDB DEFAULT CHARSET=utf8 ROW_FORMAT=DYNAMIC;

# 新增轮播图数据

INSERT INTO 'tb_newbee_mall_carousel' ('carousel_id', 'carousel_url',
'redirect_url', 'carousel_rank', 'is_deleted', 'create_time', 'create_user',
'update_time', 'update_user')
VALUES
(1,'https://newbee-ma**.oss-cn-beijing.aliyuncs.com/images/banner2.jpg',
'##',200,1,'2021-08-23 17:50:45',0,'2021-11-10 00:23:01',0),
(2,'https://newbee-ma**.oss-cn-beijing.aliyuncs.com/images/banner1.png',
'https://juej**.cn/book/7085254558678515742',13,0,'2021-11-29 00:00:00',0,'2021-11-29 00:00:00',0),
(3,'https://newbee-ma**.oss-cn-beijing.aliyuncs.com/images/banner3.jpg',
'##',0,1,'2021-09-18 18:26:38',0,'2021-11-10 00:23:01',0),
(5,'https://newbee-ma**.oss-cn-beijing.aliyuncs.com/images/banner2.png',
```

```
'https://juej**.cn/book/7085254538678515742',0,0,'2021-11-29
00:00:00',0,'2021-11-29 00:00:00',0),
(6,'https://newbee-ma**.oss-cn-beijing.aliyuncs.com/images/banner1.png',
'##',101,1,'2021-09-19 23:37:40',0,'2021-11-07 00:15:52',0),
(7,'https://newbee-ma**.oss-cn-beijing.aliyuncs.com/images/banner2.png',
'##',99,1,'2021-09-19 23:37:58',0,'2021-10-22 00:15:01',0);
```

轮播图表和首页推荐表的字段和每个字段对应的含义在上面的 SQL 语句中都有介绍，读者可以对照理解，并正确地把建表 SQL 语句导入数据库。如果有需要，读者也可以自行根据该 SQL 语句进行扩展。

5.2 创建推荐微服务编码

前文主要介绍功能模块的由来及表结构字段设计，接下来讲解具体的代码改造过程。本节的源代码是在 newbee-mall-cloud-dev-step07 工程的基础上改造的，将工程命名为 newbee-mall-cloud-dev-step08。

在工程中新增一个 newbee-mall-cloud-recommend-service 模块，并在 pom.xml 主文件中增加该模块的配置，代码如下：

```
<modules>
  <!-- 新增推荐微服务 -->
  <module>newbee-mall-cloud-recommend-service</module>
  <module>newbee-mall-cloud-goods-service</module>
  <module>newbee-mall-cloud-user-service</module>
  <module>newbee-mall-cloud-gateway-admin</module>
  <module>newbee-mall-cloud-common</module>
</modules>
```

该模块的目录结构设置与其他的功能模块的目录结构设置类似，如下所示：

```
newbee-mall-cloud-recommend-service          // 推荐微服务
    ├── newbee-mall-cloud-recommend-api      // 存放商品推荐管理模块中暴露的用于
                                                远程调用的 FeignClient 类
    └── newbee-mall-cloud-recommend-web      // 商品推荐管理模块 API 的代码及逻辑
```

在新增推荐微服务时，主要参考了当时新增商品微服务时的步骤。笔者直接将 newbee-mall-cloud-dev-step05 源代码下 newbee-mall-cloud-goods-service 模块中的代码复制了过来，并修改了模块名称和目录名称。

最终，子节点 newbee-mall-cloud-recommend-service 模块的 pom.xml 文件代码如下：

```
<?xml version="1.0" encoding="UTF-8"?>
```

```xml
<project xmlns="http://maven.apa***.org/POM/4.0.0" xmlns:xsi=
"http://www.w*.org/2001/XMLSchema-instance"
        xsi:schemaLocation="http://maven.apa***.org/POM/4.0.0
https://maven.apa***.org/xsd/maven-4.0.0.xsd">
    <modelVersion>4.0.0</modelVersion>
    <groupId>ltd.newbee.cloud</groupId>
    <artifactId>newbee-mall-cloud-recommend-service</artifactId>
    <version>0.0.1-SNAPSHOT</version>
    <packaging>pom</packaging>
    <name>newbee-mall-cloud-recommend-service</name>
    <description>商品推荐管理模块</description>

    <parent>
        <groupId>ltd.newbee.cloud</groupId>
        <artifactId>newbee-mall-cloud</artifactId>
        <version>0.0.1-SNAPSHOT</version>
    </parent>

    <properties>
        <java.version>1.8</java.version>
    </properties>

    <modules>
        <module>newbee-mall-cloud-recommend-web</module>
        <module>newbee-mall-cloud-recommend-api</module>
    </modules>

    <dependencies>

    </dependencies>

</project>
```

主要的配置项是模块名称、模块的打包方式及模块间的父子关系。其中，重点是配置父模块 newbee-mall-cloud，配置的两个子模块分别是 newbee-mall-cloud-recommend-api 和 newbee-mall-cloud-recommend-web。

newbee-mall-cloud-recommend-api 模块的 pom.xml 文件代码修改如下：

```xml
<?xml version="1.0" encoding="UTF-8"?>
<project xmlns="http://maven.apache.org/POM/4.0.0" xmlns:xsi=
"http://www.w3.org/2001/XMLSchema-instance"
        xsi:schemaLocation="http://maven.apache.org/POM/4.0.0
https://maven.apache.org/xsd/maven-4.0.0.xsd">
```

```xml
    <modelVersion>4.0.0</modelVersion>
    <groupId>ltd.recommend.newbee.cloud</groupId>
    <artifactId>newbee-mall-cloud-recommend-api</artifactId>
    <packaging>jar</packaging>
    <version>0.0.1-SNAPSHOT</version>
    <name>newbee-mall-cloud-recommend-api</name>
    <description>推荐微服务 openfeign</description>

    <parent>
        <groupId>ltd.newbee.cloud</groupId>
        <artifactId>newbee-mall-cloud-recommend-service</artifactId>
        <version>0.0.1-SNAPSHOT</version>
    </parent>

    <properties>
        <java.version>1.8</java.version>
    </properties>

    <dependencies>

        <dependency>
            <groupId>org.springframework.cloud</groupId>
            <artifactId>spring-cloud-starter-openfeign</artifactId>
        </dependency>

        <dependency>
            <groupId>ltd.newbee.cloud</groupId>
            <artifactId>newbee-mall-cloud-common</artifactId>
            <version>0.0.1-SNAPSHOT</version>
        </dependency>

    </dependencies>
</project>
```

这里定义了与父模块 newbee-mall-cloud-recommend-service 的关系,打包方式 packaging 配置项的值为 jar。同时,由于后续可能要将接口暴露,因此这里添加 FeignClient 类,引入了 OpenFeign 的依赖项。

newbee-mall-cloud-recommend-web 模块的 pom.xml 文件代码修改如下:

```
<?xml version="1.0" encoding="UTF-8"?>
<project xmlns="http://maven.apa***.org/POM/4.0.0" xmlns:xsi=
"http://www.w*.org/2001/XMLSchema-instance"
    xsi:schemaLocation="http://maven.apa***.org/POM/4.0.0
```

```xml
https://maven.apa***.org/xsd/maven-4.0.0.xsd">
    <modelVersion>4.0.0</modelVersion>
    <groupId>ltd.recommend.newbee.cloud</groupId>
    <artifactId>newbee-mall-cloud-recommend-web</artifactId>
    <version>0.0.1-SNAPSHOT</version>
    <name>newbee-mall-cloud-recommend-web</name>
    <description>推荐微服务</description>

    <parent>
        <groupId>ltd.newbee.cloud</groupId>
        <artifactId>newbee-mall-cloud-recommend-service</artifactId>
        <version>0.0.1-SNAPSHOT</version>
    </parent>

    <properties>
        <java.version>1.8</java.version>
    </properties>

    <dependencies>
        <dependency>
            <groupId>org.springframework.boot</groupId>
            <artifactId>spring-boot-starter-web</artifactId>
        </dependency>

        <dependency>
            <groupId>org.springframework.boot</groupId>
            <artifactId>spring-boot-starter-test</artifactId>
            <scope>test</scope>
        </dependency>

        <dependency>
            <groupId>com.alibaba.cloud</groupId>
            <artifactId>spring-cloud-starter-alibaba-nacos-discovery</artifactId>
        </dependency>

        <dependency>
            <groupId>org.springframework.boot</groupId>
            <artifactId>spring-boot-starter-validation</artifactId>
        </dependency>

        <dependency>
            <groupId>org.mybatis.spring.boot</groupId>
```

```xml
        <artifactId>mybatis-spring-boot-starter</artifactId>
</dependency>

<dependency>
    <groupId>org.projectlombok</groupId>
    <artifactId>lombok</artifactId>
    <version>${lombok.version}</version>
    <scope>provided</scope>
</dependency>

<dependency>
    <groupId>io.springfox</groupId>
    <artifactId>springfox-boot-starter</artifactId>
</dependency>

<dependency>
    <groupId>mysql</groupId>
    <artifactId>mysql-connector-java</artifactId>
    <scope>runtime</scope>
</dependency>

<dependency>
    <groupId>ltd.newbee.cloud</groupId>
    <artifactId>newbee-mall-cloud-common</artifactId>
    <version>0.0.1-SNAPSHOT</version>
</dependency>

<dependency>
    <groupId>org.springframework.cloud</groupId>
    <artifactId>spring-cloud-starter-openfeign</artifactId>
</dependency>

<dependency>
    <groupId>org.springframework.cloud</groupId>
    <artifactId>spring-cloud-starter-loadbalancer</artifactId>
</dependency>

<dependency>
    <groupId>ltd.user.newbee.cloud</groupId>
    <artifactId>newbee-mall-cloud-user-api</artifactId>
    <version>0.0.1-SNAPSHOT</version>
</dependency>
```

```
            </dependencies>
</project>
```

这里定义了与父模块 newbee-mall-cloud-recommend-service 的关系，同时将相关的业务依赖项移到该配置文件中，因为轮播图管理和商品推荐管理模块的主要业务代码都写在这个模块中。

因为改造过程中直接复制了 newbee-mall-cloud-dev-step05 源代码下 newbee-mall-cloud-goods-service 模块中的代码，所以在修改完依赖配置后，就要修改包名，把 ltd.goods.cloud.××× 的名称修改为 ltd.recommend.cloud.×××，之后修改 config 包中的代码，包括全局异常处理配置类、Swagger 配置类、自定义 MVC 配置类，主要修改了这些类的类名。这样就完成了一个推荐微服务的初始构建工作，此时的 newbee-mall-cloud- recommend-service 模块中并没有业务代码，目录结构如图 5-6 所示。

```
v  newbee-mall-cloud-dev-step08 [newbee-mall-cloud]
   >  .idea
   >  newbee-mall-cloud-common
   >  newbee-mall-cloud-gateway-admin
   >  newbee-mall-cloud-goods-service
   v  newbee-mall-cloud-recommend-service
      v  newbee-mall-cloud-recommend-api
            m pom.xml
      v  newbee-mall-cloud-recommend-web
         v  src
            v  main
               v  java
                  v  ltd.recommend.cloud.newbee
                     >  config
                     >  entity
                        c NewBeeMallCloudRecommendServiceApplication
            >  resources
         m pom.xml
      m pom.xml
   >  newbee-mall-cloud-user-service
   m pom.xml
```

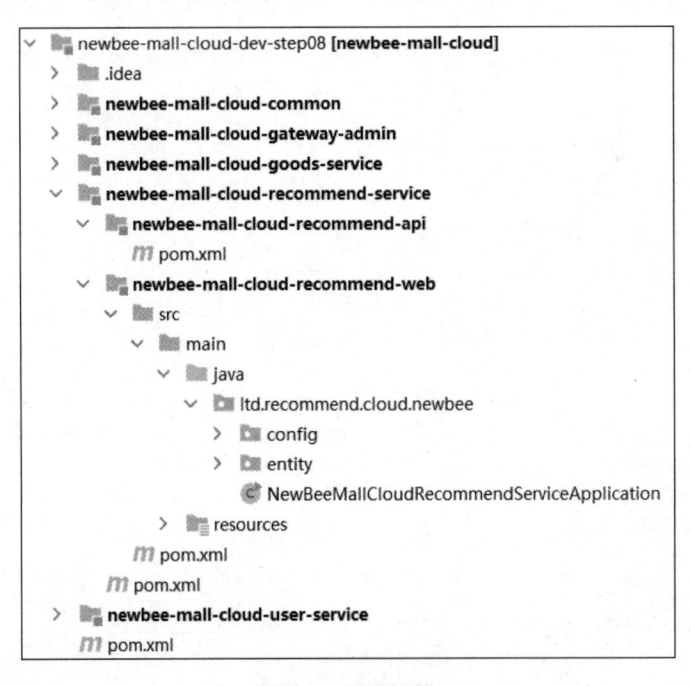

图 5-6　目录结构

本节中并未涉及具体的业务编码，主要介绍推荐微服务的模块功能和表结构设计，以及在项目中完成推荐微服务的初始化构建，后续关于推荐微服务改造的实战章节都是基于当前项目完成的。读者可以根据这些代码自己动手完成服务化编码，如拿到这份代码之后，以此为基础，自行完成推荐微服务模块所有代码的功能。如果自己实现耗费时间，也可以按照笔者给出的步骤完成服务化拆分。

5.3 推荐微服务编码

推荐微服务的源代码是在 newbee-mall-cloud-dev-step08 工程的基础上改造的，将工程命名为 newbee-mall-cloud-dev-step09。

5.3.1 推荐微服务调用用户微服务编码实践

与商品微服务一样，在推荐微服务中也需要对请求中的 token 字段进行处理，因此需要远程调用用户微服务。

第一步，增加配置，启用 OpenFeign 并使 FeignClient 类生效。

由于已经引入 LoadBalancer 依赖和 user-api 依赖，因此这里可以直接通过 OpenFeign 来调用用户微服务中的接口用于鉴权。

打开 newbee-mall-cloud-recommend-web 工程，在项目的启动类 NewBeeMallCloudRecommendServiceApplication 上添加@EnableFeignClients 注解，并配置相关的 FeignClient 类，代码如下：

```
@EnableFeignClients(basePackageClasses = {ltd.user.cloud.newbee.openfeign.
NewBeeCloudAdminUserServiceFeign.class})
```

这里使用 basePackageClasses 配置了需要的 FeignClient 类，即 NewBeeCloudAdminUserServiceFeign 类。接下来就可以直接使用 NewBeeCloudAdmin UserServiceFeign 与用户微服务进行远程通信了。

第二步，修改 token 处理类中的逻辑代码。

打开 newbee-mall-cloud-recommend-web 工程，修改 TokenToAdminUserMethodArgumentResolver 类中对 token 字段处理的逻辑代码，主要引入 NewBeeCloudAdmin UserServiceFeign 类，通过调用用户微服务来获取管理员用户的数据。

修改后的代码如下：

```
@Component
public class TokenToAdminUserMethodArgumentResolver implements
HandlerMethodArgumentResolver {

    @Autowired
    private NewBeeCloudAdminUserServiceFeign newBeeCloudAdminUserService;
```

```java
    public TokenToAdminUserMethodArgumentResolver() {
    }

    public boolean supportsParameter(MethodParameter parameter) {
        if (parameter.hasParameterAnnotation(TokenToAdminUser.class)) {
            return true;
        }
        return false;
    }

    public Object resolveArgument(MethodParameter parameter, ModelAndViewContainer mavContainer, NativeWebRequest webRequest, WebDataBinderFactory binderFactory) {
        if (parameter.getParameterAnnotation(TokenToAdminUser.class) instanceof TokenToAdminUser) {
            String token = webRequest.getHeader("token");
            if (null != token && !"".equals(token) && token.length() == 32) {
                // 通过用户微服务获取用户信息
                Result result = newBeeCloudAdminUserService.getAdminUserByToken(token);

                if (result == null || result.getResultCode() != 200 || result.getData() == null) {
                    NewBeeMallException.fail("ADMIN_NOT_LOGIN_ERROR");
                }

                LinkedHashMap resultData = (LinkedHashMap) result.getData();

                // 将返回的字段封装到 LoginAdminUser 对象中
                LoginAdminUser loginAdminUser = new LoginAdminUser();
                loginAdminUser.setAdminUserId(Long.valueOf(resultData.get("adminUserId").toString()));
                loginAdminUser.setLoginUserName((String) resultData.get("loginUserName"));
                loginAdminUser.setNickName((String) resultData.get("nickName"));
                loginAdminUser.setLocked(Byte.valueOf(resultData.get("locked").toString()));
                return loginAdminUser;
            } else {
                NewBeeMallException.fail("ADMIN_NOT_LOGIN_ERROR");
            }
        }
        return null;
    }
}
```

这里的编码和代码逻辑与商品微服务中的 token 处理类的编码和代码逻辑一致。如此，便完成了在推荐微服务中通过远程通信获取当前登录用户的功能。

5.3.2 推荐微服务编码

下面补充推荐微服务中的业务代码，主要把原来单体项目中的功能模块整合到推荐微服务中。

打开推荐微服务 newbee-mall-cloud-recommend-web 的工程目录，在 ltd.recommend.cloud.newbee 包下依次创建 config 包、dao 包、entity 包、service 包，在 resources 目录下新增 mapper 文件夹用于存放 Mapper 文件。直接将原单体 API 项目中与轮播图管理和商品推荐管理相关的业务代码及 Mapper 文件依次复制过来，如图 5-7 所示。

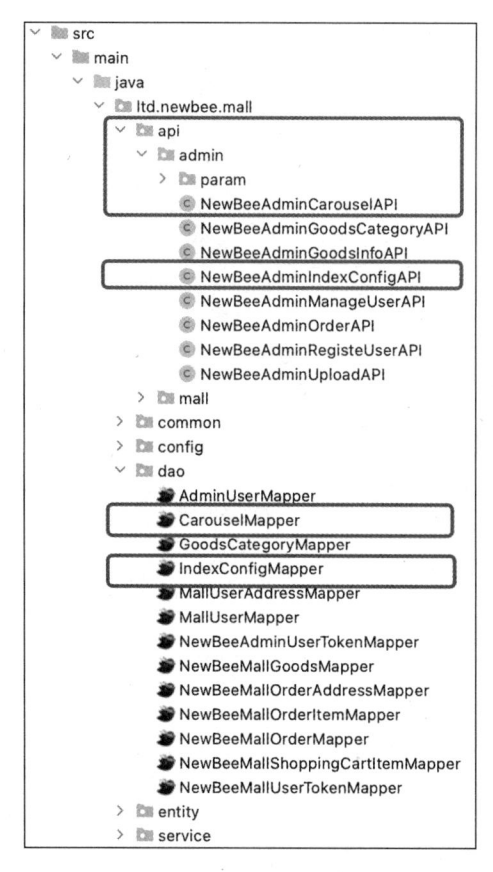

图 5-7　原单体 API 项目中与轮播图管理和商品推荐管理相关的业务代码及 Mapper 文件

由于代码量较大，这里就不一一介绍了，按照对应的文件目录复制过来即可。上述步骤完成后，最终的目录结构如图 5-8 所示。

```
v  newbee-mall-cloud-dev-step09 [newbee-mall-cloud]
   >  .idea
   >  newbee-mall-cloud-common
   >  newbee-mall-cloud-gateway-admin
   >  newbee-mall-cloud-goods-service
   v  newbee-mall-cloud-recommend-service
      >  newbee-mall-cloud-recommend-api
      v  newbee-mall-cloud-recommend-web
         v  src
            v  main
               v  java
                  v  ltd.recommend.cloud.newbee
                     >  config
                     v  controller
                        >  param
                           NewBeeAdminCarouselController
                           NewBeeAdminIndexConfigController
                     v  dao
                        CarouselMapper
                        IndexConfigMapper
                     >  entity
                     v  service
                        >  impl
                        NewBeeMallCarouselService
                        NewBeeMallIndexConfigService
                        NewBeeMallCloudRecommendServiceApplication
            >  resources
         newbee-mall-cloud-recommend-web.iml
         pom.xml
      newbee-mall-cloud-recommend-service.iml
      pom.xml
   >  newbee-mall-cloud-user-service
   newbee-mall-cloud.iml
   pom.xml
```

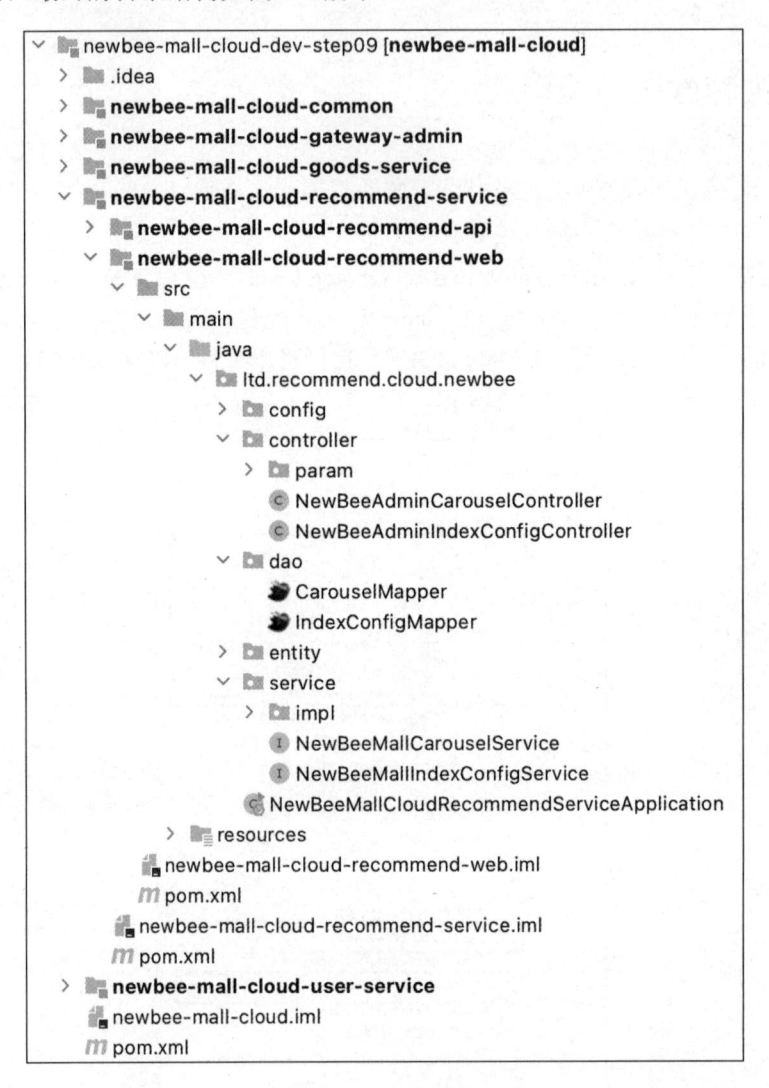

图 5-8　目录结构

最后，修改 newbee-mall-cloud-recommend-web 工程的 application.properties 配置文件，主要是数据库连接参数及 MyBatis 扫描配置，代码如下：

```
# datasource config (MySQL)
spring.datasource.name=newbee-mall-cloud-recommend-datasource
spring.datasource.driverClassName=com.mysql.cj.jdbc.Driver
```

```
spring.datasource.url=jdbc:mysql://localhost:3306/newbee_mall_cloud_reco
mmend_db?useUnicode=true&serverTimezone=Asia/Shanghai&characterEncoding=
utf8&autoReconnect=true&useSSL=false
spring.datasource.username=root
spring.datasource.password=123456
spring.datasource.hikari.minimum-idle=5
spring.datasource.hikari.maximum-pool-size=15
spring.datasource.hikari.auto-commit=true
spring.datasource.hikari.idle-timeout=60000
spring.datasource.hikari.pool-name=hikariCP
spring.datasource.hikari.max-lifetime=600000
spring.datasource.hikari.connection-timeout=30000
spring.datasource.hikari.connection-test-query=SELECT 1

# mybatis config
mybatis.mapper-locations=classpath:mapper/*Mapper.xml
```

本步骤中的源代码涉及的数据库为 newbee_mall_cloud_recommend_db，数据库表为 tb_newbee_mall_index_config 和 tb_newbee_mall_carousel。

推荐微服务中虽然创建了 recommend-api 模块，但是该微服务中暂时没有需要暴露的接口，因此并没有在该模块中编码。

最后，打开 newbee-mall-cloud-gateway-admin 项目中的 application.properties 文件，新增关于推荐微服务的路由信息，配置项为 spring.cloud.gateway.routes.*，新增内容如下：

```
# 推荐微服务的路由配置-1
spring.cloud.gateway.routes[3].id=carousels-admin-service-route
spring.cloud.gateway.routes[3].uri=lb://newbee-mall-cloud-recommend-serv
ice
spring.cloud.gateway.routes[3].order=1
spring.cloud.gateway.routes[3].predicates[0]=Path=/carousels/admin/**

# 推荐微服务的路由配置-2
spring.cloud.gateway.routes[4].id=indexConfigs-admin-service-route
spring.cloud.gateway.routes[4].uri=lb://newbee-mall-cloud-recommend-serv
ice
spring.cloud.gateway.routes[4].order=1
spring.cloud.gateway.routes[4].predicates[0]=Path=/indexConfigs/admin/**
```

这里主要配置 newbee-mall-cloud-gateway-admin 到推荐微服务的路由信息。如果访问网关项目的路径是以/carousels/admin 或/indexConfigs/admin 开头的，就会被路由到推荐微服务实例。

5.3.3 推荐微服务远程调用商品微服务编码实践

按照上述步骤对推荐微服务进行编码后,代码中依然会标红。被标红的代码及注释如下:

```java
@Override
// 添加首页配置数据
public String saveIndexConfig(IndexConfig indexConfig) {
  // 根据参数 goodsId 查询数据库中对应的商品数据是否存在
  if (goodsMapper.selectByPrimaryKey(indexConfig.getGoodsId()) == null) {
    // 若商品数据不存在,则直接返回错误提示
    return ServiceResultEnum.GOODS_NOT_EXIST.getResult();
  }
  if (indexConfigMapper.selectByTypeAndGoodsId(indexConfig.getConfigType(), indexConfig.getGoodsId()) != null) {
    return ServiceResultEnum.SAME_INDEX_CONFIG_EXIST.getResult();
  }
  if (indexConfigMapper.insertSelective(indexConfig) > 0) {
    return ServiceResultEnum.SUCCESS.getResult();
  }
  return ServiceResultEnum.DB_ERROR.getResult();
}

@Override
// 修改首页配置数据
public String updateIndexConfig(IndexConfig indexConfig) {
  // 根据参数 goodsId 查询数据库中对应的商品数据是否存在
  if (goodsMapper.selectByPrimaryKey(indexConfig.getGoodsId()) == null) {
    // 若商品数据不存在,则直接返回错误提示
    return ServiceResultEnum.GOODS_NOT_EXIST.getResult();
  }
  IndexConfig temp = indexConfigMapper.selectByPrimaryKey(indexConfig.getConfigId());
  if (temp == null) {
    return ServiceResultEnum.DATA_NOT_EXIST.getResult();
  }
  IndexConfig temp2 = indexConfigMapper.selectByTypeAndGoodsId(indexConfig.getConfigType(), indexConfig.getGoodsId());
```

```
    if (temp2 != null
&& !temp2.getConfigId().equals(indexConfig.getConfigId())) {
    //若goodsId相同但id不同，则不能继续修改
    return ServiceResultEnum.SAME_INDEX_CONFIG_EXIST.getResult();
}
indexConfig.setUpdateTime(new Date());
if (indexConfigMapper.updateByPrimaryKeySelective(indexConfig) > 0) {
    return ServiceResultEnum.SUCCESS.getResult();
}
return ServiceResultEnum.DB_ERROR.getResult();
}
```

在新增或修改一条 IndexConfig 数据时，需要对所传送的参数 goodsId 进行验证，即判断需要入库的 IndexConfig 数据中的商品是否存在，如果不存在，则标识所传送的参数不正确。因为推荐微服务未连接商品表所在的数据库，无法直接通过 GoodsMapper 去查询对应的商品数据，所以这部分代码会被标红。

这里必须进行商品数据判断逻辑的修改，由原本直接查询商品表改为远程调用商品微服务中的接口来完成这个逻辑。推荐微服务不仅要与用户微服务通信，还要与商品微服务通信。这里的代码改造步骤如下：

第一步，引入 goods-api 依赖。

打开 newbee-mall-cloud-recommend-web 工程下的 pom.xml 文件，增加与商品微服务远程通信所需的 newbee-mall-cloud-goods-api 模块，新增依赖配置如下：

```
<dependency>
  <groupId>ltd.goods.newbee.cloud</groupId>
  <artifactId>newbee-mall-cloud-goods-api</artifactId>
  <version>0.0.1-SNAPSHOT</version>
</dependency>
```

第二步，增加关于商品微服务中 FeignClient 的配置。

打开 newbee-mall-cloud-goods-web 工程，对启动类 NewBeeMallCloudRecommendServiceApplication 中的 @EnableFeignClients 注解进行修改，增加对 NewBeeCloudGoodsServiceFeign 的声明，代码如下：

```
@EnableFeignClients(basePackageClasses =
{ltd.user.cloud.newbee.openfeign.NewBeeCloudAdminUserServiceFeign.class,
        ltd.goods.cloud.newbee.openfeign.NewBeeCloudGoodsServiceFeign.class})
```

接下来就可以直接使用 NewBeeCloudGoodsServiceFeign 与商品微服务进行远程通信了。

第三步，修改商品数据判断逻辑的代码。

打开 newbee-mall-cloud-recommend-web 工程，修改 NewBeeMallIndexConfigServiceImpl 类中商品数据判断逻辑的代码。删除原本直接查询商品数据库的代码，之后注入 NewBeeCloudGoodsServiceFeign 类，并通过调用商品微服务来获取商品数据，代码如下：

```
@Autowired
private NewBeeCloudGoodsServiceFeign goodsService;

@Override
// 添加首页配置数据
public String saveIndexConfig(IndexConfig indexConfig) {
  // 根据参数 goodsId 调用商品微服务中的接口查询对应的商品数据是否存在
  Result<NewBeeMallGoodsDTO> goodsDetailResult =
goodsService.getGoodsDetail(indexConfig.getGoodsId());
  if (goodsDetailResult == null || goodsDetailResult.getResultCode() != 200) {
    // 若商品数据不存在，则直接返回错误提示
    return ServiceResultEnum.GOODS_NOT_EXIST.getResult();
  }
  if (indexConfigMapper.selectByTypeAndGoodsId
(indexConfig.getConfigType(), indexConfig.getGoodsId()) != null) {
    return ServiceResultEnum.SAME_INDEX_CONFIG_EXIST.getResult();
  }
  if (indexConfigMapper.insertSelective(indexConfig) > 0) {
    return ServiceResultEnum.SUCCESS.getResult();
  }
  return ServiceResultEnum.DB_ERROR.getResult();
}

@Override
// 修改首页配置数据
public String updateIndexConfig(IndexConfig indexConfig) {
  // 根据参数 goodsId 调用商品微服务中的接口查询对应的商品数据是否存在
  Result<NewBeeMallGoodsDTO> goodsDetailResult =
goodsService.getGoodsDetail(indexConfig.getGoodsId());
  if (goodsDetailResult == null || goodsDetailResult.getResultCode() != 200)
{
    // 若商品数据不存在，则直接返回错误提示
    return ServiceResultEnum.GOODS_NOT_EXIST.getResult();
  }
  IndexConfig temp = indexConfigMapper.selectByPrimaryKey(indexConfig.
getConfigId());
  if (temp == null) {
    return ServiceResultEnum.DATA_NOT_EXIST.getResult();
  }
```

```
        IndexConfig temp2 = indexConfigMapper.selectByTypeAndGoodsId(indexConfig.
getConfigType(), indexConfig.getGoodsId());
        if (temp2 != null && !temp2.getConfigId().equals(indexConfig.
getConfigId())) {
            //若goodsId相同但id不同，则不能继续修改
            return ServiceResultEnum.SAME_INDEX_CONFIG_EXIST.getResult();
        }
        indexConfig.setUpdateTime(new Date());
        if (indexConfigMapper.updateByPrimaryKeySelective(indexConfig) > 0) {
            return ServiceResultEnum.SUCCESS.getResult();
        }
        return ServiceResultEnum.DB_ERROR.getResult();
    }
```

这里调用的就是在商品微服务中所编写并暴露的/goods/admin/goodsDetail接口，根据参数goodsId查询对应的商品数据。

5.3.4 功能测试

代码修改完成后，测试步骤是不能漏掉的。一定要验证项目是否能正常启动、接口是否能正常调用，防止在代码移动过程中出现一些小问题，导致项目无法启动或代码报错。在项目启动前需要分别启动Nacos Server和Redis Server，之后依次启动newbee-mall-cloud-recommend-web工程、newbee-mall-cloud-user-web工程、newbee-mall-cloud-goods-web工程和newbee-mall-cloud-gateway-admin工程下的主类。启动成功后，就可以进行本节的功能测试。

打开用户微服务的Swagger页面，在浏览器中输入如下网址：http://localhost:29000/swagger-ui/index.html。

在该页面使用登录接口获取一个token值用于后续的功能测试。比如，笔者在测试时获取到一个值为"2a3d6aa70da7a1bbeb8a83ad74addb93"的token字段。

接下来打开推荐微服务的Swagger页面，在浏览器中输入如下网址：http://localhost:29020/swagger-ui/index.html。

这时就可以在Swagger提供的UI页面进行推荐微服务的接口测试了，推荐微服务接口文档显示内容如图5-9所示。

下面演示新增首页配置项接口。单击"新增首页配置项"→"Try it out"按钮，在参数栏中输入分类名称、分类等级等字段，在登录认证token输入框中输入管理员用户登录接口返回的token值，如图5-10所示。

图 5-9 推荐微服务接口文档显示内容

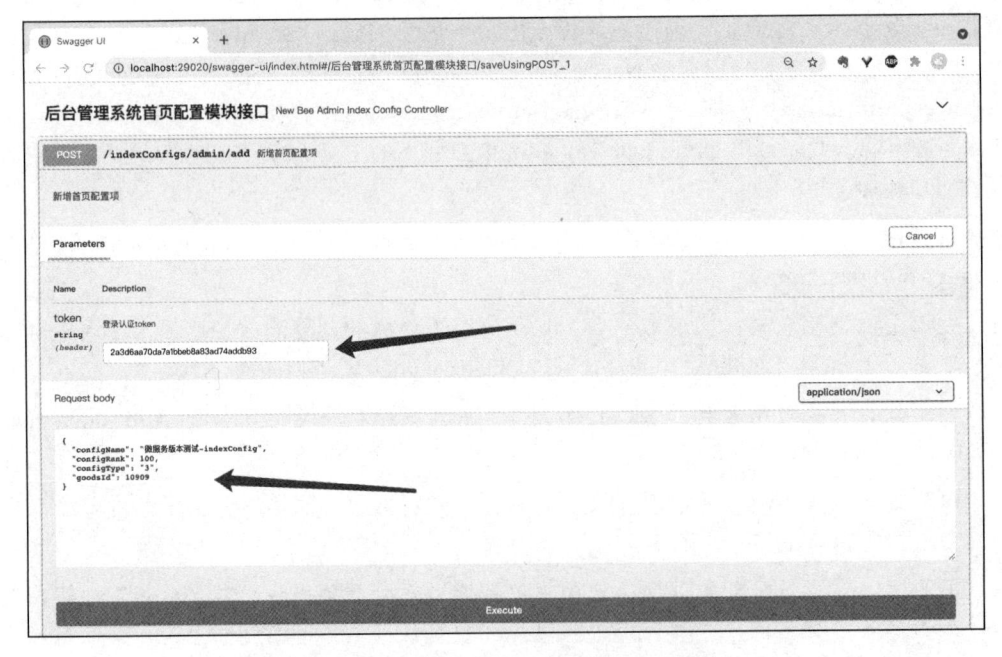

图 5-10 新增首页配置项接口的测试过程

单击"Execute"按钮,接口的响应结果如图 5-11 所示。

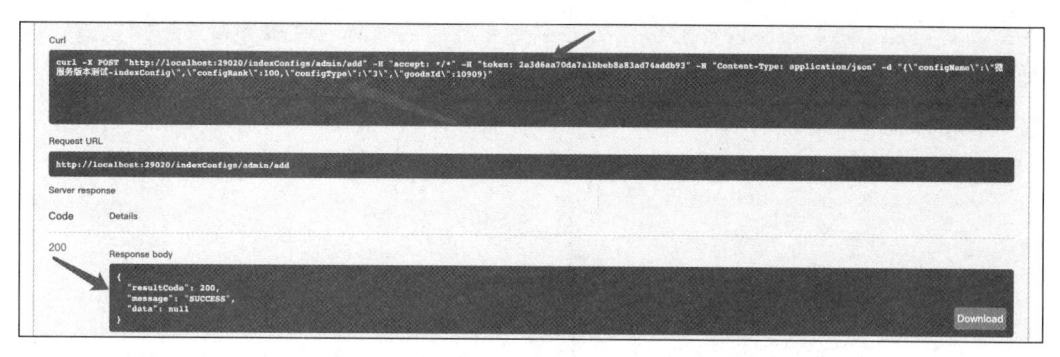

图 5-11 新增首页配置项接口的响应结果

若后端接口的测试结果中有"SUCCESS",则表示添加成功。查看推荐微服务的数据库,首页配置项表中已经新增了一条数据。笔者在测试时,输入的字段都是符合规范的。如果输入的参数没有通过基本的验证判断,就会报出对应的错误提示,如商品不存在或重复配置,提示内容如下,读者在测试时需要注意。

```
{
  "resultCode": 500,
  "message": "已存在相同的首页配置项!",
  "data": null
}

{
  "resultCode": 500,
  "message": "商品不存在!",
  "data": null
}
```

由于篇幅有限,笔者只演示了一个接口的测试过程,读者在测试时可以看看其他接口。

功能测试完成,并且接口响应一切正常,表示推荐微服务本身的功能编码完成,并且远程调用用户微服务、商品微服务也一切正常。在测试时,读者也可以通过 debug 模式启动项目,打上几个断点来查看接口测试时的完整过程。

不知不觉中已经基本完成 3 个微服务模块和 1 个网关模块的编码、配置和测试,在这些实战章节中,我们不断地运用所学到的微服务知识来"添砖加瓦",再结合一些实际的业务代码让微服务知识的应用更加生动，微服务知识加上具体的业务功能使得整个实战项目更加丰富。希望读者能够在实战中更加透彻地理解与微服务相关的知识点、组件，更加得心应手地在实战项目中应用这些知识。本章主要介绍了首页配置项管理功能接口和轮播图管理功能的编码改造，希望读者能够根据笔者提供的开发步骤顺利地完成本章的项目改造。

第 6 章

用户微服务及商品微服务功能完善

本章继续讲解微服务架构项目的改造内容，接下来先对商城端的功能模块进行编码改造，要引入用户体系中的商城用户及对应的网关服务，然后对部分商城端的接口进行编码实现。本章主要涉及用户微服务和商品微服务两个工程。

6.1 增加商城用户的相关功能

6.1.1 商城用户模块介绍

新蜂商城包含后台管理系统和商城端。在后台管理系统中可以对各个功能模块进行配置和操作，如轮播图管理、商品管理、订单管理等，对应的用户是管理员用户。管理员用户在后台管理系统的登录页面登录后，才能够在对应的页面进行操作。商城端则是用于完成整个购物流程的系统，该系统包含商城类系统的主要功能，如商品搜索、添加购物车、下单、支付等，对应的用户是商城用户。商城用户需要在商城端自行注册，登录之后就可以完成购物流程了。

新蜂商城有两种用户，分别是后台管理系统的管理员用户和商城端的商城用户，如图 6-1 所示。前面实战章节中讲解的都是后台管理系统相关的功能实现，如商品管理、轮播图管理等，代码中用到的用户是管理员用户。后续章节中将继续完善该实战项目，主要涉及商城端的功能模块，因此需要实现与商城用户相关的代码。

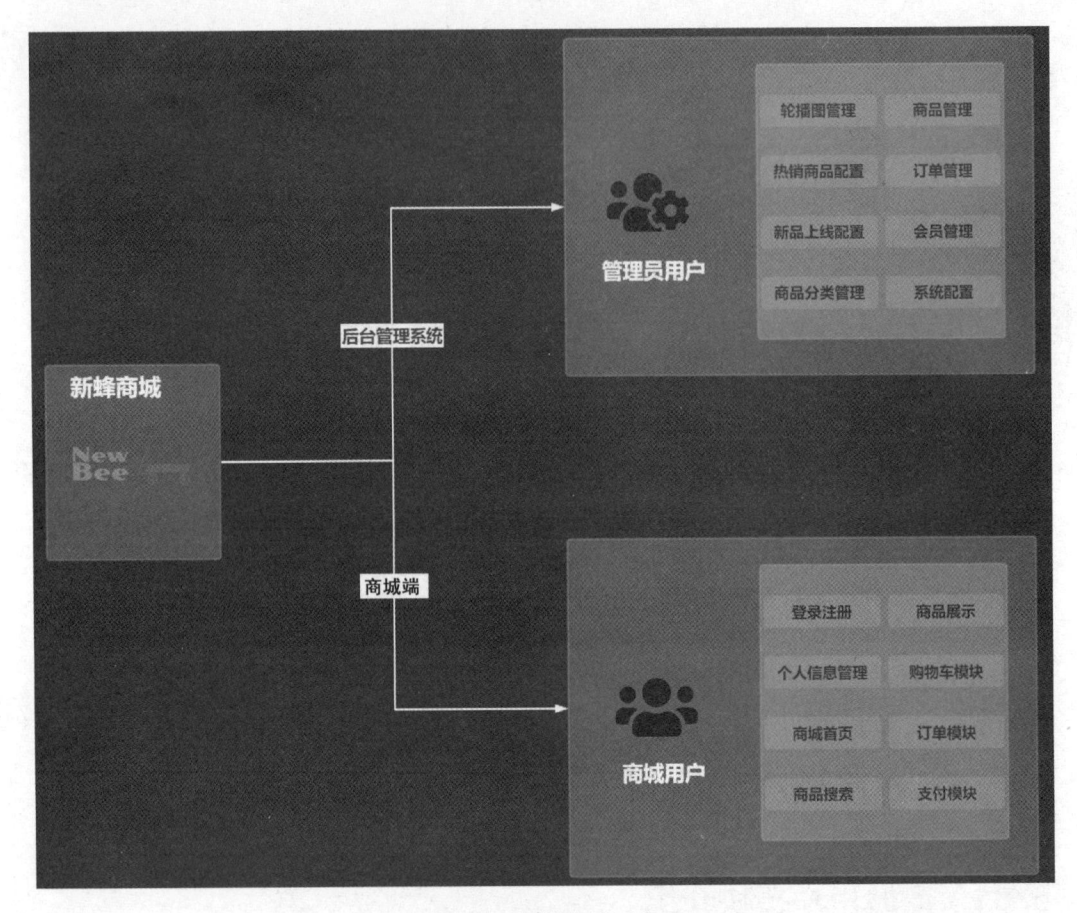

图 6-1 新蜂商城用户

笔者的做法是将两种用户的相关编码实现都放在用户微服务中，编码是直接在 newbee-mall-cloud-user-service 模块中进行的。这是一种实现形式，如果有读者想把用户微服务拆分得更细一些，完全可以拆分为商城用户微服务和管理员用户微服务。

6.1.2 商城用户功能模块编码

本节的源代码是在 newbee-mall-cloud-dev-step09 工程的基础上改造的，名称为 newbee- mall-cloud-dev-step10。打开用户微服务 newbee-mall-cloud-user-web 的工程目录，将原单体 API 项目中与商城用户相关的业务代码和 Mapper 文件依次复制过来，如图 6-2 所示。

第 6 章 用户微服务及商品微服务功能完善

```
▼ ■ newbee-mall-cloud-dev-step10 [newbee-mall-cloud]
  > ■ .idea
  > ■ newbee-mall-cloud-common
  > ■ newbee-mall-cloud-gateway-admin
  > ■ newbee-mall-cloud-goods-service
  > ■ newbee-mall-cloud-recommend-service
  ▼ ■ newbee-mall-cloud-user-service
    > ■ newbee-mall-cloud-user-api
    ▼ ■ newbee-mall-cloud-user-web
      ▼ ■ src
        ▼ ■ main
          ▼ ■ java
            ▼ ■ ltd.user.cloud.newbee
              > ■ config
              ▼ ■ controller
                > ■ param
                > ■ vo
                  ⓒ NewBeeMallCloudAdminUserController
                  ⓒ NewBeeMallCloudPersonalController  ←
              ▼ ■ dao
                  🍃 AdminUserMapper
                  🍃 MallUserMapper  ←
              > ■ entity
              ▼ ■ service
                > ■ impl
                  ⓘ AdminUserService
                  ⓘ NewBeeMallUserService  ←
              ⓒ NewBeeMallCloudUserServiceApplication
          > ■ resources
      𝘮 pom.xml
    𝘮 pom.xml
  𝘮 pom.xml
```

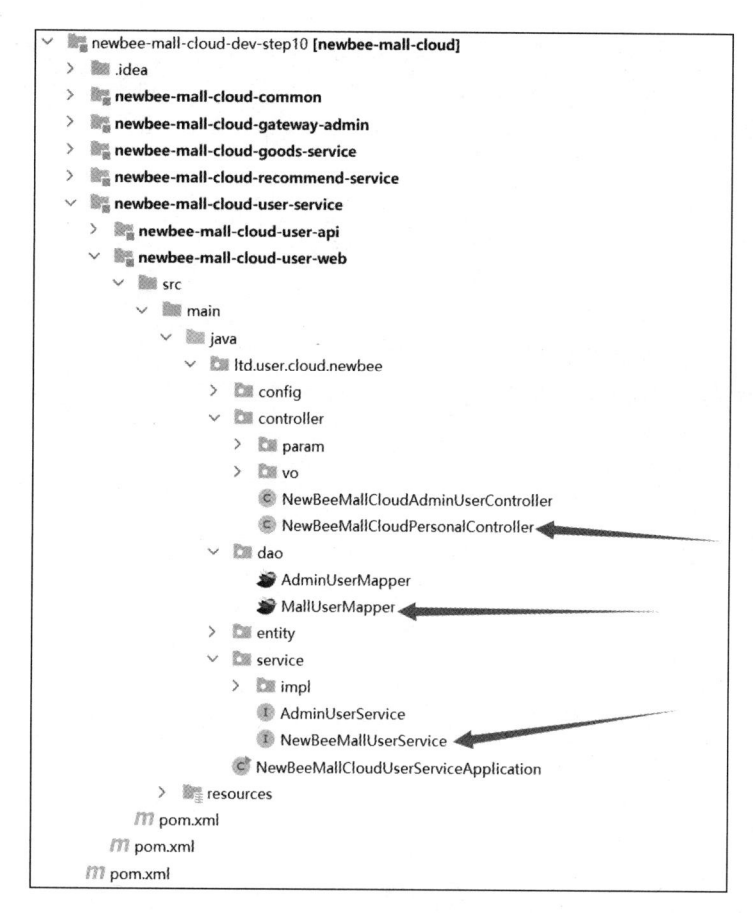

图 6-2 原单体 API 项目中与商城用户相关的业务代码和 Mapper 文件

本步骤中的源代码涉及的数据库为 newbee_mall_cloud_user_db，数据库表为 tb_newbee_mall_user。

商城用户的表结构和建表语句如下：

```
USE 'newbee_mall_cloud_user_db';

# 创建商城用户表

CREATE TABLE 'tb_newbee_mall_user' (
    'user_id' bigint(20) NOT NULL AUTO_INCREMENT COMMENT '用户主键id',
    'nick_name' varchar(50) NOT NULL DEFAULT '' COMMENT '用户昵称',
    'login_name' varchar(11) NOT NULL DEFAULT '' COMMENT '登录名称(默认为手机号)',
```

```sql
    'password_md5' varchar(32) NOT NULL DEFAULT '' COMMENT 'MD5加密后的密码',
    'introduce_sign' varchar(100) NOT NULL DEFAULT '' COMMENT '个性签名',
    'is_deleted' tinyint(4) NOT NULL DEFAULT '0' COMMENT '注销标识字段(0-正常 1-已注销)',
    'locked_flag' tinyint(4) NOT NULL DEFAULT '0' COMMENT '锁定标识字段(0-未锁定 1-已锁定)',
    'create_time' timestamp NOT NULL DEFAULT CURRENT_TIMESTAMP COMMENT '注册时间',
    PRIMARY KEY ('user_id') USING BTREE
) ENGINE=InnoDB DEFAULT CHARSET=utf8 ROW_FORMAT=DYNAMIC;

# 新增商城用户数据

INSERT INTO 'tb_newbee_mall_user' ('user_id', 'nick_name', 'login_name', 'password_md5', 'introduce_sign', 'is_deleted', 'locked_flag', 'create_time')
VALUES
(1,'十三','13700002703','e10adc3949ba59abbe56e057f20f883e','我不怕千万人阻挡,只怕自己投降',0,0,'2022-05-22 08:44:57'),
(6,'陈尼克','13711113333','e10adc3949ba59abbe56e057f20f883e','测试用户陈尼克',0,0,'2022-05-22 08:44:57');
```

除此之外,与用户微服务相关的还有 Redis 数据库。

不过,仅仅复制过来是不够的,还需要对代码进行一些微调。有些工具类已经被移到公用模块 newbee-mall-cloud-common 中,代码中使用这些工具类的地方也需要处理一下引用路径。Controller 类中的接口地址都做了微调,与之前单体项目中定义的 URL 略有不同。

6.1.3 商城用户模块代码完善

商城用户的鉴权实现逻辑也要修改。与管理员用户的处理方式类似,需要对微服务架构下用户鉴权的编码进行改造。实现思路与处理管理员用户的实现思路一致,将原来存储在 token 表中的数据改为存储在 Redis 数据库中,这样在网关层就可以通过读取 Redis 数据库中的数据实现商城用户的鉴权操作。

由于用户微服务模块中已经引入了 Redis 组件,也配置了相关内容,因此在改造商城用户相关逻辑实现时,只处理代码即可。

1. 修改登录和退出登录方法

原来的登录逻辑中会生成 MallUserToken 的相关数据并被保存到 MySQL 数据库中，现在把这部分数据保存到 Redis 数据库中，修改后的登录方法代码如下：

```
public String login(String loginName, String passwordMD5) {
    MallUser user = mallUserMapper.selectByLoginNameAndPasswd(loginName, passwordMD5);
    if (user != null) {
        if (user.getLockedFlag() == 1) {
            return ServiceResultEnum.LOGIN_USER_LOCKED_ERROR.getResult();
        }
        //登录后执行修改 token 值的操作
        String token = getNewToken(System.currentTimeMillis() + "", user.getUserId());
        MallUserToken mallUserToken = new MallUserToken();
        mallUserToken.setUserId(user.getUserId());
        mallUserToken.setToken(token);
        ValueOperations<String, MallUserToken> setToken = redisTemplate.opsForValue();
        setToken.set(token, mallUserToken, 7 * 24 * 60 * 60, TimeUnit.SECONDS); //过期时间为 7 天
        return token;

    }
    return ServiceResultEnum.LOGIN_ERROR.getResult();
}
```

商城用户不仅有登录方法，还有退出登录方法，所以还要对 logout()方法进行修改，调整后的代码如下：

```
public Boolean logout(String token) {
    redisTemplate.delete(token);
    return true;
}
```

修改的类是 newbee-mall-cloud-user-web 模块下的 ltd.user.cloud.newbee.service.impl.MallUserServiceImpl 类。

在 login()方法中，删除原有保存和修改 MallUserToken 数据到 MySQL 数据库中的逻辑代码，之后使用 RedisTemplate 对象将 MallUserToken 数据存储到 Redis 数据库中，存储时的 key 为登录成功后生成的 token 值。

在 logout()方法中，删除操作 MySQL 数据库的逻辑代码，实现方式改为使用 RedisTemplate 对象将 Redis 数据库中对应的数据删除，使用的 key 为当前的 token 值。

2. 修改商城用户 token 值处理的逻辑

修改完登录方法的逻辑后，还要修改对应的鉴权逻辑。原来的逻辑是获取请求头中的 token 值，之后根据这个值查询 MySQL 数据库是否存在及是否过期。现在不需要查询 MySQL 数据库了，直接读取 Redis 数据库中是否有对应的 token 值即可，修改后的商城用户的鉴权代码如下：

```
@Component
public class TokenToMallUserMethodArgumentResolver implements
HandlerMethodArgumentResolver {

    @Autowired
    private RedisTemplate redisTemplate;
    @Autowired
    private MallUserMapper mallUserMapper;

    public TokenToMallUserMethodArgumentResolver() {
    }

    public boolean supportsParameter(MethodParameter parameter) {
        if (parameter.hasParameterAnnotation(TokenToMallUser.class)) {
            return true;
        }
        return false;
    }

    public Object resolveArgument(MethodParameter parameter,
ModelAndViewContainer mavContainer, NativeWebRequest webRequest,
WebDataBinderFactory binderFactory) {
        if (parameter.getParameterAnnotation(TokenToMallUser.class)
instanceof TokenToMallUser) {
            String token = webRequest.getHeader("token");
            if (null != token && !"".equals(token) && token.length() == 32) {
                ValueOperations<String, MallUserToken> opsForMallUserToken =
redisTemplate.opsForValue();
                MallUserToken mallUserToken = opsForMallUserToken.get(token);
                if (mallUserToken == null ) {
                    NewBeeMallException.fail(ServiceResultEnum.TOKEN_EXPIRE_
ERROR.getResult());
                }
                MallUser mallUser = mallUserMapper.selectByPrimaryKey
(mallUserToken.getUserId());
                if (mallUser == null) {
                    NewBeeMallException.fail(ServiceResultEnum.USER_NULL_
ERROR.getResult());
```

```
            if (mallUser.getLockedFlag().intValue() == 1) {
                NewBeeMallException.fail(ServiceResultEnum.LOGIN_USER_
LOCKED_ERROR.getResult());
            }
            return mallUserToken;
        } else {
            NewBeeMallException.fail(ServiceResultEnum.NOT_LOGIN_
ERROR.getResult());
        }
    }
    return null;
    }
}
```

修改的类是 newbee-mall-cloud-user-web 模块下的 ltd.user.cloud.newbee.service.config.handler.TokenToMallUserMethodArgumentResolver 类。

删除原有查询 MySQL 数据库中 MallUserToken 数据的逻辑代码，之后添加从 Redis 数据库中读取 MallUserToken 数据的代码，后续逻辑并没有更改。

接下来在 WebMVC 配置类中增加对 TokenToMallUserMethodArgumentResolver 的配置并使其生效。增加商城用户功能模块后，WebMVC 配置类就不只是处理管理员用户了，所以将该类的名称由 AdminUserWebMvcConfigurer 改成 UserServiceWebMvcConfigurer。最终的代码如下：

```
public class UserServiceWebMvcConfigurer extends WebMvcConfigurationSupport {

    @Autowired
    private TokenToAdminUserMethodArgumentResolver tokenToAdminUserMethodArgumentResolver;

    @Autowired
    private TokenToMallUserMethodArgumentResolver tokenToMallUserMethodArgumentResolver;

    /**
     * @param argumentResolvers
     * @tip @TokenToAdminUser 注解处理方法
     */
    public void addArgumentResolvers(List<HandlerMethodArgumentResolver> argumentResolvers) {
        argumentResolvers.add(tokenToAdminUserMethodArgumentResolver);
        argumentResolvers.add(tokenToMallUserMethodArgumentResolver);
    }
```

```
    @Override
    public void addResourceHandlers(ResourceHandlerRegistry registry) {
        registry.
            addResourceHandler("/swagger-ui/**")
            .addResourceLocations("classpath:/META-INF/resources/webjars/springfox-swagger-ui/")
            .resourceChain(false);
    }
}
```

6.1.4　OpenFeign 编码暴露远程接口

在暴露远程接口前，将需要暴露的接口代码编写完成。笔者在 NewBeeMallCloudPersonalController 类中新建了一个根据 token 值查询商城用户信息的接口，并将其暴露，代码如下：

```
@RequestMapping(value = "/getDetailByToken", method = RequestMethod.GET)
public Result getMallUserByToken(@RequestParam("token") String token) {
    MallUser userDetailByToken = newBeeMallUserService.getUserDetailByToken(token);
    if (userDetailByToken != null) {
        Result result = ResultGenerator.genSuccessResult();
        result.setData(userDetailByToken);
        return result;
    }
    return ResultGenerator.genFailResult("无此用户数据");
}
```

该方法对应的业务层方法实现代码如下：

```
public MallUser getUserDetailByToken(String token) {
    ValueOperations<String, MallUserToken> opsForMallUserToken = redisTemplate.opsForValue();
    MallUserToken mallUserToken = opsForMallUserToken.get(token);
    if (mallUserToken != null) {
        MallUser mallUser = mallUserMapper.selectByPrimaryKey(mallUserToken.getUserId());
        if (mallUser == null) {
            NewBeeMallException.fail(ServiceResultEnum.DATA_NOT_EXIST.getResult());
        }
```

```
        if (mallUser.getLockedFlag().intValue() == 1) {
            NewBeeMallException.fail(ServiceResultEnum.LOGIN_USER_LOCKED_
ERROR.getResult());
        }
        return mallUser;
    }
    NewBeeMallException.fail(ServiceResultEnum.DATA_NOT_EXIST.getResult());
    return null;
}
```

代码逻辑如下。

① 根据 token 值查询 Redis 数据库中对应的 MallUserToken 对象。

② 如果不为空，则根据获取的 MallUserToken 对象中的 userId 字段查询 MySQL 数据库中的商城用户信息；如果为空，则返回错误提示。

③ 如果查询到正确的商城用户信息，则返回给调用端。

接下来，在 newbee-mall-cloud-user-api 模块中新增商城用户需要暴露的接口 FeignClient，后续在改造商城端的功能模块时需要用到。

之前管理员用户中的远程接口配置在 ltd.user.cloud.newbee.openfeign.NewBeeCloud AdminUserServiceFeign 类中，增加商城用户功能模块后，NewBeeCloudAdminUserService Feign 类就不只处理管理员用户了，因此将该类的名称由 NewBeeCloudAdminUser ServiceFeign 改为 NewBeeCloudUserServiceFeign。注意，修改类名后，引用了该类的代码都需要同步改动。新增对 users/mall/getDetailByToken 接口的配置，最终的代码如下：

```
package ltd.user.cloud.newbee.openfeign;

import ltd.user.cloud.newbee.dto.MallUserDTO;
import org.springframework.cloud.openfeign.FeignClient;
import org.springframework.web.bind.annotation.GetMapping;
import org.springframework.web.bind.annotation.PathVariable;
import ltd.common.cloud.newbee.dto.Result;
import org.springframework.web.bind.annotation.RequestParam;

@FeignClient(value = "newbee-mall-cloud-user-service", path = "/users")
public interface NewBeeCloudUserServiceFeign {

    @GetMapping(value = "/admin/{token}")
    Result getAdminUserByToken(@PathVariable(value = "token") String token);

    @GetMapping(value = "/mall/getDetailByToken")
    Result<MallUserDTO> getMallUserByToken(@RequestParam(value = "token") String token);
}
```

这个方法的作用就是根据 token 值获取商城用户的信息，响应的实体结构为 MallUserDTO，其中的字段都是商城用户的属性。调用的接口是 newbee-mall-cloud-user-web 模块下 NewBeeMallCloudPersonalController 类中的 getMallUserByToken()方法。这样，如果其他服务需要通过 token 值获取商城用户数据，就可以引入 newbee-mall-cloud-user-api 模块作为依赖，并且直接调用 NewBeeCloudUserServiceFeign 类中的 getMallUserByToken()方法。

6.1.5　商城用户鉴权功能测试

修改完代码后，测试步骤是不能漏掉的，一定要验证项目是否能正常启动、接口是否能正常调用，防止在代码移动过程中出现一些小问题，导致项目无法启动或代码报错。在项目启动前需要分别启动 Nacos Server 和 Redis Server，之后依次启动 newbee-mall-cloud-user-web 工程、newbee-mall-cloud-gateway-admin 工程下的主类。启动成功后，就可以进行本节的功能测试了。

打开用户微服务的 Swagger 页面，在浏览器中输入如下网址：http://localhost:29000/swagger-ui/index.html。

加入商城用户后的用户微服务接口文档如图 6-3 所示，页面中包括管理员用户的接口文档和商城用户的接口文档，本节在测试时使用的是与商城用户相关的接口。

图 6-3　加入商城用户后的用户微服务接口文档

1. 商城用户登录功能测试

依次单击"登录接口""Try it out"按钮，在参数栏中输入账号、密码字段，在这里笔者使用建表时的测试账号进行测试，如图 6-4 所示。

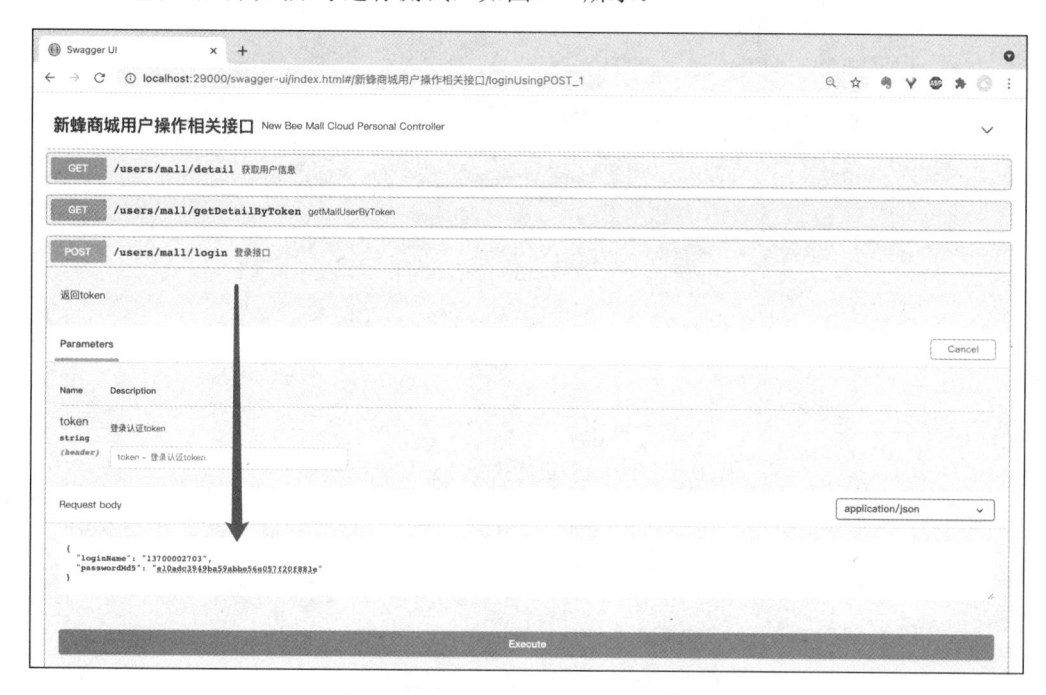

图 6-4　商城用户登录接口的测试过程

当然，也可以使用商城用户的注册接口自行注册一个商城账号用于后续的功能测试。单击"Execute"按钮，接口的测试结果如图 6-5 所示。

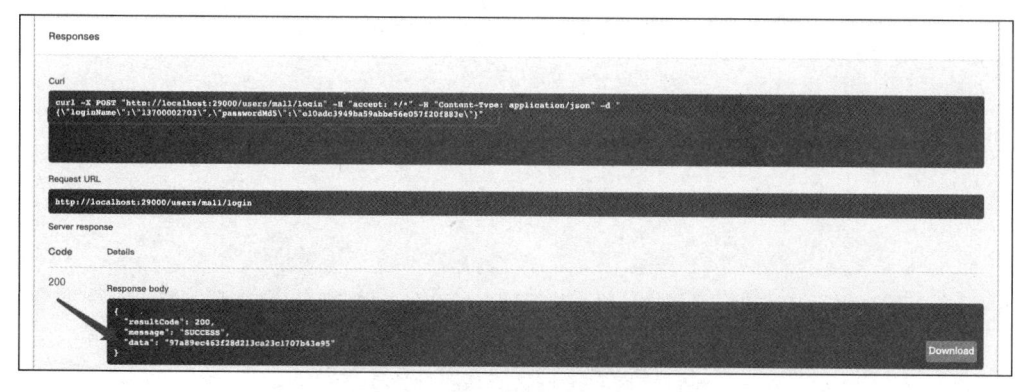

图 6-5　商城用户登录接口的测试结果

接口测试成功，使用登录接口获取的 token 值可以用于后续关于商城端的功能测试。比如，笔者在测试时获取了一个值为"97a89ec463f28d213ca23c1707b43e95"的 token 字段。

由于对 token 字段处理的逻辑做了改动，因此还需要查看 Redis 数据库中是否存在刚刚获取的 token 值，如图 6-6 所示。

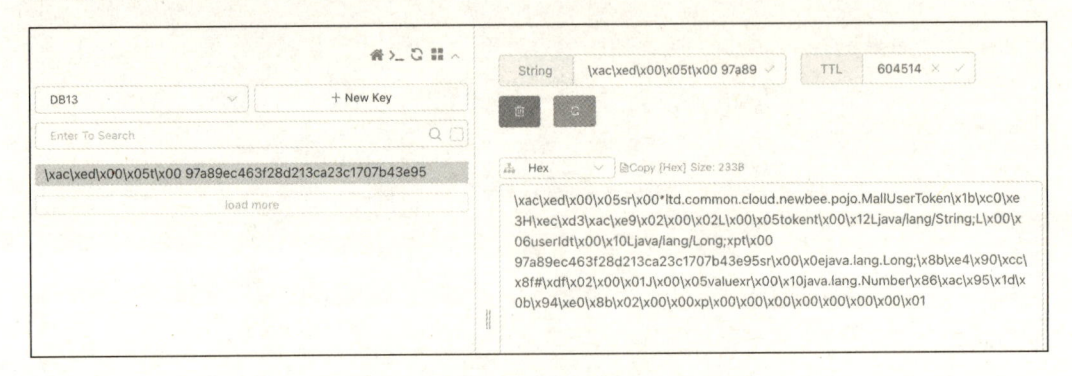

图 6-6 Redis 数据库中的用户登录数据

此时，可以查看 Redis 数据库中索引为 13 的数据。token 值被正确地存入 Redis 数据库，并且设置了过期时间。

2. 获取商城用户信息接口测试

接下来测试其他需要进行鉴权的接口，确认 token 字段处理的逻辑在改动后功能是否正常。

依次单击"获取用户信息""Try it out"按钮，在登录认证 token 的输入框中输入管理员用户登录接口返回的 token 值，如图 6-7 所示。

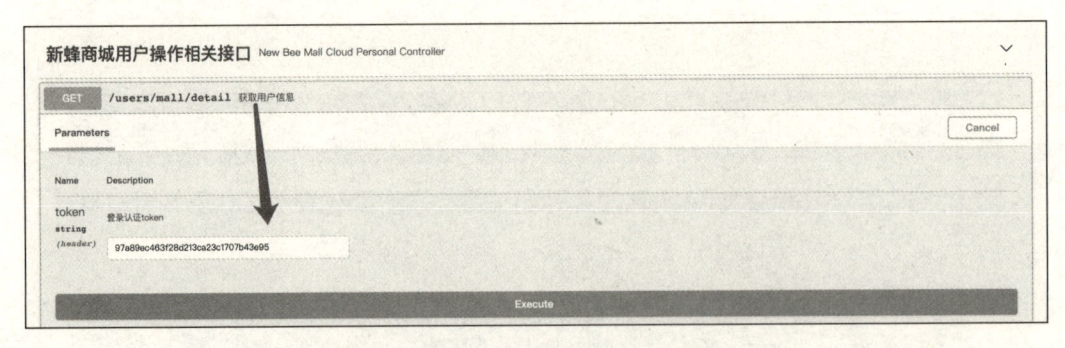

图 6-7 获取商城用户信息接口的测试过程

单击"Execute"按钮，接口的测试结果如图 6-8 所示。

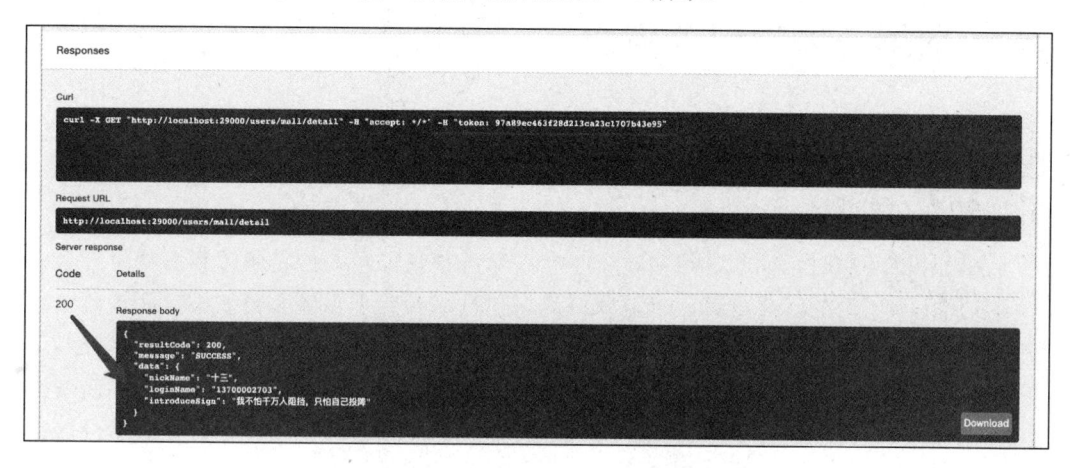

图 6-8 获取商城用户信息接口的测试结果

若后端接口的测试结果中有"SUCCESS"，则表示信息获取成功。该接口对应到实际的项目页面中，是新蜂商城系统中的个人信息页面，如图 6-9 所示。

图 6-9 新蜂商城系统中的个人信息页面

笔者在这里只演示了一个接口的测试过程，读者也可以测试其他接口。

6.2 新增商城端网关模块

6.2.1 创建商城端网关 newbee-mall-cloud-gateway-mall

在设计网关层时，笔者觉得创建一个网关服务模块即可，但是考虑到后期可能会做不同的优化和不同的逻辑处理，毕竟新蜂商城包含后台管理系统和商城端，而这两个系统分别对应不同的页面流程和不同的用户体系，所以创建了两个单独的网关服务模块。在之前的编码中已经创建了 newbee-mall-cloud-gateway-admin，在该代码目录中新增路由配置和鉴权过滤器用于后台管理系统相关接口的转发和鉴权。现在，开始改造商城端的接口，需要再创建一个网关服务 newbee-mall-cloud-gateway-mall。

本节的源代码是在 newbee-mall-cloud-dev-step10 工程的基础上改造的，将工程命名为 newbee-mall-cloud-dev-step11。

创建过程并不复杂，由于代码量不大，因此直接在工程中复制 newbee-mall-cloud-gateway-admin 即可，并将复制的项目改名为 newbee-mall-cloud-gateway-mall。

打开 newbee-mall-cloud-gateway-mall 工程中的 pom.xml 文件，修改依赖配置信息，主要修改 artifactId 和 name 两个属性，修改后的代码如下：

```xml
<modelVersion>4.0.0</modelVersion>
<groupId>ltd.newbee.cloud</groupId>
<artifactId>newbee-mall-cloud-gateway-mall</artifactId>
<version>0.0.1-SNAPSHOT</version>
<name>newbee-mall-cloud-gateway-mall</name>
<description>商城端网关模块</description>
```

在工程的 pom.xml 主文件中增加该模块的配置，将该模块纳入实战项目，代码如下：

```xml
<modules>
  <module>newbee-mall-cloud-recommend-service</module>
  <module>newbee-mall-cloud-goods-service</module>
  <module>newbee-mall-cloud-user-service</module>
  <!-- 新增商城端网关 -->
  <module>newbee-mall-cloud-gateway-mall</module>
  <module>newbee-mall-cloud-gateway-admin</module>
  <module>newbee-mall-cloud-common</module>
</modules>
```

将项目主类 NewBeeMallCloudAdminGatewayApplication 修改为 NewBeeMallCloudMallGatewayApplication，之后打开配置文件 application.properties，修改端口号和微服务名称，删除后台管理系统相关接口的路由配置，增加商城端用户微服务的路由配置，最终的配置文件代码如下：

```
server.port=29110
# 微服务名称
spring.application.name=newbee-mall-cloud-gateway-mall
# Nacos 地址
spring.cloud.nacos.discovery.server-addr=127.0.0.1:8848
# Nacos 登录用户名(默认为 nacos，在生产环境中一定要修改)
spring.cloud.nacos.username=nacos
# Nacos 登录密码(默认为 nacos，在生产环境中一定要修改)
spring.cloud.nacos.password=nacos
# 网关开启微服务注册与微服务发现
spring.cloud.gateway.discovery.locator.enabled=true
spring.cloud.gateway.discovery.locator.lower-case-service-id=true

# 用户微服务的路由配置
spring.cloud.gateway.routes[0].id=user-service-route
spring.cloud.gateway.routes[0].uri=lb://newbee-mall-cloud-user-service
spring.cloud.gateway.routes[0].order=1
spring.cloud.gateway.routes[0].predicates[0]=Path=/users/mall/**

##Redis 配置
# Redis 数据库索引（默认为 0）
spring.redis.database=13
# Redis 服务器地址
spring.redis.host=127.0.0.1
# Redis 服务器连接端口
spring.redis.port=6379
# Redis 服务器连接密码
spring.redis.password=123456789
# 连接池最大连接数（使用负值表示没有限制）
spring.redis.jedis.pool.max-active=8
# 连接池最大阻塞等待时间（使用负值表示没有限制）
spring.redis.jedis.pool.max-wait=-1
# 连接池中的最大空闲连接
spring.redis.jedis.pool.max-idle=8
# 连接池中的最小空闲连接
spring.redis.jedis.pool.min-idle=0
# 连接超时时间（毫秒）
spring.redis.timeout=5000
```

与后台管理系统的网关模块一样，商城端网关也需要进行基础的用户鉴权操作。在网关层进行统一鉴权，这样就能够避免无正确身份标识的请求直接进入微服务实例中。如果请求头中有正确的身份标识，则放行，让后方的微服务实例进行请求处理；如果没有正确的身份标识，则直接在网关层响应一个错误提示。具体的编码实现在前文中已经介绍过，使用 Spring Cloud Gateway 的全局过滤器。

由于使用复制过来的代码，因此直接将 ValidTokenGlobalFilter 类改名为 ValidMallUserTokenGlobalFilter 类，在这个类中编写对商城用户的网关层鉴权逻辑，代码如下：

```java
package ltd.gateway.cloud.newbee.filter;

import com.fasterxml.jackson.databind.ObjectMapper;
import com.fasterxml.jackson.databind.node.ObjectNode;
import ltd.common.cloud.newbee.dto.Result;
import ltd.common.cloud.newbee.dto.ResultGenerator;
import ltd.common.cloud.newbee.pojo.AdminUserToken;
import ltd.common.cloud.newbee.pojo.MallUserToken;
import org.springframework.beans.factory.annotation.Autowired;
import org.springframework.cloud.gateway.filter.GatewayFilterChain;
import org.springframework.cloud.gateway.filter.GlobalFilter;
import org.springframework.core.Ordered;
import org.springframework.core.io.buffer.DataBuffer;
import org.springframework.data.redis.core.RedisTemplate;
import org.springframework.data.redis.core.ValueOperations;
import org.springframework.http.HttpHeaders;
import org.springframework.http.HttpStatus;
import org.springframework.stereotype.Component;
import org.springframework.util.StringUtils;
import org.springframework.web.server.ServerWebExchange;
import reactor.core.publisher.Flux;
import reactor.core.publisher.Mono;

import java.nio.charset.StandardCharsets;

@Component
public class ValidMallUserTokenGlobalFilter implements GlobalFilter, Ordered {

    @Autowired
    private RedisTemplate redisTemplate;

    @Override
    public Mono<Void> filter(ServerWebExchange exchange, GatewayFilterChain
```

```java
chain) {

    // 登录注册接口，直接放行
    if (exchange.getRequest().getURI().getPath().equals("/users/mall/login") || exchange.getRequest().getURI().getPath().equals("/users/mall/register")) {
        return chain.filter(exchange);
    }

    HttpHeaders headers = exchange.getRequest().getHeaders();

    if (headers == null || headers.isEmpty()) {
        // 返回错误提示
        return wrapErrorResponse(exchange, chain);
    }

    String token = headers.getFirst("token");

    if (!StringUtils.hasText(token)) {
        // 返回错误提示
        return wrapErrorResponse(exchange, chain);
    }
    ValueOperations<String, MallUserToken> opsForMallUserToken = redisTemplate.opsForValue();
    MallUserToken tokenObject = opsForMallUserToken.get(token);
    if (tokenObject == null) {
        // 返回错误提示
        return wrapErrorResponse(exchange, chain);
    }

    return chain.filter(exchange);
}

@Override
public int getOrder() {
    return Ordered.HIGHEST_PRECEDENCE;
}

Mono<Void> wrapErrorResponse(ServerWebExchange exchange, GatewayFilterChain chain) {
    Result result = ResultGenerator.genErrorResult(416, "无权限访问");
    ObjectMapper mapper = new ObjectMapper();
    ObjectNode resultNode = mapper.valueToTree(result);
```

```
        byte[] bytes = resultNode.toString().getBytes(StandardCharsets.UTF_8);
        DataBuffer dataBuffer = exchange.getResponse().bufferFactory().
wrap(bytes);
        exchange.getResponse().setStatusCode(HttpStatus.OK);
        return exchange.getResponse().writeWith(Flux.just(dataBuffer));
    }
}
```

主要的代码逻辑就在 filter()方法中，处理步骤如下。

① 判断请求路径，如果是登录接口或注册接口，则直接放行；如果不是登录接口或注册接口，则进行后续处理。

② 获取请求头对象，如果为空，则直接返回错误提示，不会将请求转发到后方的微服务实例中。

③ 如果请求头不为空，则取出其中的 token 值。如果该值不存在，则直接返回错误提示，不会将请求转发到后方的微服务实例中。

④ 根据 token 值查询 Redis 数据库中是否存在对应的数据。如果 Redis 数据库中不存在对应的数据，则直接返回错误提示，不会将请求转发到后方的微服务实例中。

⑤ 如果 Redis 数据库中存在对应的数据，则表示对商城用户的鉴权成功，将请求转发到后方的微服务实例中去处理请求。

至此，商城端网关层鉴权功能所需的基本配置与功能代码已经完备，接下来进行功能测试。

6.2.2　商城端网关功能测试

测试内容主要包括商城端网关的路由配置是否正确，以及网关层鉴权是否正常。

编码完成后，准备好数据库和表就可以启动项目了。当然，在项目启动前需要分别启动 Nacos Server 和 Redis Server，之后依次启动 newbee-mall-cloud-goods-web 工程和 newbee-mall-cloud-gateway-mall 工程下的主类。启动成功后，就可以测试通过网关访问商城端接口的功能与网关层鉴权功能了。

笔者使用 Postman 工具进行接口请求和功能测试，由于通过网关层访问，因此这里的请求地址需要修改，用户微服务的端口号为 29000，商城端网关的端口号为 29110。

在 Postman 工具的地址栏中输入如下网址：http://localhost:29110/users/mall/detail。

选择请求方法为 GET，测试结果如图 6-10 所示。

图 6-10　未登录状态下通过网关获取商城用户信息的测试结果

　　这里通过直接访问网关层地址来获取商城用户的信息。因为没有在请求头中添加 token 值，所以网关层的 token 过滤器直接拦截了这个请求，并返回错误提示。这个请求根本没有进入用户微服务实例，商城端网关层鉴权生效了。

　　上述测试过程演示的是鉴权失败的情况，接下来通过请求登录接口获取正确的 token 值，演示一下鉴权成功的情况。

　　在地址栏中输入如下网址：http://localhost:29110/users/mall/detail。

　　设置请求方法为 GET，并且添加一个请求头参数，Key 为 "token"，Value 为之前登录成功后获取的 token 值，相关请求配置及测试结果如图 6-11 所示。

　　最终成功获取了正确的数据，商城端网关的路由配置正常且网关层鉴权正常，功能测试完成。当然，读者在测试时也可以选择 debug 模式启动两个项目，并打上几个断点，发起请求后看一下代码的执行步骤，会理解得更透彻一些。这里主要引入了商城用户相关功能模块，以及对商城用户登录后的 token 值处理和商城端网关鉴权。后续章节将继续完善该实战项目，主要涉及商城端的功能模块，因此需要实现与商城用户相关的代码。功能及编码都不复杂，很多知识点都在管理员用户功能模块改造时介绍过了，如果有问题，可以参考管理员用户的功能改造来理解。希望读者能够根据笔者提供的开发步骤顺利地完成本节的项目改造。

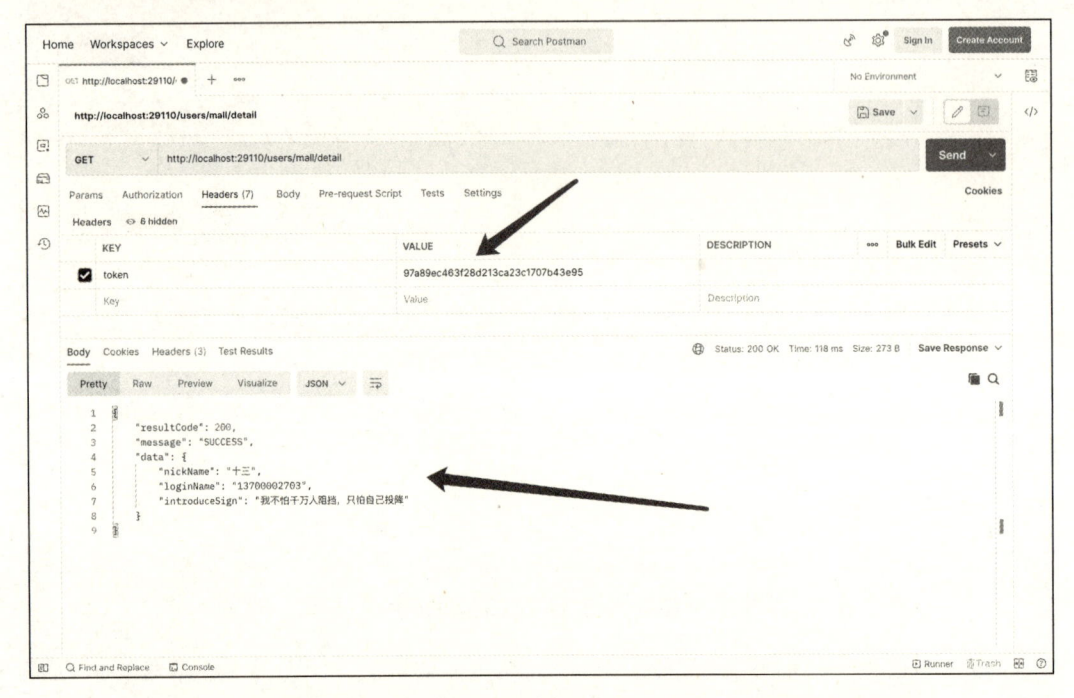

图 6-11 正常登录状态下通过网关获取商城用户信息的请求配置及测试结果

接下来继续讲解微服务架构下商城端接口的编码改造，主要包括商城首页数据的接口、商城分类页面的接口、商品列表和商品详情页面的接口。它们的源代码是在 newbee-mall-cloud-dev-step11 工程的基础上改造的，笔者根据开发步骤整理了 3 份源代码文件，分别是 newbee-mall-cloud-dev-step12、newbee-mall-cloud-dev-step13 和 newbee-mall-cloud-dev-step13-2。

6.3 商城首页数据的接口实现

商城首页数据的接口在原单体 API 项目中已经实现，改造过程并不复杂，将相关的代码复制过来并进行一些修改即可。由于该接口主要涉及轮播图和首页推荐配置两个功能点，因此该接口是定义在推荐微服务中的。本节的源代码是在 newbee-mall-cloud-dev-step11 工程的基础上改造的，将工程命名为 newbee-mall-cloud-dev-step12。

6.3.1 首页的排版设计

新蜂商城的商城首页布局和排版效果如图 6-12 所示。

图 6-12 新蜂商城的商城首页布局和排版效果

商城首页的整个设计版面被分成 6 个部分，总结如下。

① 商城 Logo 及搜索框：可以放置 Logo 图片、商品搜索输入框。

② 轮播图：以轮播的形式展示后台配置的轮播图。

③ 热销商品：展示后台配置的热销商品数据。

④ 新品推荐：展示后台配置的新品数据。

⑤ 推荐商品：展示后台配置的推荐商品数据。

⑥ 导航栏：固定导航栏，放置新蜂商城几个重要的功能模块页面。

当然，以上版面设计只是针对新蜂商城这个项目，在开源社区或企业开发中还有许多其他的商城系统项目。前端的设计和实现灵活多变，不同的商城系统可能有不同的页面样式和页面布局。

6.3.2 首页接口的响应结果设计

因为首页展示的数据是多个功能模块的数据，包括轮播图数据、首页配置推荐数据、商品数据，所以需要重新抽象出一个首页数据视图层对象，并对数据格式进行规范和定义。

关于轮播图数据结构，需要将轮播图的图片地址和单击轮播图后的跳转路径返回给前端，因此定义了视图层对象 NewBeeMallIndexCarouselVO，字段定义如下：

```
/**
 * 首页轮播图 VO
 */
@Data
public class NewBeeMallIndexCarouselVO implements Serializable {

    @ApiModelProperty("轮播图图片地址")
    private String carouselUrl;

    @ApiModelProperty("单击轮播图后的跳转路径")
    private String redirectUrl;
}
```

在商品推荐管理模块中，通过首页效果图可以看到，每个推荐商品的展示区域都有 3 个字段，分别是商品名称、商品图片、商品价格。另外，因为单击后会跳转至商品详情页面，所以需要加上商品 id 字段，视图层对象 NewBeeMallIndexConfigGoodsVO 的字段定义如下：

```
/**
 * 首页配置商品VO
 */
@Data
public class NewBeeMallIndexConfigGoodsVO implements Serializable {

    @ApiModelProperty("商品 id")
    private Long goodsId;
    @ApiModelProperty("商品名称")
    private String goodsName;
    @ApiModelProperty("商品图片地址")
```

```
    private String goodsCoverImg;
    @ApiModelProperty("商品价格")
    private Integer sellingPrice;
}
```

上述两个 VO 对象都是单项,只能表示一个轮播图对象和一个推荐商品对象。而在首页展示时,需要展示多条数据,即轮播图需要返回多张,推荐商品有 3 种类型,每种类型也有多条商品数据。因此需要返回 List 类型的数据,最终抽象出一个首页信息对象 IndexInfoVO,字段定义如下:

```
import io.swagger.annotations.ApiModelProperty;
import lombok.Data;

import java.io.Serializable;
import java.util.List;

@Data
public class IndexInfoVO implements Serializable {

    @ApiModelProperty("轮播图(列表)")
    private List<NewBeeMallIndexCarouselVO> carousels;

    @ApiModelProperty("首页热销商品(列表)")
    private List<NewBeeMallIndexConfigGoodsVO> hotGoodses;

    @ApiModelProperty("首页新品推荐(列表)")
    private List<NewBeeMallIndexConfigGoodsVO> newGoodses;

    @ApiModelProperty("首页推荐商品(列表)")
    private List<NewBeeMallIndexConfigGoodsVO> recommendGoodses;
}
```

打开 newbee-mall-cloud-recommend-web 工程,新建 ltd.recommend.cloud.newbee.controller.vo 包,并创建上述的 3 个 VO 类。

VO 对象就是视图层使用的对象,与 entity 对象有一些区别。entity 对象中的字段与数据库表字段逐一对应,VO 对象里的字段是视图层需要哪些字段就设置哪些字段。在编码时,可以不去额外新增 VO 对象而直接返回 entity 对象,这取决于开发者的编码习惯。

这样,首页展示时后端需要返回的数据格式就定义完成了。如果读者对该项目进行了功能的删减,则可以参考笔者定义的视图层对象灵活地增减字段。

6.3.3 业务层代码的实现

因为获取首页的数据只涉及查询操作,所以业务层方法中分别定义了 getCarouselsForIndex()方法和 getConfigGoodsesForIndex()方法用于查询首页接口中所需要的数据,查询数据后进行首页 VO 对象的封装。

getCarouselsForIndex()方法的代码如下:

```
public List<NewBeeMallIndexCarouselVO> getCarouselsForIndex(int number) {
  List<NewBeeMallIndexCarouselVO> newBeeMallIndexCarouselVOS = new ArrayList<>(number);
  // 读取 MySQL 语句,查询固定数量的轮播图数据
  List<Carousel> carousels = carouselMapper.findCarouselsByNum(number);
  if (!CollectionUtils.isEmpty(carousels)) {
    // 将数据转换为需要的 VO 类型
    newBeeMallIndexCarouselVOS = BeanUtil.copyList(carousels, NewBeeMallIndexCarouselVO.class);
  }
  return newBeeMallIndexCarouselVOS;
}
```

getCarouselsForIndex()方法的作用是返回固定数量的轮播图对象供首页数据渲染,执行逻辑如下:先查询固定数量的轮播图数据,参数为 number,再进行非空判断,然后将查询出来的轮播图对象转换为视图层对象 NewBeeMallIndexCarouselVO,最终返回给调用端。

getConfigGoodsesForIndex()方法的代码如下:

```
public List<NewBeeMallIndexConfigGoodsVO> getConfigGoodsesForIndex(int configType, int number) {
  List<NewBeeMallIndexConfigGoodsVO> newBeeMallIndexConfigGoodsVOS = new ArrayList<>(number);
  List<IndexConfig> indexConfigs = indexConfigMapper.findIndexConfigsByTypeAndNum(configType, number);
  if (!CollectionUtils.isEmpty(indexConfigs)) {
    //取出所有的 goodsId
    List<Long> goodsIds = indexConfigs.stream().map(IndexConfig::getGoodsId).collect(Collectors.toList());
    // 读取 MySQL 语句,根据商品 id 列表查询对应的商品数据
    List<NewBeeMallGoods> newBeeMallGoods = goodsMapper.selectByPrimaryKeys(goodsIds);
```

```
    newBeeMallIndexConfigGoodsVOS = BeanUtil.copyList(newBeeMallGoods,
NewBeeMallIndexConfigGoodsVO.class);
    // 转换为 VO 对象
    for (NewBeeMallIndexConfigGoodsVO newBeeMallIndexConfigGoodsVO :
newBeeMallIndexConfigGoodsVOS) {
      String goodsName = newBeeMallIndexConfigGoodsVO.getGoodsName();
      // 字符串过长导致文字超出的问题
      if (goodsName.length() > 30) {
        goodsName = goodsName.substring(0, 30) + "...";
        newBeeMallIndexConfigGoodsVO.setGoodsName(goodsName);
      }
    }
  return newBeeMallIndexConfigGoodsVOS;
}
}
```

getConfigGoodsesForIndex()方法的作用是返回一定数量的配置项对象供首页数据渲染，执行逻辑如下：先根据 configType 参数读取固定数量的首页配置数据，再获取配置项中关联的商品 id 列表，然后查询商品表并将首页展示时所需的几个字段依次读取并封装到 VO 对象里，最后返回给调用端。

打开 newbee-mall-cloud-recommend-web 工程，在 service.impl 包的 NewBeeMall-CarouselServiceImpl 类和 NewBeeMallIndexConfigServiceImpl 类中添加上述两个方法即可。

6.3.4 调用商品微服务进行数据的查询与封装

按照上述步骤对推荐微服务进行编码后，代码依然会被标红，getConfigGoodses-ForIndex()方法中需要根据商品 id 列表查询对应的商品数据，才能进行后续的 VO 数据封装。而推荐微服务未连接商品表所在的数据库，无法直接通过 GoodsMapper 去查询对应的商品数据，因此这部分代码会被标红。

所以，这里必须进行商品数据查询逻辑的改造，由原本直接查询商品表改为远程调用商品微服务中的接口来完成这个逻辑。在之前的改造步骤中，推荐微服务已经引入了 goods-api 依赖并增加了关于商品微服务中 FeignClient 的配置，因此这部分配置可以省略，直接在商品微服务中编写一个根据 goodsId 查询商品列表数据的接口并暴露即可，改造步骤如下。

第一步，增加根据 goodsId 查询商品列表数据的接口。

打开 newbee-mall-cloud-goods-web 工程,在 NewBeeAdminGoodsInfoController 类中新增接口代码:

```java
/**
 * 根据goodsId查询商品列表
 */
@GetMapping("/listByGoodsIds")
@ApiOperation(value = "根据goodsId查询商品列表", notes = "根据goodsId查询")
public Result getNewBeeMallGoodsByIds(@RequestParam("goodsIds") List<Long> goodsIds) {
    List<NewBeeMallGoods> newBeeMallGoods = newBeeMallGoodsService.getNewBeeMallGoodsByIds(goodsIds);
    return ResultGenerator.genSuccessResult(newBeeMallGoods);
}
```

打开 newbee-mall-cloud-goods-api 工程,在 NewBeeCloudGoodsServiceFeign 类中定义该接口,使得其他微服务可以调用,新增代码如下:

```java
@GetMapping(value = "/admin/listByGoodsIds")
Result<List<NewBeeMallGoodsDTO>> listByGoodsIds(@RequestParam(value = "goodsIds") List<Long> goodsIds);
```

第二步,修改推荐微服务中的商品查询代码。

修改 NewBeeMallIndexConfigServiceImpl 类中商品列表数据查询的代码。删除原本直接查询商品数据库的代码,通过调用商品微服务来获取商品的数据,修改后的 getConfigGoodsesForIndex()方法代码如下:

```java
public List<NewBeeMallIndexConfigGoodsVO> getConfigGoodsesForIndex(int configType, int number) {
    List<NewBeeMallIndexConfigGoodsVO> newBeeMallIndexConfigGoodsVOS = new ArrayList<>(number);
    List<IndexConfig> indexConfigs = indexConfigMapper.findIndexConfigsByTypeAndNum(configType, number);
    if (!CollectionUtils.isEmpty(indexConfigs)) {
        // 取出所有的 goodsId
        List<Long> goodsIds = indexConfigs.stream().map(IndexConfig::getGoodsId).collect(Collectors.toList());
        // 调用商品微服务来获取商品的数据
        Result<List<NewBeeMallGoodsDTO>> newBeeMallGoodsDTOResult = goodsService.listByGoodsIds(goodsIds);
        if (newBeeMallGoodsDTOResult == null || newBeeMallGoodsDTOResult.getResultCode() != 200 || CollectionUtils.isEmpty(newBeeMallGoodsDTOResult.getData())) {
```

```
        // 未查询到数据，返回空链表（也可以直接在这里抛出异常）
        return newBeeMallIndexConfigGoodsVOS;
    }
    newBeeMallIndexConfigGoodsVOS =
BeanUtil.copyList(newBeeMallGoods DTOResult.getData(),
NewBeeMallIndexConfigGoodsVO.class);
    for (NewBeeMallIndexConfigGoodsVO newBeeMallIndexConfigGoodsVO :
newBeeMallIndexConfigGoodsVOS) {
      String goodsName = newBeeMallIndexConfigGoodsVO.getGoodsName();
      String goodsIntro = newBeeMallIndexConfigGoodsVO.getGoodsIntro();
      // 字符串过长导致文字超出的问题
      if (goodsName.length() > 30) {
        goodsName = goodsName.substring(0, 30) + "...";
        newBeeMallIndexConfigGoodsVO.setGoodsName(goodsName);
      }
      if (goodsIntro.length() > 22) {
        goodsIntro = goodsIntro.substring(0, 22) + "...";
        newBeeMallIndexConfigGoodsVO.setGoodsIntro(goodsIntro);
      }
    }
    return newBeeMallIndexConfigGoodsVOS;
}
```

这里调用的就是在商品微服务中暴露的 /goods/admin/listByGoodsIds 接口，根据 goodsId 列表查询对应的商品数据。

6.3.5 首页接口控制层代码的实现

增加首页接口的定义与编码。打开 newbee-mall-cloud-recommend-web 工程，选择 ltd.recommend.cloud.newbee.controller 包并单击鼠标右键，在弹出的快捷菜单中选择 "New→Java Class" 选项，之后在弹出的窗口中输入 "NewBeeMallIndexAPI"，对首页数据请求进行处理，在 NewBeeMallIndexAPI 类中新增如下代码：

```
package ltd.recommend.cloud.newbee.controller;

import io.swagger.annotations.Api;
import io.swagger.annotations.ApiOperation;
import ltd.common.cloud.newbee.dto.Result;
import ltd.common.cloud.newbee.dto.ResultGenerator;
import ltd.common.cloud.newbee.enums.IndexConfigTypeEnum;
```

```java
import ltd.recommend.cloud.newbee.controller.vo.IndexInfoVO;
import ltd.recommend.cloud.newbee.controller.vo.NewBeeMallIndexCarouselVO;
import ltd.recommend.cloud.newbee.controller.vo.NewBeeMallIndexConfigGoodsVO;
import ltd.recommend.cloud.newbee.service.NewBeeMallCarouselService;
import ltd.recommend.cloud.newbee.service.NewBeeMallIndexConfigService;
import org.springframework.web.bind.annotation.GetMapping;
import org.springframework.web.bind.annotation.RequestMapping;
import org.springframework.web.bind.annotation.RestController;

import javax.annotation.Resource;
import java.util.List;

@RestController
@Api(value = "v1", tags = "新蜂商城首页接口")
@RequestMapping("/mall/index")
public class NewBeeMallIndexController {

    @Resource
    private NewBeeMallCarouselService newBeeMallCarouselService;

    @Resource
    private NewBeeMallIndexConfigService newBeeMallIndexConfigService;

    @GetMapping("/recommondInfos")
    @ApiOperation(value = "获取首页数据", notes = "轮播图、新品、推荐等")
    public Result<IndexInfoVO> indexInfo() {
        IndexInfoVO indexInfoVO = new IndexInfoVO();
        List<NewBeeMallIndexCarouselVO> carousels =
newBeeMallCarouselService.getCarouselsForIndex(5);
        List<NewBeeMallIndexConfigGoodsVO> hotGoodses =
newBeeMallIndexConfigService.getConfigGoodsesForIndex(IndexConfigTypeEnum.
INDEX_GOODS_HOT.getType(), 4);
        List<NewBeeMallIndexConfigGoodsVO> newGoodses =
newBeeMallIndexConfigService.getConfigGoodsesForIndex(IndexConfigTypeEnum.
INDEX_GOODS_NEW.getType(), 6);
        List<NewBeeMallIndexConfigGoodsVO> recommendGoodses =
newBeeMallIndexConfigService.getConfigGoodsesForIndex(IndexConfigTypeEnum.
INDEX_GOODS_RECOMMOND.getType(), 10);
        indexInfoVO.setCarousels(carousels);
        indexInfoVO.setHotGoodses(hotGoodses);
        indexInfoVO.setNewGoodses(newGoodses);
        indexInfoVO.setRecommendGoodses(recommendGoodses);
```

```
        return ResultGenerator.genSuccessResult(indexInfoVO);
    }
}
```

处理首页数据请求的方法名称为 indexInfo()，请求类型为 GET，映射的路径为 /mall/index/recommondInfos，响应结果类型为统一的响应对象 Result，实际的 data 属性类型为 IndexInfoVO 视图层对象。

实现逻辑是分别调用轮播图业务实现类 NewBeeMallCarouselService 中的查询方法和首页配置业务实现类 NewBeeMallIndexConfigService 中的查询方法，查询首页所需的数据并逐一设置到 IndexInfoVO 对象中。最终响应给前端，因为商品推荐有热销商品、新品上线、推荐商品 3 种类型，所以首页配置业务实现类中的 getConfigGoodsesForIndex() 方法在此处被调用了 3 次，但是每次传入的参数并不相同。

还有一点需要注意，首页上的内容是任何访问者都能够直接查看的，并不需要登录认证，因此首页数据接口中并没有使用权限认证的注解@TokenToMallUser。

由于无须登录认证，因此需要在商城端网关的全局过滤器中添加配置，在鉴权时忽略这个接口，直接放行。代码修改如下：

```
// 登录接口、注册接口、首页接口，直接放行
if ("/users/mall/login".equals(uri) || "/users/mall/register".equals(uri)
|| "/mall/index/recommondInfos".equals(uri)) {
  return chain.filter(exchange);
}
```

6.3.6 首页接口网关配置

打开 newbee-mall-cloud-gateway-mall 项目中的 application.properties 文件，新增关于推荐微服务的路由信息，配置项为 spring.cloud.gateway.routes.*，新增代码如下：

```
# 首页接口的路由配置
spring.cloud.gateway.routes[1].id=recommend-service-route
spring.cloud.gateway.routes[1].uri=lb://newbee-mall-cloud-recommend-service
spring.cloud.gateway.routes[1].order=1
spring.cloud.gateway.routes[1].predicates[0]=Path=/mall/index/**
```

这里配置 newbee-mall-cloud-gateway-mall 到推荐微服务的路由信息，主要配置"获取首页数据"的接口。如果访问商城端网关项目的路径是以/mall/index/开头的，就路由到推荐微服务实例。

6.4 商城分类页面的接口实现

本节的源代码是在 newbee-mall-cloud-dev-step12 工程的基础上改造的，将工程命名为 newbee-mall-cloud-dev-step13。

6.4.1 分类页面的接口响应数据

分类页面的页面布局和交互依然由前端代码来实现，后端只要将页面所需的分类数据通过接口响应回去即可。接下来定义分类接口中返回数据的格式，这里结合新蜂商城的商城分类页面的布局和排版来讲解，如图 6-13 所示。

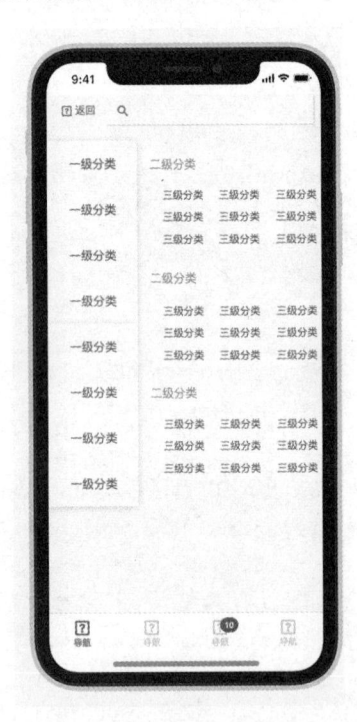

图 6-13 新蜂商城的商城分类页面的布局和排版

由图 6-13 可知，分类页面中需要返回分类的列表数据。同时，分类信息有层级关系，分别是一级分类、二级分类和三级分类，并且三者的展示位置不同。一级分类固定在页面的左侧，由上至下平铺显示。二级分类和三级分类展示在页面右侧，每个二级分类下展示对应三级分类的列表，三级分类由左至右平铺展示，二级分类则分开展示。同时，二级分类数据和三级分类数据的展示区域会随着一级分类的切换动态变化。

在分类接口的返回数据格式定义中，需要返回一级分类列表，还包括每个一级分类的二级分类列表，而二级分类列表中的每个二级分类也有一个三级分类列表。因此，后端 API 项目中定义了 3 个视图层的分类 VO 对象，并做了层级的定义和关联。

因为都是分类信息，所以 3 个 VO 对象中都有分类级别和分类名称，三级分类的视图层对象 ThirdLevelCategoryVO 字段定义如下：

```
/**
 * 分类数据VO(第三级)
 */
@Data
public class ThirdLevelCategoryVO implements Serializable {

    @ApiModelProperty("当前三级分类id")
    private Long categoryId;

    @ApiModelProperty("当前分类级别")
    private Byte categoryLevel;

    @ApiModelProperty("当前三级分类名称")
    private String categoryName;
}
```

一级分类的视图层对象 NewBeeMallIndexCategoryVO 和二级分类的视图层对象 SecondLevelCategoryVO 字段定义如下：

```
/**
 * 分类数据VO(第二级)
 */
@Data
public class SecondLevelCategoryVO implements Serializable {

    @ApiModelProperty("当前二级分类id")
    private Long categoryId;

    @ApiModelProperty("父级分类id")
    private Long parentId;

    @ApiModelProperty("当前分类级别")
    private Byte categoryLevel;
```

```java
    @ApiModelProperty("当前二级分类名称")
    private String categoryName;

    @ApiModelProperty("三级分类列表")
    private List<ThirdLevelCategoryVO> thirdLevelCategoryVOS;
}
/**
 * 分类数据VO
 */
@Data
public class NewBeeMallIndexCategoryVO implements Serializable {

    @ApiModelProperty("当前一级分类 id")
    private Long categoryId;

    @ApiModelProperty("当前分类级别")
    private Byte categoryLevel;

    @ApiModelProperty("当前一级分类名称")
    private String categoryName;

    @ApiModelProperty("二级分类列表")
    private List<SecondLevelCategoryVO> secondLevelCategoryVOS;
}
```

在分类的 VO 对象中都定义了分类层级字段，并且在一级分类 VO 对象和二级分类 VO 对象中定义了下级分类的列表字段。比如，在二级分类中，不仅包含二级分类的信息，还包含该二级分类下所有的三级分类的信息。

打开 newbee-mall-cloud-goods-web 工程，新建 ltd.goods.cloud.newbee.controller.vo 包，创建上述 3 个 VO 对象即可。

6.4.2 业务层代码的实现

因为获取分类页面的数据只涉及查询操作，所以业务层方法中定义了 getCategoriesForIndex() 方法用于查询分类页面中所需要的数据，查询到数据后进行首页 VO 对象的封装。代码及注释信息如下：

```java
public List<NewBeeMallIndexCategoryVO> getCategoriesForIndex() {
    List<NewBeeMallIndexCategoryVO> newBeeMallIndexCategoryVOS = new
```

```java
ArrayList<>();
    //获取一级分类的固定数量的数据
    List<GoodsCategory> firstLevelCategories = goodsCategoryMapper.selectByLevelAndParentIdsAndNumber(Collections.singletonList(0L), NewBeeMallCategoryLevelEnum.LEVEL_ONE.getLevel(), Constants.INDEX_CATEGORY_NUMBER);
    if (!CollectionUtils.isEmpty(firstLevelCategories)) {
        List<Long> firstLevelCategoryIds = firstLevelCategories.stream().map(GoodsCategory::getCategoryId).collect(Collectors.toList());
        //获取二级分类的数据
        List<GoodsCategory> secondLevelCategories = goodsCategoryMapper.selectByLevelAndParentIdsAndNumber(firstLevelCategoryIds, NewBeeMallCategoryLevelEnum.LEVEL_TWO.getLevel(), 0);
        if (!CollectionUtils.isEmpty(secondLevelCategories)) {
            List<Long> secondLevelCategoryIds = secondLevelCategories.stream().map(GoodsCategory::getCategoryId).collect(Collectors.toList());
            //获取三级分类的数据
            List<GoodsCategory> thirdLevelCategories = goodsCategoryMapper.selectByLevelAndParentIdsAndNumber(secondLevelCategoryIds, NewBeeMallCategoryLevelEnum.LEVEL_THREE.getLevel(), 0);
            if (!CollectionUtils.isEmpty(thirdLevelCategories)) {
                //根据 parentId 将 thirdLevelCategories 分组
                Map<Long, List<GoodsCategory>> thirdLevelCategoryMap = thirdLevelCategories.stream().collect(groupingBy(GoodsCategory::getParentId));
                List<SecondLevelCategoryVO> secondLevelCategoryVOS = new ArrayList<>();
                //处理二级分类
                for (GoodsCategory secondLevelCategory : secondLevelCategories) {
                    SecondLevelCategoryVO secondLevelCategoryVO = new SecondLevelCategoryVO();
                    BeanUtil.copyProperties(secondLevelCategory, secondLevelCategoryVO);
                    //如果该二级分类下有数据, 则放入 secondLevelCategoryVOS 对象
                    if (thirdLevelCategoryMap.containsKey(secondLevelCategory.getCategoryId())) {
                        //根据二级分类的 id 取出 thirdLevelCategoryMap 分组中的三级分类列表
                        List<GoodsCategory> tempGoodsCategories = thirdLevelCategoryMap.get(secondLevelCategory.getCategoryId());
                        secondLevelCategoryVO.setThirdLevelCategoryVOS((BeanUtil.copyList(tempGoodsCategories, ThirdLevelCategoryVO.class)));
                        secondLevelCategoryVOS.add(secondLevelCategoryVO);
                    }
                }
```

```
    //处理一级分类
      if (!CollectionUtils.isEmpty(secondLevelCategoryVOS)) {
        //根据 parentId 将 thirdLevelCategories 分组
        Map<Long, List<SecondLevelCategoryVO>> secondLevelCategoryVOMap =
secondLevelCategoryVOS.stream().collect(groupingBy(SecondLevelCategoryVO
::getParentId));
        for (GoodsCategory firstCategory : firstLevelCategories) {
          NewBeeMallIndexCategoryVO newBeeMallIndexCategoryVO = new
NewBeeMallIndexCategoryVO();
          BeanUtil.copyProperties(firstCategory, newBeeMallIndexCategoryVO);
          //如果该一级分类下有数据,则放入 newBeeMallIndexCategoryVOS 对象
          if (secondLevelCategoryVOMap.containsKey(firstCategory.
getCategoryId())) {
            //根据一级分类的 id 取出 secondLevelCategoryVOMap 分组中的二级分类列表
            List<SecondLevelCategoryVO> tempGoodsCategories =
secondLevelCategoryVOMap.get(firstCategory.getCategoryId());
            newBeeMallIndexCategoryVO.setSecondLevelCategoryVOS
(tempGoodsCategories);
            newBeeMallIndexCategoryVOS.add(newBeeMallIndexCategoryVO);
          }
        }
      }
    }
    return newBeeMallIndexCategoryVOS;
  } else {
    return null;
  }
}
```

getCategoriesForIndex()方法的作用是返回已配置完成的分类数据并响应给前端,实现思路如下:先读取固定数量的一级分类数据,再获取二级分类数据并设置到对应的一级分类下,然后获取和设置每个二级分类下的三级分类数据,接着将所有的分类列表数据都读取出来并根据层级进行划分和封装,最后将视图层对象返回给调用端。

下面结合代码具体讲解。查询一级分类列表的代码如下:

```
//获取一级分类的固定数量的数据
List<GoodsCategory> firstLevelCategories =
goodsCategoryMapper.selectByLevelAndParentIdsAndNumber(Collections.singl
etonList(0L), NewBeeMallCategoryLevelEnum.LEVEL_ONE.getLevel(),
Constants.INDEX_CATEGORY_NUMBER);
```

因为一级分类是没有父类的,即父级分类的 id 为默认值 0,同时,parentIds 参数为 List 类型,所以这里 parentIds 参数传的是 Collections.singletonList(0L),分类级别传的是 1,用的是枚举类 NewBeeMallCategoryLevelEnum.LEVEL_ONE。查询数量是 10 条,用的也是一个常量 Constants.INDEX_CATEGORY_NUMBER,该值默认为 10。当然,这里直接传数字 10 也是可以的。

查询二级分类列表的代码如下:

```
//获取二级分类的数据
List<GoodsCategory> secondLevelCategories =
goodsCategoryMapper.selectByLevelAndParentIdsAndNumber(firstLevelCategor
yIds, NewBeeMallCategoryLevelEnum.LEVEL_TWO.getLevel(), 0);
```

因为上一步已经获取了所有的一级分类列表数据,所以把其中的 id 字段全部取出来并放到一个 List 对象 firstLevelCategoryIds 中,之后作为查询二级分类列表的 parentIds 参数。分类级别传的是 2,用的是枚举类 NewBeeMallCategoryLevelEnum.LEVEL_TWO。数量 number 参数传的是 0,表示查询所有当前一级分类下的二级分类数据,并不是代表查询 0 条数据。三级分类查询方式与二级分类查询方式类似,这里不再赘述。

打开 newbee-mall-cloud-goods-web 工程,在 service.impl 包的 NewBeeMallCategoryServiceImpl 类中添加上述方法即可。

6.4.3 分类页面数据接口控制层代码的实现

增加分类页面接口的定义与编码。打开 newbee-mall-cloud-goods-web 工程,选择 ltd.goods.cloud.newbee.controller 包并单击鼠标右键,在弹出的快捷菜单中选择"New→Java Class"选项,在弹出的窗口中输入"NewBeeMallGoodsCategoryController",对分类页面的数据请求进行处理,在 NewBeeMallGoodsCategoryController 类中新增如下代码:

```
package ltd.goods.cloud.newbee.controller;

import io.swagger.annotations.Api;
import io.swagger.annotations.ApiOperation;
import ltd.common.cloud.newbee.ServiceResultEnum;
import ltd.common.cloud.newbee.dto.Result;
import ltd.common.cloud.newbee.dto.ResultGenerator;
import ltd.common.cloud.newbee.exception.NewBeeMallException;
import ltd.goods.cloud.newbee.controller.vo.NewBeeMallIndexCategoryVO;
import ltd.goods.cloud.newbee.service.NewBeeMallCategoryService;
import org.springframework.util.CollectionUtils;
import org.springframework.web.bind.annotation.GetMapping;
```

```java
import org.springframework.web.bind.annotation.RequestMapping;
import org.springframework.web.bind.annotation.RestController;

import javax.annotation.Resource;
import java.util.List;

@RestController
@Api(value = "v1", tags = "新蜂商城分类页面接口")
@RequestMapping("/categories/mall")
public class NewBeeMallGoodsCategoryController {

    @Resource
    private NewBeeMallCategoryService newBeeMallCategoryService;

    @GetMapping("/listAll")
    @ApiOperation(value = "获取分类数据", notes = "分类页面使用")
    public Result<List<NewBeeMallIndexCategoryVO>> getCategories() {
        List<NewBeeMallIndexCategoryVO> categories =
newBeeMallCategoryService.getCategoriesForIndex();
        if (CollectionUtils.isEmpty(categories)) {
            NewBeeMallException.fail(ServiceResultEnum.DATA_NOT_EXIST.
getResult());
        }
        return ResultGenerator.genSuccessResult(categories);
    }
}
```

处理分类列表数据请求的方法名称为 getCategories()，请求类型为 GET，映射的路径为 /categories/mall/listAll，响应结果类型为 Result，实际的 data 属性类型为 List 对象，即分类列表数据。

实现逻辑是调用分类业务实现类 NewBeeMallCategoryService 中的查询方法，查询所需的数据并响应给前端，所有的实现逻辑都是在业务实现类中处理的，包括查询和字段设置，在控制层代码中只是将获得的数据结果设置给 Result 对象并返回。

还有一点需要注意，分类页面的内容是任何访问者都能够直接查看的，并不需要登录认证，因此首页数据接口中并没有使用权限认证的注解@TokenToMallUser。

由于无须登录认证，因此需要在商城端网关的全局过滤器中添加配置，在鉴权时忽略这个接口，直接放行。代码修改如下：

```java
final List<String> ignoreURLs = new ArrayList<>();
ignoreURLs.add("/users/mall/login");
ignoreURLs.add("/users/mall/register");
ignoreURLs.add("/categories/mall/listAll");
```

```
ignoreURLs.add("/mall/index");

// 登录接口、注册接口、首页接口、分类接口，直接放行
if (ignoreURLs.contains(exchange.getRequest().getURI().getPath())) {
    return chain.filter(exchange);
}
```

在当前版本中，分类数据的获取是一个单独的接口。换一种思路来设计该接口，可以做成三个接口。第一个接口返回所有的一级分类数据，第二个接口根据选择的一级分类 id 查询所有的二级分类数据，第三个接口则根据选择的二级分类 id 查询所有的三级分类数据。笔者认为，一次性全部查出来的设计更好一些，这种接口设计方式可以让前端开发人员一次性处理和渲染这些数据，而不是分多次查询、渲染页面。

做成单独一个接口，前端方便处理，既不用因为用户切换了不同的分类而重新改变页面 DOM，也不用根据页面选项卡的切换多次发送请求，在一定程度上可以节省网络开销。不过，在做接口设计时，依然要因地制宜、灵活变通，并不是说笔者的想法就是正确的，读者要结合自己实际开发的项目灵活地分析和设计。

6.4.4 分类接口网关配置

打开 newbee-mall-cloud-gateway-mall 项目中的 application.properties 文件，新增关于商品微服务的路由信息，配置项为 spring.cloud.gateway.routes.*，新增内容如下：

```
# 分类接口的路由配置
spring.cloud.gateway.routes[2].id=goods-service-route
spring.cloud.gateway.routes[2].uri=lb://newbee-mall-cloud-goods-service
spring.cloud.gateway.routes[2].order=1
spring.cloud.gateway.routes[2].predicates[0]=Path=/categories/mall/**
```

以上是配置 newbee-mall-cloud-gateway-mall 到商品微服务的路由信息，主要配置"获取分类页面数据"的接口。如果访问商城端网关项目的路径是以/categories/mall/开头的，就路由到商品微服务实例。

6.5 商品列表和商品详情页面的接口实现

首页接口和分类接口已经介绍完毕，接下来进行商品列表和商品详情页面的功能完善，需要实现商品搜索和商品详情页面的后端接口，本节的源代码是在 newbee-mall-cloud-dev-step13 工程的基础上改造的，将工程命名为 newbee-mall-cloud-dev-step13-2。

6.5.1 接口传参解析及返回字段定义

获取用户输入的搜索关键字或分类 id，跳转到搜索结果页面，以及搜索结果页面渲染，这些都是前端开发人员来实现的。前端页面在实现这些效果时，需要获取当前页面要渲染的数据，也就是在跳转到搜索结果页面后，组装商品搜索的请求参数并向后端接口发送搜索请求，获得后端返回的数据后进行页面渲染。后端开发人员需要定义接口参数、接口地址、接口返回字段，后端接口需要接收前端传过来的参数，先根据参数查询数据，再组装数据并返回给前端，最后由前端进行渲染。

1. 接口传参

页面 UI 还原和样式编写是由前端开发人员来实现的，前端开发人员包括 Web 开发人员和移动端原生 App 开发人员，负责页面及交互。后端开发人员则需要实现一个列表数据的返回接口，根据用户输入的关键字或用户单击的分类来返回数据，即商品搜索接口。因此，搜索接口传参的两个重要参数为搜索关键字和商品的三级分类 id，将其分别定义为 keyword 和 goodsCategoryId，类型分别为 String 和 Long。另外，还有两个参数是排序方式 orderBy 和分页参数 pageNumber。

商品详情页面是让用户看到更多的商品信息，以便更好地进行选择和比较。不过，获取商品详情信息这个接口并不是一个复杂的逻辑，实现逻辑就是根据商品 id 查询商品表中的记录并返回给前端。

2. 分页逻辑

虽然只是返回一个列表，但是有一个隐藏的知识点，那就是该接口返回的数据条数可能很多，需要加入分页的逻辑。在展示搜索内容后，商品列表页面可以不断往上拉，数据会不断地加载进来，在移动端的实现效果就是人们常说的"上拉加载"，这里就用到了分页逻辑。虽然没有分页页码和翻页按钮，但依然是分页展示的逻辑，移动端页面通常的做法就是如此，毕竟人们在移动端的操作习惯与在 PC 端的操作习惯不同，移动端页面的面积也比 PC 端页面的面积小了很多，不可能完全做成 PC 端的分页效果。

分页数据还做了一个封装，由于所有的分页结果基本上都包括以下几个属性，因此分页结果集的数据格式定义如下（注：完整代码位于 ltd.common.cloud.newbee.dto.PageResult）：

```java
//分页的通用结果类
public class PageResult implements Serializable {
    //总记录数
    private int totalCount;
    //每页记录数
    private int pageSize;
```

```
    //总页码
    private int totalPage;
    //当前页的页码
    private int currPage;
    //当前页的所有数据列表
    private List<?> list;
}
```

在实现分页功能的返回对象 PageResult 中定义了以下 4 个参数，即当前页的所有数据列表、当前页的页码、总页码、总记录数，并将它们放入 Result 返回结果的 data 属性中，在商品搜索接口中我们最终得到的返回对象为 Result<PageResult<Lis <NewBeeMallSearchGoodsVO>>>，最外层是 Result 对象，这个不用多说，所有的接口返回类型都是 Result 类型，里面一层是 PageResult 对象，因为搜索接口需要返回分页信息，PageResult 对象中则是具体的当前页所需要的商品列表信息 List<NewBeeMallSearchGoodsVO>。

之后由前端开发人员直接读取对应的参数并对这些数据进行处理，这就是前后端进行数据交互时分页数据的格式定义，希望读者能够结合代码及实际的分页效果进行理解和学习。

3. 返回数据格式

图 6-14 所示为商品列表页面和商品详情页面中需要渲染的内容。商品分页列表是一个 List 对象，还有分页功能，需要返回分页字段，因此最终接收的结果返回格式为 PageResult 对象，而列表的单项对象中的字段则需要通过图 6-14 中的内容进行确认。

图 6-14　商品列表页面和商品详情页面中需要渲染的内容

通过图片可以看到商品预览图字段、商品标题字段、商品简介字段、商品价格字段。当然，这里通常会设计成可跳转的形式，即单击标题或预览图会跳转到对应的商品详情页面中，因此还需要一个商品实体的 id 字段。这样，返回数据的格式就得出来了，代码如下：

```java
package ltd.goods.cloud.newbee.controller.vo;

import io.swagger.annotations.ApiModelProperty;
import lombok.Data;

import java.io.Serializable;

/**
 * 搜索列表页面商品VO
 */
@Data
public class NewBeeMallSearchGoodsVO implements Serializable {

    @ApiModelProperty("商品id")
    private Long goodsId;

    @ApiModelProperty("商品名称")
    private String goodsName;

    @ApiModelProperty("商品简介")
    private String goodsIntro;

    @ApiModelProperty("商品图片地址")
    private String goodsCoverImg;

    @ApiModelProperty("商品价格")
    private Integer sellingPrice;

}
```

商品详情页面需要展示的字段更多一些，数据格式如下：

```java
/**
 * 商品详情页面VO
 */
@Data
public class NewBeeMallGoodsDetailVO implements Serializable {

    @ApiModelProperty("商品id")
```

```
    private Long goodsId;

    @ApiModelProperty("商品名称")
    private String goodsName;

    @ApiModelProperty("商品简介")
    private String goodsIntro;

    @ApiModelProperty("商品图片地址")
    private String goodsCoverImg;

    @ApiModelProperty("商品价格")
    private Integer sellingPrice;

    @ApiModelProperty("商品标签")
    private String tag;

    @ApiModelProperty("商品图片")
    private String[] goodsCarouselList;

    @ApiModelProperty("商品原价")
    private Integer originalPrice;

    @ApiModelProperty("商品详情字段")
    private String goodsDetailContent;
}
```

打开 newbee-mall-cloud-goods-web 工程，在 ltd.goods.cloud.newbee.controller.vo 包中增加上述两个 VO 类即可。

6.5.2 业务层代码的实现

获取商品列表的数据只涉及查询操作，因此业务层方法中定义了 searchNewBeeMallGoods() 方法用于查询分类页面中所需要的数据，查询到数据后进行分页对象的封装。查询商品详情数据则更加简单，根据主键查询对应的数据库记录即可，代码如下：

```
public PageResult searchNewBeeMallGoods(PageQueryUtil pageUtil) {
    List<NewBeeMallGoods> goodsList = goodsMapper.findNewBeeMallGoodsListBySearch(pageUtil);
    int total = goodsMapper.getTotalNewBeeMallGoodsBySearch(pageUtil);
    List<NewBeeMallSearchGoodsVO> newBeeMallSearchGoodsVOS = new
```

```
ArrayList<>();
    if (!CollectionUtils.isEmpty(goodsList)) {
        newBeeMallSearchGoodsVOS = BeanUtil.copyList(goodsList,
NewBeeMallSearchGoodsVO.class);
        for (NewBeeMallSearchGoodsVO newBeeMallSearchGoodsVO :
newBeeMallSearchGoodsVOS) {
            String goodsName = newBeeMallSearchGoodsVO.getGoodsName();
            String goodsIntro = newBeeMallSearchGoodsVO.getGoodsIntro();
            // 字符串过长导致文字超出的问题
            if (goodsName.length() > 28) {
                goodsName = goodsName.substring(0, 28) + "...";
                newBeeMallSearchGoodsVO.setGoodsName(goodsName);
            }
            if (goodsIntro.length() > 30) {
                goodsIntro = goodsIntro.substring(0, 30) + "...";
                newBeeMallSearchGoodsVO.setGoodsIntro(goodsIntro);
            }
        }
    }
    PageResult pageResult = new PageResult(newBeeMallSearchGoodsVOS, total,
pageUtil.getLimit(), pageUtil.getPage());
    return pageResult;
}

public NewBeeMallGoods getNewBeeMallGoodsById(Long id) {
    NewBeeMallGoods newBeeMallGoods = goodsMapper.selectByPrimaryKey(id);
    if (newBeeMallGoods == null) {
        NewBeeMallException.fail(ServiceResultEnum.GOODS_NOT_EXIST.getResult());
    }
    return newBeeMallGoods;
}
```

searchNewBeeMallGoods()方法使用 PageQueryUtil 对象作为参数，商品类目 id 、关键字 keyword 字段、分页所需的 page 字段、排序字段等都作为属性放在了这个对象中，关键字或商品类目 id 用来过滤想要的商品列表，page 字段用于确定查询第几页的数据，之后通过 SQL 语句查询对应的分页数据，并填充数据。某些字段太长导致页面上的展示效果不好，所以对这些字段进行了简单的字符串处理并设置到 NewBeeMallSearchGoodsVO 对象中，最终返回的数据是 PageResult 对象，包含当前页返回的商品列表数据和分页信息。

打开 newbee-mall-cloud-goods-web 工程，在 service.impl 包的 NewBeeMallCategoryServiceImpl 类中添加上述方法即可。

具体执行的 SQL 语句如下（注：完整代码位于 resources/mapper/ NewBeeMallGoodsMapper.xml 文件中）：

```
<select id="findNewBeeMallGoodsListBySearch" parameterType="Map" resultMap=
```

```xml
"BaseResultMap">
    select
    <include refid="Base_Column_List"/>
    from tb_newbee_mall_goods_info
    <where>
        <if test="keyword!=null and keyword!=''">
            and (goods_name like CONCAT('%',#{keyword},'%') or goods_intro like CONCAT('%',#{keyword},'%'))
        </if>
        <if test="goodsCategoryId!=null and goodsCategoryId!=''">
            and goods_category_id = #{goodsCategoryId}
        </if>
    </where>
    <if test="orderBy!=null and orderBy!=''">
        <choose>
            <when test="orderBy == 'new'">
                <!-- 按照发布时间倒序排列 -->
                order by goods_id desc
            </when>
            <when test="orderBy == 'price'">
                <!-- 按照售价从便宜到贵排列 -->
                order by selling_price asc
            </when>
            <otherwise>
                <!-- 默认按照库存数量从多到少排列 -->
                order by stock_num desc
            </otherwise>
        </choose>
    </if>
    <if test="start!=null and limit!=null">
        limit #{start},#{limit}
    </if>
</select>

<select id="getTotalNewBeeMallGoodsBySearch" parameterType="Map" resultType="int">
    select count(*) from tb_newbee_mall_goods_info
    <where>
        <if test="keyword!=null and keyword!=''">
            and (goods_name like CONCAT('%',#{keyword},'%') or goods_intro like CONCAT('%',#{keyword},'%'))
        </if>
        <if test="goodsCategoryId!=null and goodsCategoryId!=''">
```

```
            and goods_category_id = #{goodsCategoryId}
        </if>
    </where>
</select>
```

根据前端传过来的关键字和商品类目 id 对商品记录进行检索，使用 MySQL 数据库的 LIKE 语法对关键字进行过滤，或者根据 goods_category_id 字段对商品类目进行过滤，之后根据 orderBy 字段进行商品搜索分页结果的排序，最后使用 start 和 limit 两个分页所必需的关键字。

6.5.3 控制层代码的实现

增加商品列表接口的定义与编码。打开 newbee-mall-cloud-goods-web 工程，选择 ltd.goods.cloud.newbee.controller 包并单击鼠标右键，在弹出的快捷菜单中选择"New→Java Class"选项，在弹出的窗口中输入"NewBeeMallGoodsController"，在 NewBeeMallGoodsController 类中新增如下代码：

```
package ltd.goods.cloud.newbee.controller;

import io.swagger.annotations.Api;
import io.swagger.annotations.ApiOperation;
import io.swagger.annotations.ApiParam;
import ltd.common.cloud.newbee.enums.ServiceResultEnum;
import ltd.common.cloud.newbee.dto.PageQueryUtil;
import ltd.common.cloud.newbee.dto.PageResult;
import ltd.common.cloud.newbee.dto.Result;
import ltd.common.cloud.newbee.dto.ResultGenerator;
import ltd.common.cloud.newbee.exception.NewBeeMallException;
import ltd.common.cloud.newbee.pojo.MallUserToken;
import ltd.common.cloud.newbee.util.BeanUtil;
import ltd.goods.cloud.newbee.config.annotation.TokenToMallUser;
import ltd.goods.cloud.newbee.controller.vo.NewBeeMallGoodsDetailVO;
import ltd.goods.cloud.newbee.controller.vo.NewBeeMallSearchGoodsVO;
import ltd.goods.cloud.newbee.entity.NewBeeMallGoods;
import ltd.goods.cloud.newbee.service.NewBeeMallGoodsService;
import org.slf4j.Logger;
import org.slf4j.LoggerFactory;
import org.springframework.util.StringUtils;
import org.springframework.web.bind.annotation.*;

import javax.annotation.Resource;
```

```java
import java.util.HashMap;
import java.util.List;
import java.util.Map;

@RestController
@Api(value = "v1", tags = "新蜂商城商品相关接口")
@RequestMapping("/goods/mall")
public class NewBeeMallGoodsController {

    private static final Logger logger = LoggerFactory.getLogger(NewBeeMall
GoodsController.class);

    @Resource
    private NewBeeMallGoodsService newBeeMallGoodsService;

    @GetMapping("/search")
    @ApiOperation(value = "商品搜索接口", notes = "根据关键字和分类id进行搜索")
    public Result<PageResult<List<NewBeeMallSearchGoodsVO>>>
search(@RequestParam(required = false) @ApiParam(value = "搜索关键字") String
keyword,@RequestParam(required = false) @ApiParam(value = "分类id") Long
goodsCategoryId,@RequestParam(required = false) @ApiParam(value = "orderBy")
String orderBy,@RequestParam(required = false) @ApiParam(value = "页码")
Integer pageNumber,@TokenToMallUser MallUserToken loginMallUserToken) {

        logger.info("goods search api,keyword={},goodsCategoryId={},
orderBy={},pageNumber={},userId={}", keyword, goodsCategoryId, orderBy,
pageNumber, loginMallUserToken.getUserId());

        Map params = new HashMap(8);
        //两个搜索参数都为空，直接返回异常
        if (goodsCategoryId == null && !StringUtils.hasText(keyword)) {
            NewBeeMallException.fail("非法的搜索参数");
        }
        if (pageNumber == null || pageNumber < 1) {
            pageNumber = 1;
        }
        params.put("goodsCategoryId", goodsCategoryId);
        params.put("page", pageNumber);
        params.put("limit", 10);
        //对keyword做过滤，去掉空格
        if (StringUtils.hasText(keyword)) {
            params.put("keyword", keyword);
        }
```

```java
        if (StringUtils.hasText(orderBy)) {
            params.put("orderBy", orderBy);
        }
        //搜索上架状态的商品
        params.put("goodsSellStatus", 0);
        //封装商品数据
        PageQueryUtil pageUtil = new PageQueryUtil(params);
        return ResultGenerator.genSuccessResult (newBeeMallGoodsService.
searchNewBeeMallGoods(pageUtil));
    }

    @GetMapping("/detail/{goodsId}")
    @ApiOperation(value = "商品详情接口", notes = "传参为商品id")
    public Result<NewBeeMallGoodsDetailVO> goodsDetail(@ApiParam(value = "
商品id") @PathVariable("goodsId") Long goodsId, @TokenToMallUser
MallUserToken loginMallUserToken) {
        logger.info("goods detail api,goodsId={},userId={}", goodsId,
loginMallUserToken.getUserId());
        if (goodsId < 1) {
            return ResultGenerator.genFailResult("参数异常");
        }
        NewBeeMallGoods goods = newBeeMallGoodsService. getNewBeeMall
GoodsById(goodsId);
        if (0 != goods.getGoodsSellStatus()) {
            NewBeeMallException.fail(ServiceResultEnum.GOODS_PUT_DOWN.
getResult());
        }
        NewBeeMallGoodsDetailVO goodsDetailVO = new NewBeeMallGoodsDetailVO();
        BeanUtil.copyProperties(goods, goodsDetailVO);
        goodsDetailVO.setGoodsCarouselList(goods.getGoodsCarousel().split(","));
        return ResultGenerator.genSuccessResult(goodsDetailVO);
    }
}
```

商品列表接口的方法名称为 search()，请求类型为 GET，路径映射为/goods/mall/search，所有的传参都用@RequestParam 注解进行接收，前端传过来的参数主要有：

① keyword；

② goodsCategoryId；

③ orderBy；

④ pageNumber。

参数 pageNumber 是分页所必需的字段，如果不传，则默认为第 1 页。参数 keyword

是关键字,用来过滤商品名称和商品简介。参数 goodsCategoryId 是用来过滤商品分类 id 的字段。参数 orderBy 是排序字段,传过来不同的排序方式,返回的数据也会不同。另外,还有一个参数是当前登录用户的信息,已经用@TokenToMallUser 注解来接收,相关逻辑已经介绍过,这里不再赘述。根据以上字段进行查询参数的封装,之后通过 SQL 语句查询对应的分页数据 pageResult。响应结果类型为 Result,实际的 data 属性类型为 PageResult<List>对象,即商品搜索结果的分页列表数据。

实现逻辑就是调用商品业务实现类 NewBeeMallGoodsService 中的查询方法,将所需的数据进行查询并响应给前端,所有的实现逻辑都是在业务实现类中处理的,包括查询和返回字段的内容设置。在控制层代码中,主要进行参数判断和参数的封装,以及将获得的数据结果设置给 Result 对象并返回。

商品详情接口的方法名称为 goodsDetail(),请求类型为 GET,路径映射为 /goods/mall/detail/{goodsId},参数 goodsId 就是商品主键 id,通过@PathVariable 注解读取路径中的这个字段值,并根据这个字段值调用 NewBeeMallGoodsService 中的 getNewBeeMallGoodsById()方法,获取 NewBeeMallGoods 对象。getNewBeeMallGoodsById()方法的实现方式是根据主键 id 查询数据库中的商品表并返回商品实体数据,将查询到的商品详情数据转换为 NewBeeMallGoodsDetailVO 对象并返回给前端。因为在商品表中并不是所有的字段都需要返回,所以这里做了一次对象的转换。

6.5.4 商品接口网关配置

打开 newbee-mall-cloud-gateway-mall 项目中的 application.properties 文件,新增关于商品微服务的路由信息,配置项为 spring.cloud.gateway.routes.*,新增内容如下:

```
# 商品接口的路由配置
spring.cloud.gateway.routes[3].id=goods-service-route2
spring.cloud.gateway.routes[3].uri=lb://newbee-mall-cloud-goods-service
spring.cloud.gateway.routes[3].order=1
spring.cloud.gateway.routes[3].predicates[0]=Path=/goods/mall/**
```

这里配置 newbee-mall-cloud-gateway-mall 到商品微服务的路由信息,主要配置商品列表和商品详情页面的接口。如果访问商城端网关项目的路径是以/goods/mall/开头的,就路由到商品微服务实例。

6.6 商城端部分接口的功能测试

代码修改完成后，测试步骤是不能漏掉的。一定要验证项目是否能正常启动、接口是否能正常调用，防止在代码移动过程中出现一些小问题，导致项目无法启动或代码报错。在项目启动前，需要分别启动 Nacos Server 和 Redis Server，之后依次启动 newbee-mall-cloud-user-web 工程、newbee-mall-cloud-goods-web 工程和 newbee-mall-cloud-gateway-mall 工程下的主类。启动成功后，就可以进行本节的功能测试了。

6.6.1 获取首页数据的接口测试

打开推荐微服务的 Swagger 页面，在浏览器中输入如下网址：http://localhost:29020/swagger-ui/index.html。

接下来就可以在 Swagger 提供的 UI 页面中进行推荐微服务的接口测试了，接口文档显示内容如图 6-15 所示。

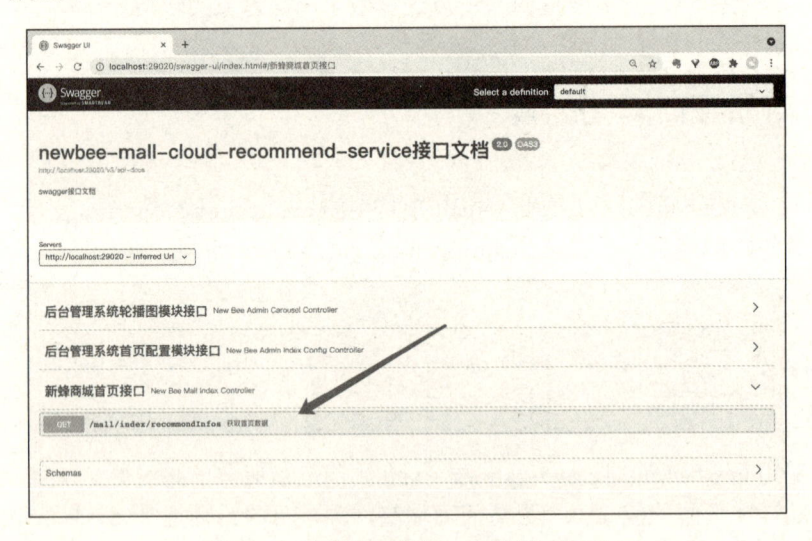

图 6-15 推荐微服务接口文档显示内容

单击"获取首页数据"选项，因为不需要身份验证，所以在 token 输入框中不输入 token 值也可以获取首页数据，之后单击"Execute"按钮，即可获取首页展示所需要的数据，测试结果如图 6-16 所示。

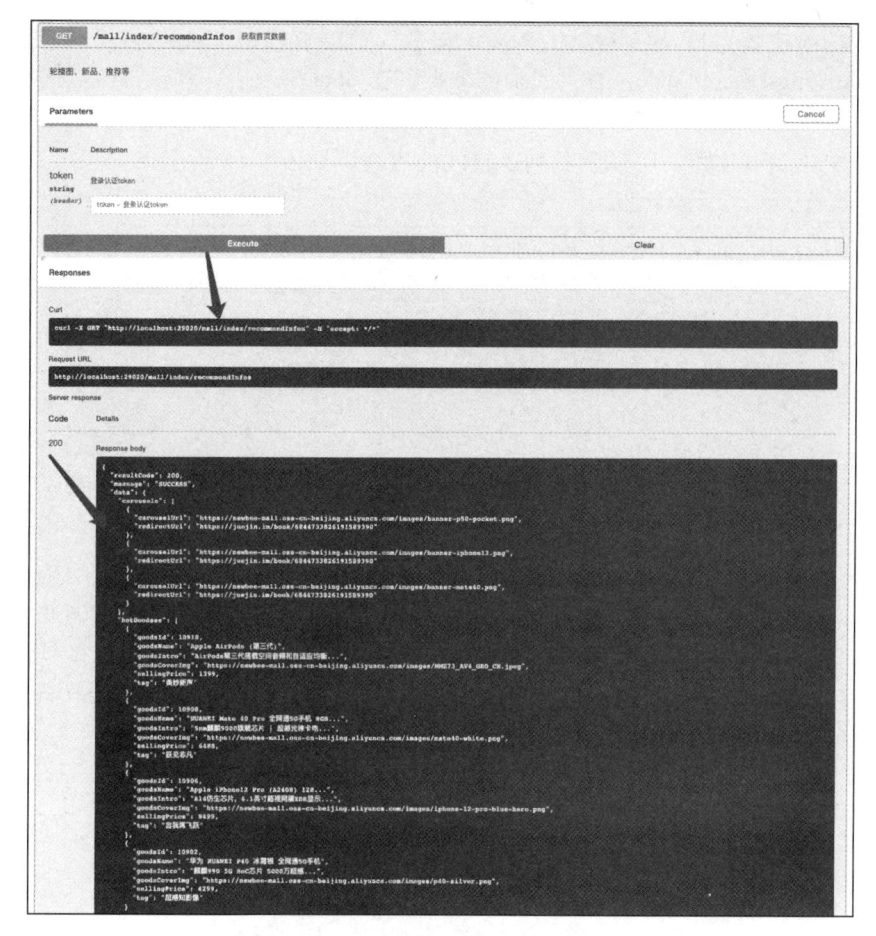

第 6 章　用户微服务及商品微服务功能完善

图 6-16　首页接口的测试结果

接下来分析一下返回结果，在当前页面上单击鼠标右键，在弹出的快捷菜单中选择"检查"选项，或按 F12 快捷键打开浏览器控制台，查看该接口返回的数据格式，如图 6-17 所示。

图 6-17　查看接口返回的数据格式

接收的响应数据是一个标准的 Result 对象，前端解析为 JSON 格式，字段分别为 resultCode、message 和 data，首页所需的数据都在 data 字段中，有轮播图数据和推荐商品数据。

依次单击这些字段，可以看到列表格式的内容，如图 6-18 所示。

图 6-18　列表格式的内容

接口响应的数据与预期一致，首页接口编码完成。

由于在商城端网关层做了路由配置，因此还需要通过网关层访问该接口确认通信是否正常。这里，笔者使用 Postman 工具进行测试，由于通过网关层访问，因此这里的请求网址需要进行修改，推荐微服务的端口号为 29020，商城端网关的端口号为 29110。

在 Postman 工具的地址栏中输入如下网址：http://localhost:29110/mall/index/recommondInfos。

设置请求方法为 GET，测试结果如图 6-19 所示。

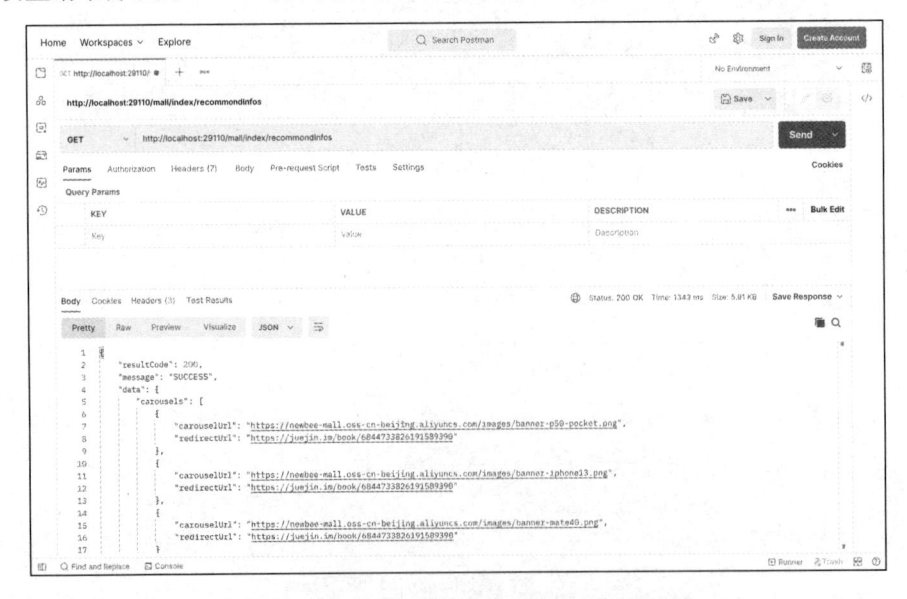

图 6-19　通过网关层请求首页接口的测试结果

最终，在网关层也成功获取了正确的数据响应。

6.6.2 获取分类页面的数据接口测试

打开商品微服务的 Swagger 页面，在浏览器中输入如下网址：http://localhost:29010/swagger-ui/index.html。

接着就可以在 Swagger 提供的 UI 页面中进行商品微服务的接口测试了，接口文档显示内容如图 6-20 所示。

图 6-20　商品微服务接口文档显示内容

依次单击"获取分类数据""Try it out"按钮，因为不需要身份验证，所以在 token 输入框中不输入 token 值也可以获取数据，之后单击"Execute"按钮，即可获取分类页面展示所需要的数据，测试结果如图 6-21 所示。

由于数据过多，因此截图无法截全。打开浏览器控制台，通过控制台 Network 面板中的内容可以分析返回的数据结构。单击刚刚请求的/categories/mall/listAll 链接，再单击"Preview"选项卡，就可以看到返回的数据的基本结构，如图 6-22 所示。

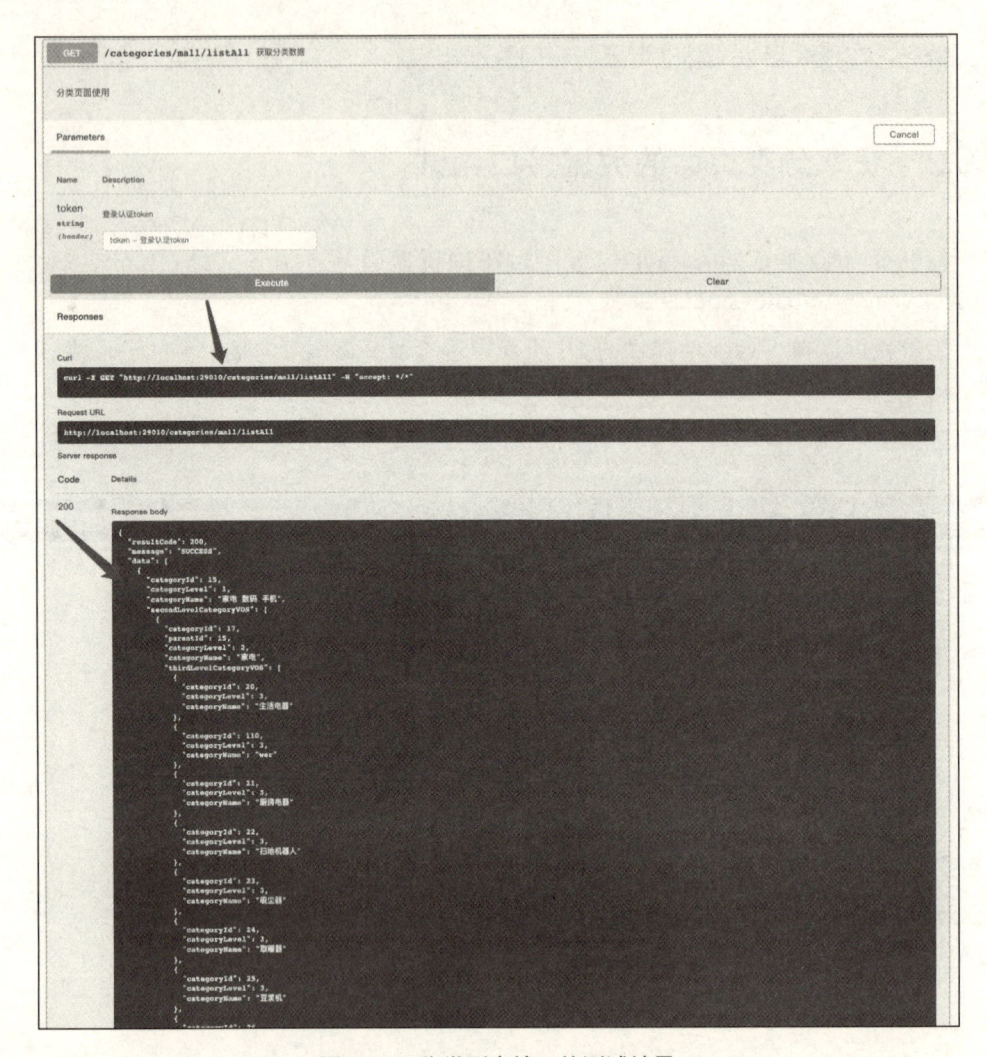

图 6-21 分类列表接口的测试结果

图 6-22 分类接口响应数据的基本结构

前端接收的响应数据是一个标准的 Result 对象，前端解析为 JSON 格式，字段分别为 resultCode、message 和 data，所有的分类数据在 data 字段中，图 6-23 中线框里的数据就是所有的一级分类数据，逐一点开每条一级分类，可以看到一级分类下还有二级分类列表，每个二级分类列表下还有三级分类列表。

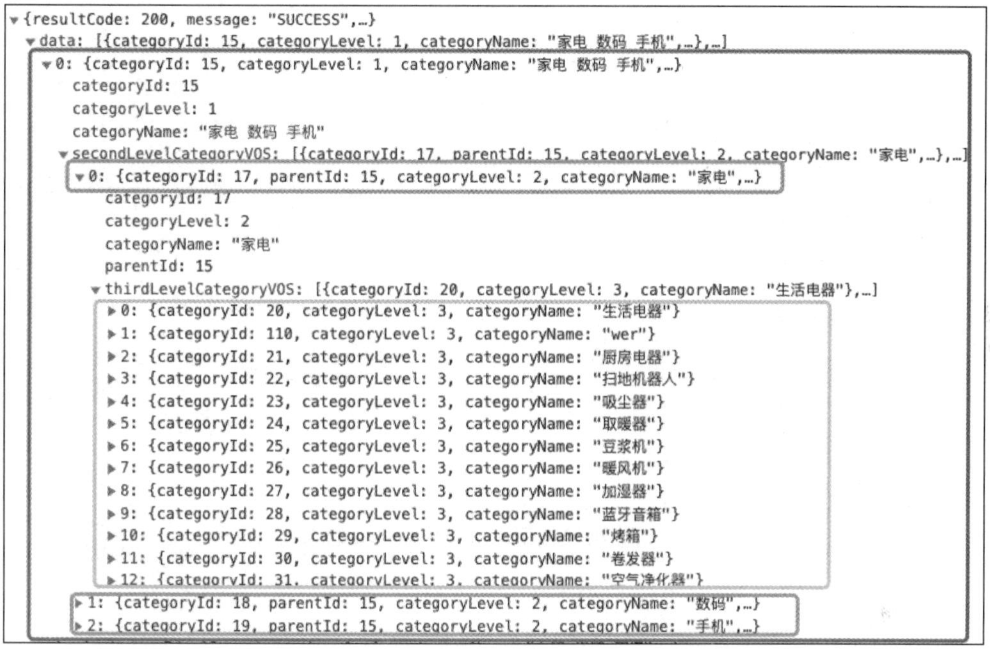

```
▼{resultCode: 200, message: "SUCCESS",…}
  ▼data: [{categoryId: 15, categoryLevel: 1, categoryName: "家电 数码 手机",…},…]
    ▼0: {categoryId: 15, categoryLevel: 1, categoryName: "家电 数码 手机",…}
        categoryId: 15
        categoryLevel: 1
        categoryName: "家电 数码 手机"
      ▼secondLevelCategoryVOS: [{categoryId: 17, parentId: 15, categoryLevel: 2, categoryName: "家电",…},…]
        ▼0: {categoryId: 17, parentId: 15, categoryLevel: 2, categoryName: "家电",…}
            categoryId: 17
            categoryLevel: 2
            categoryName: "家电"
            parentId: 15
          ▼thirdLevelCategoryVOS: [{categoryId: 20, categoryLevel: 3, categoryName: "生活电器"},…]
            ▶0: {categoryId: 20, categoryLevel: 3, categoryName: "生活电器"}
            ▶1: {categoryId: 110, categoryLevel: 3, categoryName: "wer"}
            ▶2: {categoryId: 21, categoryLevel: 3, categoryName: "厨房电器"}
            ▶3: {categoryId: 22, categoryLevel: 3, categoryName: "扫地机器人"}
            ▶4: {categoryId: 23, categoryLevel: 3, categoryName: "吸尘器"}
            ▶5: {categoryId: 24, categoryLevel: 3, categoryName: "取暖器"}
            ▶6: {categoryId: 25, categoryLevel: 3, categoryName: "豆浆机"}
            ▶7: {categoryId: 26, categoryLevel: 3, categoryName: "暖风机"}
            ▶8: {categoryId: 27, categoryLevel: 3, categoryName: "加湿器"}
            ▶9: {categoryId: 28, categoryLevel: 3, categoryName: "蓝牙音箱"}
            ▶10: {categoryId: 29, categoryLevel: 3, categoryName: "烤箱"}
            ▶11: {categoryId: 30, categoryLevel: 3, categoryName: "卷发器"}
            ▶12: {categoryId: 31, categoryLevel: 3, categoryName: "空气净化器"}
        ▶1: {categoryId: 18, parentId: 15, categoryLevel: 2, categoryName: "数码",…}
        ▶2: {categoryId: 19, parentId: 15, categoryLevel: 2, categoryName: "手机",…}
```

图 6-23　分类列表接口响应数据的详细结构

前端获取这些数据后，就可以渲染到页面中进行显示，分类接口编码完成。

由于在商城端网关层做了路由配置，因此还需要通过网关层访问该接口确认通信是否正常。这里，笔者使用 Postman 工具进行测试，由于通过网关层访问，因此这里的请求网址需要进行修改，推荐微服务的端口号为 29010，商城端网关的端口号为 29110。

在 Postman 工具的地址栏中输入如下网址：http://localhost:29110/categories/mall/listAll。

设置请求方法为 GET，测试结果如图 6-24 所示。

最终，在网关层也成功获取了正确的数据响应。

至此，增加商城用户相关逻辑、搭建商城端网关及完善用户微服务和商品微服务的工作就完成了，希望读者能够根据笔者提供的开发步骤顺利地完成本章的项目改造。

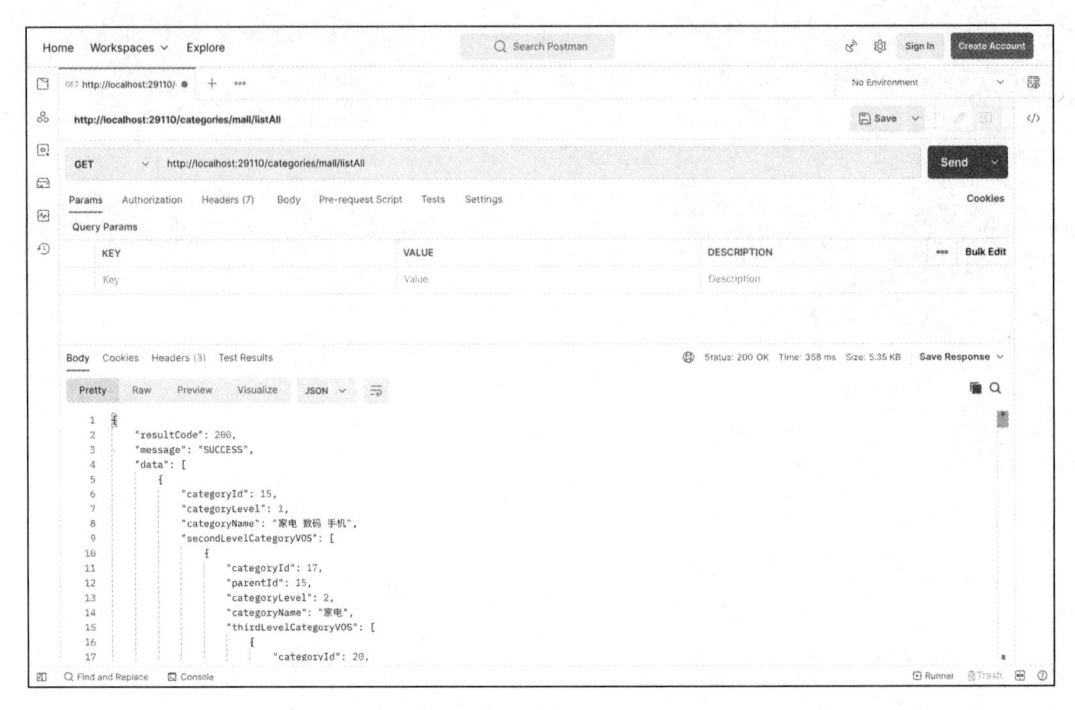

图 6-24 通过网关层请求分类列表接口的测试结果

第 7 章
购物车微服务编码实践及功能讲解

在商品模块和订单模块之间有一个模块,即购物车模块,它处于整个购物环节由商品到订单转换过程的中间状态,负责打通商品和订单这两个模块。本章继续讲解微服务架构项目的改造过程,主要介绍购物车模块的功能及开发步骤。

7.1 购物车微服务主要功能介绍

7.1.1 购物车模块介绍

为了让读者更好地理解购物车模块,这里先介绍顾客在线下超市的购物流程。
① 进入超市。
② 在入口处取一辆购物车或一个购物篮。
③ 在超市中逛。
④ 在不同的区域挑选不同的商品(水果、蔬菜、熟食、衣服、家居用品等)。
⑤ 经过一番筛选,将想要购买的商品放入购物车或购物篮。
⑥ 某些商品需要称重或做其他处理。
⑦ 到收银台清点商品并计算价格。
⑧ 付款(用微信/用支付宝/刷卡/用现金)。
⑨ 离开超市。

简单梳理一下购物流程：筛选商品、放入购物车或购物篮、清点结算、付款。这是日常生活中大家都有的体验。

大部分线上商城的购物车模块都是将线下购物流程进行抽象而开发的，新蜂商城也是如此。与线下实体的购物车不同，线上购物车模块的作用是存放商城用户挑选的商品数据。

购物车模块主要有以下 4 个功能：

① 将商品添加到购物车。

② 查询购物车中的购物项表。

③ 修改购物项。

④ 删除购物车中的某个购物项。

对应地，在设计后端接口时，需要设计并实现上述 4 个功能的接口。这 4 个接口的功能非常典型，分别是增、查、改、删，即程序员常挂在嘴边的"增删改查"。

7.1.2　购物车模块的表结构设计

商城系统的购物车模块的表结构的主要字段如下。

① user_id：用户的 id，根据这个字段确定用户购物车中的数据。

② goods_id：关联的商品 id，根据这个字段查询对应的商品信息并显示到页面上。

③ goods_count：购物车中某件商品的数量。

④ create_time：商品被添加到购物车中的时间。

购物车模块的表结构设计如下：

```sql
# 创建购物车微服务所需数据

CREATE DATABASE /*!32312 IF NOT EXISTS*/'newbee_mall_cloud_cart_db' /*!40100 DEFAULT CHARACTER SET utf8 */;

USE 'newbee_mall_cloud_cart_db';

DROP TABLE IF EXISTS 'tb_newbee_mall_shopping_cart_item';

CREATE TABLE 'tb_newbee_mall_shopping_cart_item' (
  'cart_item_id' bigint(20) NOT NULL AUTO_INCREMENT COMMENT '购物项主键id',
  'user_id' bigint(20) NOT NULL COMMENT '用户主键id',
```

```sql
  'goods_id' bigint(20) NOT NULL DEFAULT '0' COMMENT '关联商品id',
  'goods_count' int(11) NOT NULL DEFAULT '1' COMMENT '数量(最大为5)',
  'is_deleted' tinyint(4) NOT NULL DEFAULT '0' COMMENT '删除标识字段(0-未删除 1-已删除)',
  'create_time' timestamp NOT NULL DEFAULT CURRENT_TIMESTAMP COMMENT '创建时间',
  'update_time' timestamp NOT NULL DEFAULT CURRENT_TIMESTAMP COMMENT '最新修改时间',
  PRIMARY KEY ('cart_item_id')
) ENGINE=InnoDB DEFAULT CHARSET=utf8;
```

每个字段对应的含义在上面的 SQL 语句中都有介绍，读者可以对照理解，并正确地把建表 SQL 语句导入数据库。如果有需要，读者也可以根据该设计进行扩展。

上述代码中的购物车模块用来存储用户选择的商品数据，为订单结算做准备。这也是距离结算环节最近的一个步骤和功能点，接下来笔者将讲解购物车模块在微服务架构模式下的改造过程。

7.2 创建购物车微服务模块

本节介绍具体的代码改造过程，本节的源代码是在 newbee-mall-cloud-dev-step13-2 工程的基础上改造的，将工程命名为 newbee-mall-cloud-dev-step14。

在工程中新增一个 newbee-mall-cloud-shop-cart-service 模块，并在 pom.xml 主文件中增加该模块的配置，代码如下：

```xml
<modules>
  <!-- 新增购物车微服务 -->
  <module>newbee-mall-cloud-shop-cart-service</module>
  <module>newbee-mall-cloud-recommend-service</module>
  <module>newbee-mall-cloud-goods-service</module>
  <module>newbee-mall-cloud-user-service</module>
  <module>newbee-mall-cloud-gateway-mall</module>
  <module>newbee-mall-cloud-gateway-admin</module>
  <module>newbee-mall-cloud-common</module>
</modules>
```

该模块的目录结构设置与其他功能模块的目录结构设置类似，如下所示：

```
newbee-mall-cloud-shop-cart-service          // 购物车微服务
    ├── newbee-mall-cloud-shop-cart-api      // 存放购物车模块中暴露的用于远程
                                                调用的 FeignClient 类
    └── newbee-mall-cloud-shop-cart-web      // 购物车 API 的代码及逻辑
```

在新增购物车微服务时,主要参考了新增商品微服务时的步骤。笔者直接将 newbee-mall-cloud-dev-step05 源代码下 newbee-mall-cloud-goods-service 模块中的代码复制过来,之后对模块名称和目录名称进行修改。

最终,子节点 newbee-mall-cloud-shop-cart-service 模块的 pom.xml 文件代码如下:

```xml
<?xml version="1.0" encoding="UTF-8"?>
<project xmlns="http://maven.apa***.org/POM/4.0.0" xmlns:xsi=
"http://www.w*.org/2001/XMLSchema-instance"
     xsi:schemaLocation="http://ma***.apache.org/POM/4.0.0
https://maven.apa***.org/xsd/maven-4.0.0.xsd">
    <modelVersion>4.0.0</modelVersion>
    <groupId>ltd.newbee.cloud</groupId>
    <artifactId>newbee-mall-cloud-shop-cart-service</artifactId>
    <version>0.0.1-SNAPSHOT</version>
    <packaging>pom</packaging>
    <name>newbee-mall-cloud-shop-cart-service</name>
    <description>购物车模块</description>

    <parent>
        <groupId>ltd.newbee.cloud</groupId>
        <artifactId>newbee-mall-cloud</artifactId>
        <version>0.0.1-SNAPSHOT</version>
    </parent>

    <properties>
        <java.version>1.8</java.version>
    </properties>

    <modules>
        <module>newbee-mall-cloud-shop-cart-web</module>
        <module>newbee-mall-cloud-shop-cart-api</module>
    </modules>

    <dependencies>

    </dependencies>
</project>
```

主要的配置项是模块名称、模块的打包方式及模块间的父子关系。其中,重点配置了父模块 newbee-mall-cloud,两个子模块分别是 newbee-mall-cloud-shop-cart-api 和 newbee-mall-cloud-shop-cart-web。

newbee-mall-cloud-shop-cart-api 模块的 pom.xml 文件代码修改如下：

```xml
<?xml version="1.0" encoding="UTF-8"?>
<project xmlns="http://maven.apa***.org/POM/4.0.0" xmlns:xsi=
"http://www.w*.org/2001/XMLSchema-instance"
    xsi:schemaLocation="http://maven.apa***.org/POM/4.0.0
https://maven.apa***.org/xsd/maven-4.0.0.xsd">
    <modelVersion>4.0.0</modelVersion>
    <groupId>ltd.shopcart.newbee.cloud</groupId>
    <artifactId>newbee-mall-cloud-shop-cart-api</artifactId>
    <packaging>jar</packaging>
    <version>0.0.1-SNAPSHOT</version>
    <name>newbee-mall-cloud-shop-cart-api</name>
    <description>购物车微服务 openfeign</description>

    <parent>
        <groupId>ltd.newbee.cloud</groupId>
        <artifactId>newbee-mall-cloud-shop-cart-service</artifactId>
        <version>0.0.1-SNAPSHOT</version>
    </parent>

    <properties>
        <java.version>1.8</java.version>
    </properties>

    <dependencies>

        <dependency>
            <groupId>org.springframework.cloud</groupId>
            <artifactId>spring-cloud-starter-openfeign</artifactId>
        </dependency>

        <dependency>
            <groupId>ltd.newbee.cloud</groupId>
            <artifactId>newbee-mall-cloud-common</artifactId>
            <version>0.0.1-SNAPSHOT</version>
        </dependency>

    </dependencies>
</project>
```

上述代码定义了与父模块 newbee-mall-cloud-shop-cart-service 的关系，打包方式 packaging 配置项的值为 jar。同时，由于后续可能要将接口暴露，因此这里引入了 OpenFeign 的依赖项。

newbee-mall-cloud-shop-cart-web 模块的 pom.xml 文件代码修改如下：

```xml
<?xml version="1.0" encoding="UTF-8"?>
<project xmlns="http://maven.apa***.org/POM/4.0.0" xmlns:xsi=
"http://www.w*.org/2001/XMLSchema-instance"
         xsi:schemaLocation="http://maven.apa***.org/POM/4.0.0
https://maven.apa***.org/xsd/maven-4.0.0.xsd">
    <modelVersion>4.0.0</modelVersion>
    <groupId>ltd.shopcart.newbee.cloud</groupId>
    <artifactId>newbee-mall-cloud-shop-cart-web</artifactId>
    <version>0.0.1-SNAPSHOT</version>
    <name>newbee-mall-cloud-shop-cart-web</name>
    <description>购物车微服务</description>

    <parent>
        <groupId>ltd.newbee.cloud</groupId>
        <artifactId>newbee-mall-cloud-shop-cart-service</artifactId>
        <version>0.0.1-SNAPSHOT</version>
    </parent>

    <properties>
        <java.version>1.8</java.version>
    </properties>

    <dependencies>
        <dependency>
            <groupId>org.springframework.boot</groupId>
            <artifactId>spring-boot-starter-web</artifactId>
        </dependency>

        <dependency>
            <groupId>org.springframework.boot</groupId>
            <artifactId>spring-boot-starter-test</artifactId>
            <scope>test</scope>
        </dependency>

        <dependency>
            <groupId>com.alibaba.cloud</groupId>
            <artifactId>spring-cloud-starter-alibaba-nacos-discovery</artifactId>
        </dependency>

        <dependency>
            <groupId>org.springframework.boot</groupId>
            <artifactId>spring-boot-starter-validation</artifactId>
```

```xml
</dependency>

<dependency>
    <groupId>org.mybatis.spring.boot</groupId>
    <artifactId>mybatis-spring-boot-starter</artifactId>
</dependency>

<dependency>
    <groupId>org.projectlombok</groupId>
    <artifactId>lombok</artifactId>
    <version>${lombok.version}</version>
    <scope>provided</scope>
</dependency>

<dependency>
    <groupId>io.springfox</groupId>
    <artifactId>springfox-boot-starter</artifactId>
</dependency>

<dependency>
    <groupId>mysql</groupId>
    <artifactId>mysql-connector-java</artifactId>
    <scope>runtime</scope>
</dependency>

<dependency>
    <groupId>ltd.newbee.cloud</groupId>
    <artifactId>newbee-mall-cloud-common</artifactId>
    <version>0.0.1-SNAPSHOT</version>
</dependency>

<dependency>
    <groupId>org.springframework.cloud</groupId>
    <artifactId>spring-cloud-starter-openfeign</artifactId>
</dependency>

<dependency>
    <groupId>org.springframework.cloud</groupId>
    <artifactId>spring-cloud-starter-loadbalancer</artifactId>
</dependency>

<dependency>
    <groupId>ltd.user.newbee.cloud</groupId>
    <artifactId>newbee-mall-cloud-user-api</artifactId>
```

```xml
            <version>0.0.1-SNAPSHOT</version>
        </dependency>

        <dependency>
            <groupId>ltd.goods.newbee.cloud</groupId>
            <artifactId>newbee-mall-cloud-goods-api</artifactId>
            <version>0.0.1-SNAPSHOT</version>
        </dependency>

    </dependencies>
</project>
```

上述代码定义了与父模块 newbee-mall-cloud-shop-cart-service 的关系，同时将相关的业务依赖项也移到该配置文件中，因为购物车模块的主要业务代码都写在这个模块中。

由于改造过程中直接复制了 newbee-mall-cloud-dev-step05 源代码下 newbee-mall-cloud-goods-service 模块中的代码，因此在修改完依赖配置后，需要修改包名，把 ltd.goods.cloud.×××的名称修改为 ltd.shopcart.cloud.×××，并修改 config 包中的代码，包括全局异常处理配置类、Swagger 配置类、自定义 MVC 配置类，主要修改了这些类的类名。这样就完成了一个购物车微服务的初始构建工作。

7.3 远程调用用户微服务及其他注意事项

与其他微服务一样，在购物车微服务中也需要完善 token 字段处理的逻辑，即在购物车微服务中调用用户微服务完成商城用户的鉴权及用户信息的获取。

第一步，增加配置，启用 OpenFeign 并使 FeignClient 类生效。

由于已经引入了 LoadBalancer 依赖和 user-api 依赖，因此这里可以直接通过 OpenFeign 来调用用户微服务中的接口用于商城用户的鉴权和信息获取。

打开 newbee-mall-cloud-shop-cart-web 工程，在项目的启动类 NewBeeMallCloudShopCartServiceApplication 中添加@EnableFeignClients 注解，并配置相关的 FeignClient 类，代码如下：

```
@EnableFeignClients(basePackageClasses = {ltd.user.cloud.newbee.openfeign.NewBeeCloudUserServiceFeign.class})
```

这里使用 basePackageClasses 配置需要使用的 FeignClient 类，即 NewBeeCloudUserServiceFeign 类。接下来就可以直接使用 NewBeeCloudUserServiceFeign 类与用户微服务进行远程通信了。

第二步，修改 token 字段处理类中的逻辑代码。

打开 newbee-mall-cloud-shop-cart-web 工程，修改 TokenToMallUserMethodArgumentResolver 类中对 token 字段处理的逻辑代码，主要引入 NewBeeCloudUserServiceFeign 类，通过调用用户微服务来获取商城用户的数据。

修改后的代码如下：

```
@Component
public class TokenToMallUserMethodArgumentResolver implements
HandlerMethodArgumentResolver {

    @Autowired
    private NewBeeCloudUserServiceFeign newBeeCloudUserService;

    public TokenToMallUserMethodArgumentResolver() {
    }

    public boolean supportsParameter(MethodParameter parameter) {
        if (parameter.hasParameterAnnotation(TokenToMallUser.class)) {
            return true;
        }
        return false;
    }

    public Object resolveArgument(MethodParameter parameter,
ModelAndViewContainer mavContainer, NativeWebRequest webRequest,
WebDataBinderFactory binderFactory) {
        if (parameter.getParameterAnnotation(TokenToMallUser.class)
instanceof TokenToMallUser) {
            String token = webRequest.getHeader("token");
            if (null != token && !"".equals(token) && token.length() == 32) {
                // 调用用户微服务，根据 token 字段获取商城用户的数据
                Result<MallUserDTO> result = newBeeCloudUserService.get
MallUserByToken(token);
                if (result == null || result.getResultCode() != 200 ||
result.getData() == null) {
                    NewBeeMallException.fail(ServiceResultEnum.TOKEN_EXPIRE_
ERROR.getResult());
                }
```

```
                MallUserToken mallUserToken = new MallUserToken();
                mallUserToken.setToken(token);
                mallUserToken.setUserId(result.getData().getUserId());
                return mallUserToken;
            } else {
                NewBeeMallException.fail(ServiceResultEnum.NOT_LOGIN_ERROR.getResult());
            }
        }
        return null;
    }
}
```

如此便完成了在购物车微服务中通过远程通信获取当前登录用户信息的功能。

另外，购物车模块中的所有功能都只与商城用户相关，因此只需要处理商城用户即可。在全局异常处理类 ShopCartServiceExceptionHandler 中，由于购物车模块未涉及管理员用户账户，因此在区分自定义异常时有一些修改，要判断商城用户是否正常登录，未正常登录返回特定的响应代码 416（管理员用户未正常登录返回的响应代码为 419），代码如下：

```
@ExceptionHandler(Exception.class)
public Object handleException(Exception e, HttpServletRequest req) {
  Result result = new Result();
  result.setResultCode(500);
  //区分是否为自定义异常
  if (e instanceof NewBeeMallException) {
    result.setMessage(e.getMessage());
    // 判断商城用户是否正常登录，未正常登录返回特定的响应代码 416（管理员用户未正常登录返回的响应代码为 419）
    if (e.getMessage().equals(ServiceResultEnum.NOT_LOGIN_ERROR.getResult()) || e.getMessage().equals(ServiceResultEnum.TOKEN_EXPIRE_ERROR.getResult())) {
      result.setResultCode(416);
    }
  } else {
    e.printStackTrace();
    result.setMessage("未知异常，请查看控制台日志并检查配置文件。");
  }
  return result;
}
```

此时的 newbee-mall-cloud-shop-cart-service 模块中没有业务代码，目录结构如图 7-1 所示。

```
v newbee-mall-cloud-dev-step14 [newbee-mall-cloud]
  > .idea
  > newbee-mall-cloud-common
  > newbee-mall-cloud-gateway-admin
  > newbee-mall-cloud-gateway-mall
  > newbee-mall-cloud-goods-service
  > newbee-mall-cloud-recommend-service
  v newbee-mall-cloud-shop-cart-service
    > newbee-mall-cloud-shop-cart-api
    v newbee-mall-cloud-shop-cart-web
      v src
        v main
          v java
            v ltd.shopcart.cloud.newbee
              v config
                v annotation
                  @ TokenToMallUser
                v handler
                  © TokenToMallUserMethodArgumentResolver
                © ShopCartServiceExceptionHandler
                © ShopCartServiceSwagger3Config
                © ShopCartServiceWebMvcConfigurer
              © NewBeeMallCloudShopCartServiceApplication
          > resources
      newbee-mall-cloud-shop-cart-web.iml
      pom.xml
    newbee-mall-cloud-shop-cart-service.iml
    pom.xml
  > newbee-mall-cloud-user-service
  newbee-mall-cloud.iml
  pom.xml
```

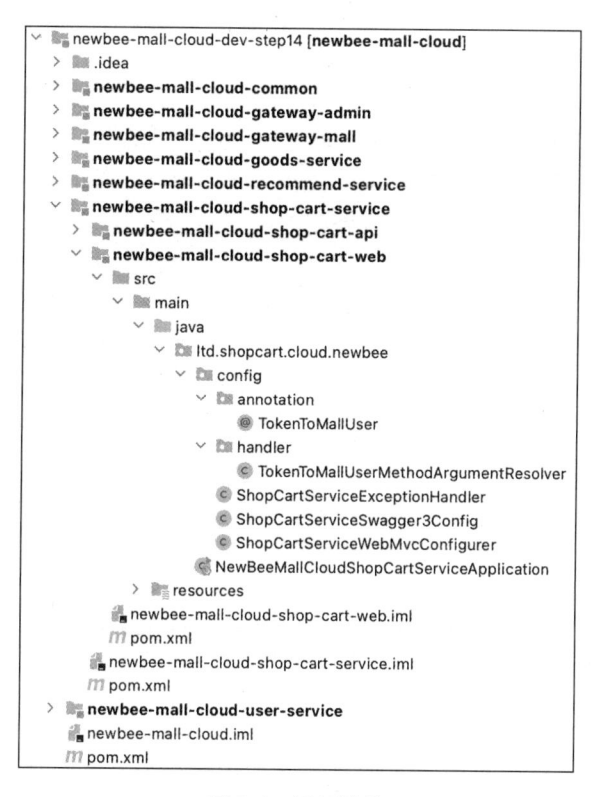

图 7-1 目录结构

这里并未涉及具体的业务代码，主要介绍购物车微服务的模块功能和表结构设计，以及在项目中完成购物车微服务的初始化构建，后续关于购物车微服务改造的实战章节都是基于当前项目来完成的。读者可以根据这些代码自己动手完成微服务编码，如获得这份代码后，以此为基础，自行完成购物车微服务模块所有代码的功能。如果自己实现耗费时间，也可以按照笔者给出的步骤完成服务化拆分。

7.4 购物车微服务编码

接下来，补充购物车微服务中的业务代码，主要把原单体 API 项目中的功能模块整合到购物车微服务中，源代码是在 newbee-mall-cloud-dev-step14 工程的基础上改造的，将工程命名为 newbee-mall-cloud-dev-step15。

7.4.1 购物车微服务代码改造

打开购物车微服务 newbee-mall-cloud-shop-cart-web 的工程目录，在 ltd.shopcart.cloud.newbee 包下依次创建 config 包、dao 包、entity 包、service 包，在 resources 目录下新增 Mapper 文件夹用于存放 Mapper 文件。接着，将原单体 API 项目中与购物车模块相关的业务代码和 Mapper 文件（如图 7-2 所示）依次复制过来。

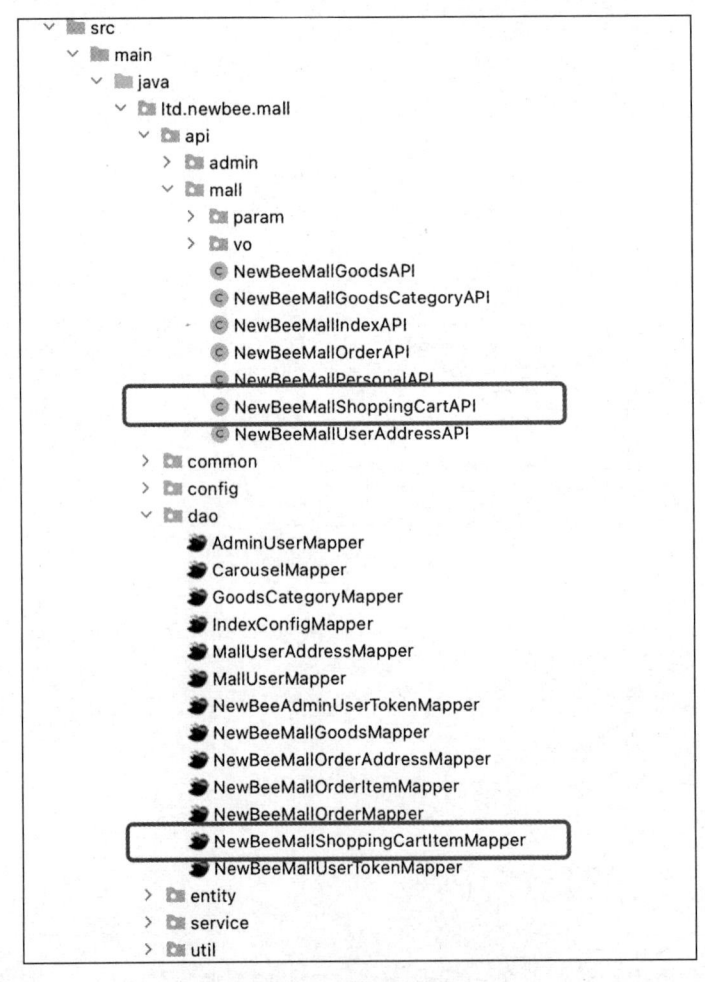

```
v src
  v main
    v java
      v ltd.newbee.mall
        > api
          > admin
          v mall
            > param
            > vo
              NewBeeMallGoodsAPI
              NewBeeMallGoodsCategoryAPI
              NewBeeMallIndexAPI
              NewBeeMallOrderAPI
              NewBeeMallPersonalAPI
             ┌─────────────────────────────┐
             │ NewBeeMallShoppingCartAPI   │
             └─────────────────────────────┘
              NewBeeMallUserAddressAPI
        > common
        > config
        v dao
            AdminUserMapper
            CarouselMapper
            GoodsCategoryMapper
            IndexConfigMapper
            MallUserAddressMapper
            MallUserMapper
            NewBeeAdminUserTokenMapper
            NewBeeMallGoodsMapper
            NewBeeMallOrderAddressMapper
            NewBeeMallOrderItemMapper
            NewBeeMallOrderMapper
           ┌─────────────────────────────────┐
           │ NewBeeMallShoppingCartItemMapper│
           └─────────────────────────────────┘
            NewBeeMallUserTokenMapper
        > entity
        > service
        > util
```

图 7-2 原单体 API 项目中与购物车模块相关的业务代码和 Mapper 文件

上述步骤完成后，最终的目录结构如图 7-3 所示。

第7章 购物车微服务编码实践及功能讲解

```
> newbee-mall-cloud-dev-step15 [newbee-mall-cloud]
  > .idea
  > newbee-mall-cloud-common
  > newbee-mall-cloud-gateway-admin
  > newbee-mall-cloud-gateway-mall
  > newbee-mall-cloud-goods-service
  > newbee-mall-cloud-recommend-service
  v newbee-mall-cloud-shop-cart-service
    > newbee-mall-cloud-shop-cart-api
    v newbee-mall-cloud-shop-cart-web
      v src
        v main
          v java
            v ltd.shopcart.cloud.newbee
              > config
              v controller
                > param
                > vo
                  NewBeeMallShoppingCartController
              v dao
                  NewBeeMallShoppingCartItemMapper
              v entity
                  NewBeeMallShoppingCartItem
              v service
                > impl
                  NewBeeMallShoppingCartService
                  NewBeeMallCloudShopCartServiceApplication
        > resources
      newbee-mall-cloud-shop-cart-web.iml
      m pom.xml
    newbee-mall-cloud-shop-cart-service.iml
    m pom.xml
  > newbee-mall-cloud-user-service
    newbee-mall-cloud.iml
    m pom.xml
```

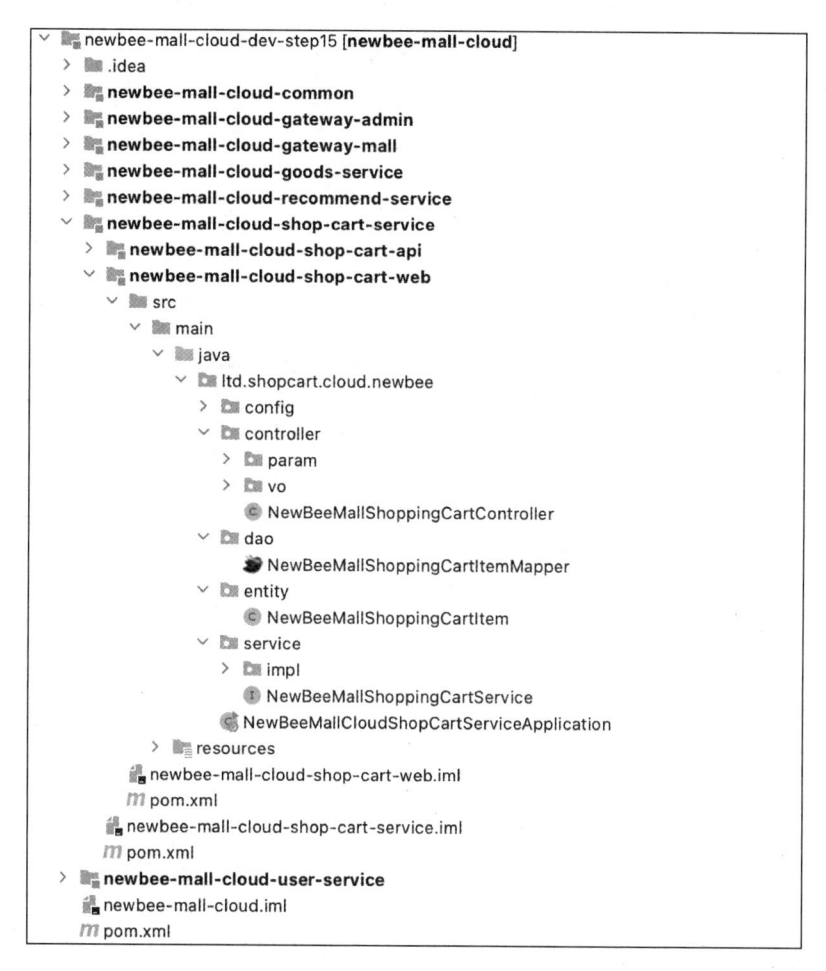

图 7-3 目录结构

修改 newbee-mall-cloud-shop-cart-web 工程中的 application.properties 配置文件，主要进行数据库连接参数及 MyBatis 扫描配置，代码如下：

```
# datasource config (MySQL)
spring.datasource.name=newbee-mall-cloud-recommend-datasource
spring.datasource.driverClassName=com.mysql.cj.jdbc.Driver
spring.datasource.url=jdbc:mysql://localhost:3306/newbee_mall_cloud_cart
_db?useUnicode=true&serverTimezone=Asia/Shanghai&characterEncoding=utf8&
autoReconnect=true&useSSL=false
spring.datasource.username=root
spring.datasource.password=123456
spring.datasource.hikari.minimum-idle=5
spring.datasource.hikari.maximum-pool-size=15
```

```
spring.datasource.hikari.auto-commit=true
spring.datasource.hikari.idle-timeout=60000
spring.datasource.hikari.pool-name=hikariCP
spring.datasource.hikari.max-lifetime=600000
spring.datasource.hikari.connection-timeout=30000
spring.datasource.hikari.connection-test-query=SELECT 1

# mybatis config
mybatis.mapper-locations=classpath:mapper/*Mapper.xml
```

本步骤中的源代码涉及的数据库为 newbeemallcloudcartdb，数据库表为 tbnewbeemallshoppingcart_item。除此之外，还会用到商品数据和用户数据，这部分数据就要使用远程调用技术了。

在调整购物车微服务代码前，有些工具类已经被移到公用模块 newbee-mall-cloud-common 中，所以在 pom.xml 文件中需要引入公用模块。同时，代码中使用这些工具类的地方也需要修改一下引用路径。

另外，购物车微服务中也有一些公用类，都放到了公用模块 newbee-mall-cloud-common 中。

Controller 类中的接口地址都做了微调，与原单体项目中定义的 URL 不同。调整的原因主要是在网关配置时方便一些。

7.4.2 网关模块配置

打开商城端网关 newbee-mall-cloud-gateway-mall 项目中的 application.properties 文件，新增关于购物车微服务的路由信息，配置项为 spring.cloud.gateway.routes.*，新增代码如下：

```
# 购物车接口的路由配置
spring.cloud.gateway.routes[4].id=shop-cart-service-route
spring.cloud.gateway.routes[4].uri=lb://newbee-mall-cloud-shop-cart-service
spring.cloud.gateway.routes[4].order=1
spring.cloud.gateway.routes[4].predicates[0]=Path=/shop-cart/**
```

这里主要配置 newbee-mall-cloud-gateway-mall 到购物车微服务的路由信息。如果访问网关项目的路径是以 /shop-cart 开头的，就路由到购物车微服务实例。

7.5 购物车微服务远程调用商品微服务编码实践

按照上述步骤对购物车微服务进行编码后,代码中依然会被标红。被标红的代码及注释如下:

```
@Override
/**
 * 添加商品至购物车
 */
public String saveNewBeeMallCartItem(SaveCartItemParam saveCartItemParam,
Long userId) {
  NewBeeMallShoppingCartItem temp = newBeeMallShoppingCartItemMapper.
selectByUserIdAndGoodsId(userId, saveCartItemParam.getGoodsId());
  if (temp != null) {
    //若已存在,则修改该记录
    NewBeeMallException.fail(ServiceResultEnum.SHOPPING_CART_ITEM_EXIST_
ERROR.getResult());
  }
  // 根据goodsId参数查询数据库中对应的商品数据
  NewBeeMallGoods newBeeMallGoods = newBeeMallGoodsMapper. selectByPrimaryKey
(saveCartItemParam.getGoodsId());
  //商品为空
  if (newBeeMallGoods == null) {
    return ServiceResultEnum.GOODS_NOT_EXIST.getResult();
  }
    省略部分代码
  //保存记录
  if (newBeeMallShoppingCartItemMapper.insertSelective (newBeeMallShopping
CartItem) > 0) {
    return ServiceResultEnum.SUCCESS.getResult();
  }
  return ServiceResultEnum.DB_ERROR.getResult();
}

/**
 * VO实体封装
 */
private List<NewBeeMallShoppingCartItemVO> getNewBeeMallShoppingCartItemVOS
(List<NewBeeMallShoppingCartItemVO> newBeeMallShoppingCartItemVOS,
List<NewBeeMallShoppingCartItem> newBeeMallShoppingCartItems) {
  if (!CollectionUtils.isEmpty(newBeeMallShoppingCartItems)) {
```

```
    //查询商品信息并进行数据转换
    List<Long> newBeeMallGoodsIds = newBeeMallShoppingCartItems.stream().
map(NewBeeMallShoppingCartItem::getGoodsId).collect(Collectors.toList());
    // 根据 goodsIds 链表查询数据库中对应的商品列表数据
    List<NewBeeMallGoods> newBeeMallGoods = newBeeMallGoodsMapper.selectBy
PrimaryKeys(newBeeMallGoodsIds);
    Map<Long, NewBeeMallGoods> newBeeMallGoodsMap = new HashMap<>();
    省略部分代码
  }
  return newBeeMallShoppingCartItemVOS;
}
```

在添加商品至购物车和查询购物车列表数据时，需要根据 goodId 参数或 goodsIds 链表查询数据库中的一条或多条商品数据，之后才能执行后续的业务逻辑。而购物车微服务未连接商品表所在的数据库，无法直接通过 GoodsMapper 去查询对应的商品数据，所以这部分代码会被标红。

因此，这里必须进行商品数据查询代码的改造，由原本直接查询商品表改为远程调用商品微服务中的接口来完成这个逻辑。购物车微服务不仅要与用户微服务通信，还要与商品微服务通信。这里的代码改造步骤如下：

第一步，引入 goods-api 依赖。

打开 newbee-mall-cloud-shop-cart-web 工程下的 pom.xml 文件，增加与商品微服务远程通信所需的 newbee-mall-cloud-goods-api 模块，新增依赖配置如下：

```xml
<dependency>
  <groupId>ltd.goods.newbee.cloud</groupId>
  <artifactId>newbee-mall-cloud-goods-api</artifactId>
  <version>0.0.1-SNAPSHOT</version>
</dependency>
```

第二步，增加关于商品微服务中 FeignClient 类的配置。

打开 newbee-mall-cloud-shop-cart-web 工程，对启动类 NewBeeMallCloudShopCartServiceApplication 中的 @EnableFeignClients 注解进行修改，增加对 NewBeeCloudGoodsServiceFeign 类的声明，代码如下：

```
@EnableFeignClients(basePackageClasses ={ltd.user.cloud.newbee.openfeign.
NewBeeCloudUserServiceFeign.class, ltd.goods.cloud.newbee.openfeign.
NewBeeCloudGoodsServiceFeign.class})
```

接下来就可以直接使用 NewBeeCloudGoodsServiceFeign 类与商品微服务进行远程通信了。

第三步，修改商品数据判断逻辑的代码。

打开 newbee-mall-cloud-shop-cart-web 工程，修改 NewBeeMallIndexConfigServiceImpl 类中商品数据判断的逻辑代码。删除原本直接查询商品数据的代码，之后注入 NewBeeCloudGoodsServiceFeign 类，并通过调用商品微服务来获取商品数据，代码如下：

```java
@Autowired
private NewBeeCloudGoodsServiceFeign goodsService;

@Override
/**
 * 添加商品至购物车
 */
public String saveNewBeeMallCartItem(SaveCartItemParam saveCartItemParam, Long userId) {
    NewBeeMallShoppingCartItem temp = newBeeMallShoppingCartItemMapper.selectByUserIdAndGoodsId(userId, saveCartItemParam.getGoodsId());
    if (temp != null) {
        // 若已存在，则修改该记录
        NewBeeMallException.fail(ServiceResultEnum.SHOPPING_CART_ITEM_EXIST_ERROR.getResult());
    }
    // 根据goodsId参数调用商品微服务中的接口，查询对应的商品数据
    Result<NewBeeMallGoodsDTO> goodsDetailResult = goodsService.getGoodsDetail(saveCartItemParam.getGoodsId());
    // 商品为空
    if (goodsDetailResult == null || goodsDetailResult.getResultCode() != 200) {
        return ServiceResultEnum.GOODS_NOT_EXIST.getResult();
    }
    省略部分代码
    // 保存记录
    if (newBeeMallShoppingCartItemMapper.insertSelective(newBeeMallShoppingCartItem) > 0) {
        return ServiceResultEnum.SUCCESS.getResult();
    }
    return ServiceResultEnum.DB_ERROR.getResult();
}

/**
 * VO实体封装
 */
```

```
private List<NewBeeMallShoppingCartItemVO> getNewBeeMallShoppingCartItemVOS
(List<NewBeeMallShoppingCartItemVO> newBeeMallShoppingCartItemVOS,
List<NewBeeMallShoppingCartItem> newBeeMallShoppingCartItems) {
    if (!CollectionUtils.isEmpty(newBeeMallShoppingCartItems)) {
        List<Long> newBeeMallGoodsIds = newBeeMallShoppingCartItems.stream().
map(NewBeeMallShoppingCartItem::getGoodsId).collect(Collectors.toList());
        // 根据goodsIds链表调用商品微服务中的接口，查询对应的商品列表数据
        Result<List<NewBeeMallGoodsDTO>> newBeeMallGoodsDTOResult =
goodsService.listByGoodsIds(newBeeMallGoodsIds);
        // 远程调用，返回的数据为空
        if (newBeeMallGoodsDTOResult == null || newBeeMallGoodsDTOResult.
getResultCode() != 200) {
            NewBeeMallException.fail(ServiceResultEnum.GOODS_NOT_EXIST.getResult());
        }
        Map<Long, NewBeeMallGoodsDTO> newBeeMallGoodsDTOMap = new HashMap<>();
        List<NewBeeMallGoodsDTO> newBeeMallGoodsDTOS = newBeeMallGoodsDTOResult.
getData();
        省略部分代码
    }
    return newBeeMallShoppingCartItemVOS;
}
```

这里调用的就是在商品微服务中暴露的 /goods/admin/goodsDetail 接口和 /goods/admin/listByGoodsIds 接口，根据参数或链表查询对应的一条或多条商品数据。

7.6 购物车微服务功能测试

代码修改完成后，测试步骤是不能漏掉的。一定要验证项目是否能正常启动、接口是否能正常调用，防止在代码移动过程中出现一些小问题，导致项目无法启动或代码报错。在项目启动前需要分别启动 Nacos Server 和 Redis Server，之后依次启动 newbee-mall-cloud-shop-cart-web 工程、newbee-mall-cloud-user-web 工程、newbee-mall-cloud-goods-web 工程和 newbee-mall-cloud-gateway-mall 工程下的主类。启动成功后，就可以进行功能测试了。

打开用户微服务的 Swagger 页面，在浏览器中输入如下网址：http://localhost:29000/swagger-ui/index.html。

在该页面使用登录接口获取一个 token 值用于后续的功能测试，如笔者在测试时获取了一个值为 "496660e70edb82437f7c56c61f5540bf" 的 token 字段。

打开购物车微服务的 Swagger 页面，在浏览器中输入如下网址：http://localhost:29030/swagger-ui/index.html。

接着就可以在 Swagger 提供的 UI 页面中进行购物车模块的接口测试了，接口文档显示内容如图 7-4 所示。

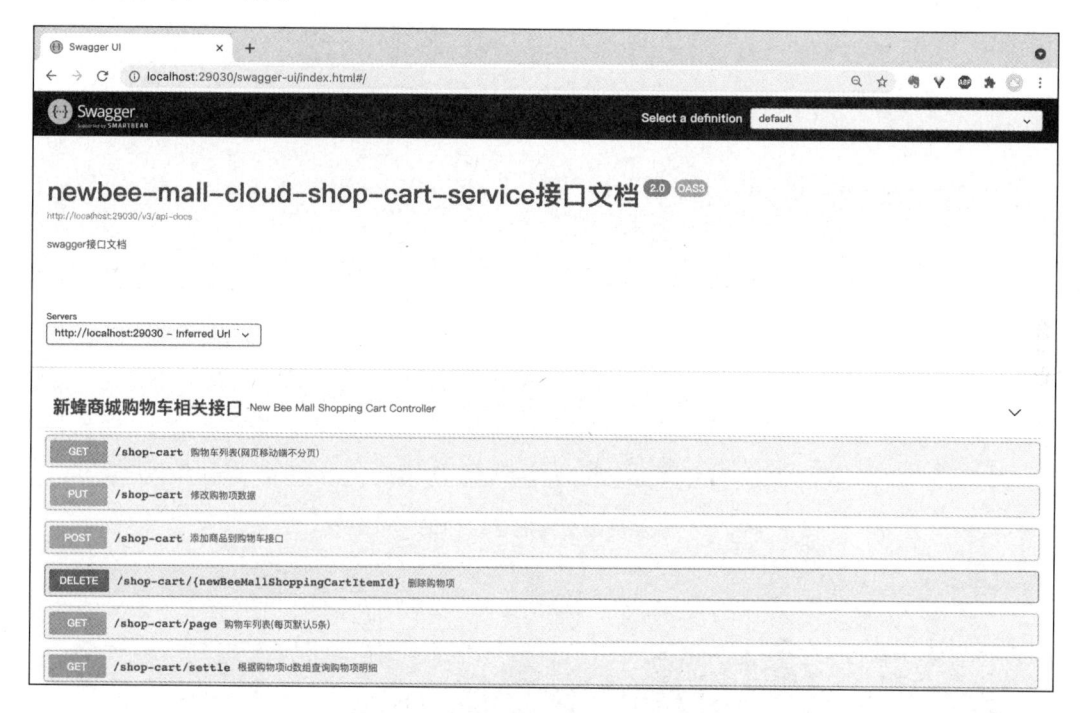

图 7-4　购物车模块接口文档显示内容

下面将演示把商品 id 分别为 10925 和 10926 的商品信息添加至购物车。依次单击"添加商品到购物车接口""Try it out"按钮，在参数栏中输入商品 id 字段和添加数量，在登录认证 token 的输入框中输入登录接口返回的 token 值，如图 7-5 所示。

单击"Execute"按钮，接口的测试结果如图 7-6 所示。

若后端接口的测试结果中有"SUCCESS"，则表示添加成功。在演示时使用的商品 id 为 10925，添加商品 id 为 10926 的商品至购物车的步骤与其相同。

笔者在测试时，输入的商品数量和商品 id 都是符合规范的且商品 id 在数据库中真实存在。如果输入的商品数量过大，则会报错"超出单个商品的最大购买数量"。如果输入的商品 id 在数据库中不存在，则会报错"商品不存在"。对于这一点，读者在测试时需要注意。

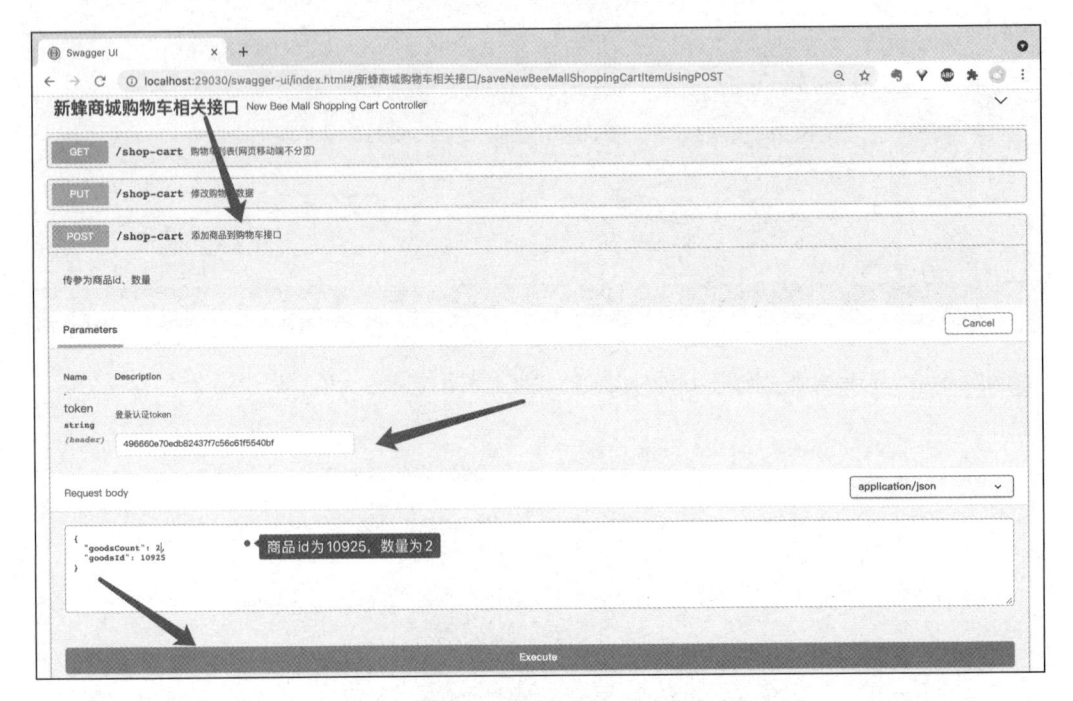

图 7-5 添加商品至购物车接口的测试过程

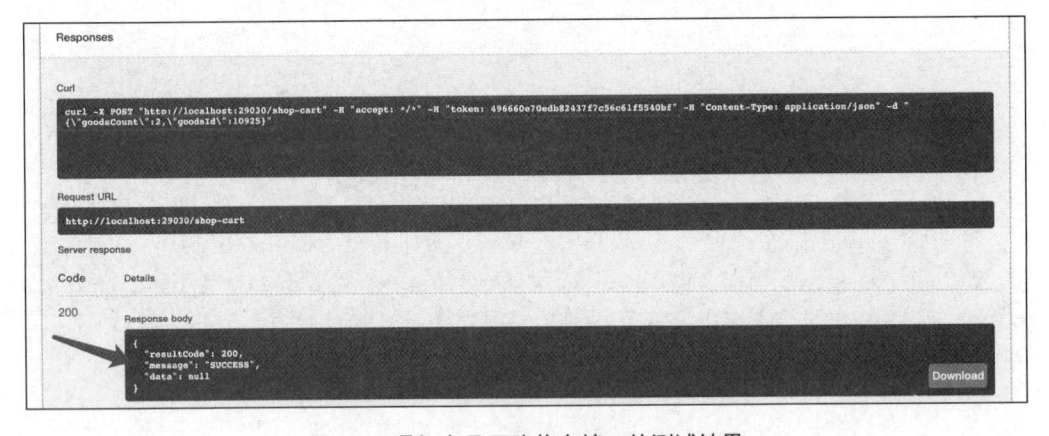

图 7-6 添加商品至购物车接口的测试结果

依次单击"购物车列表（网页移动端不分页）""Try it out"按钮，在登录认证 token 的输入框中输入登录接口返回的 token 值，如图 7-7 所示。

单击"Execute"按钮，接口的测试结果如图 7-8 所示。

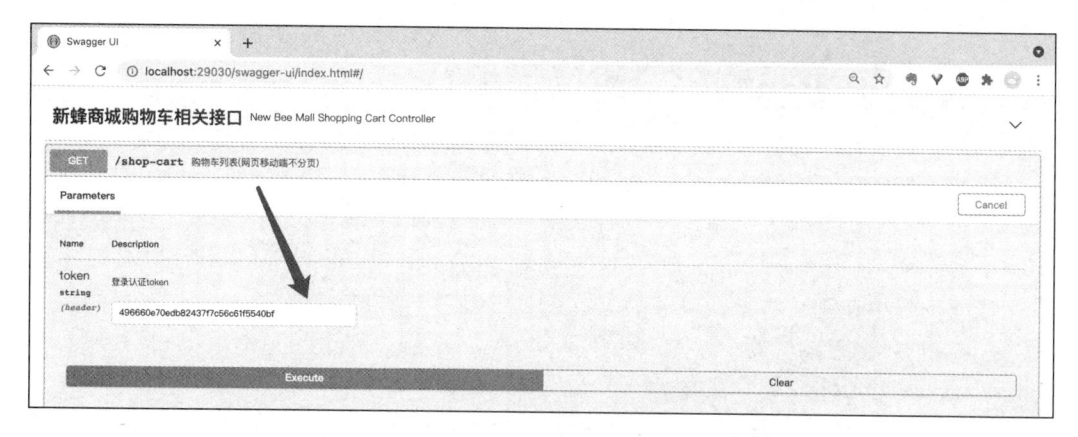

图 7-7　购物车列表接口的测试过程

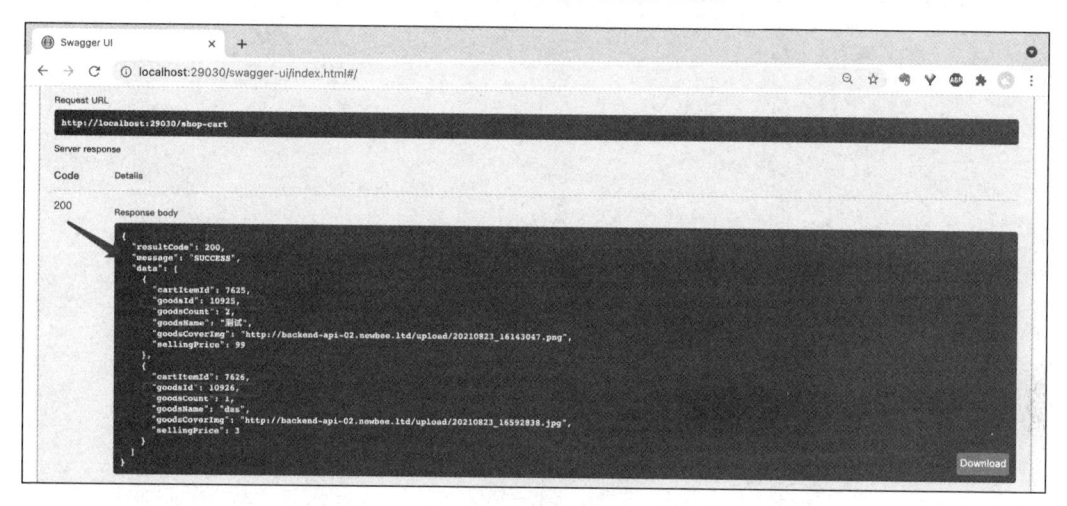

图 7-8　购物车列表接口的测试结果

若后端接口的测试结果中有"SUCCESS",则表示添加商品成功,演示时使用的商品 id 分别为 10925 和 10926 的数据都出现在响应结果中,对应的购物项 cartItemId 分别为 7625 和 7626。

功能测试完成且接口响应一切正常,表示购物车微服务本身的功能编码完成,并且远程调用用户微服务、商品微服务也一切正常。在测试时,读者也可以通过 debug 模式启动项目,之后打上几个断点来查看接口测试时的完整过程。

由于篇幅限制,笔者这里只演示了两个接口的测试过程,读者在测试时可以查看其他接口。除在购物车微服务架构项目的 Swagger 页面测试接口外,也可以通过商城端网关访问这些接口并进行功能测试。

7.7 OpenFeign编码暴露远程接口

在 newbee-mall-cloud-shop-cart-api 模块中新增需要暴露的接口 FeignClient。如果其他微服务实例需要获取购物车微服务中的资源，则可以通过调用购物车微服务下暴露的接口来实现。

当前项目中使用购物车微服务的只有订单微服务，主要有两个功能，分别是根据购物项 id 链表查询购物项表和删除多条购物项记录。因此，笔者在 NewBeeMall-ShoppingCartController 类中新建了一个商品详情接口，并将其暴露，代码如下：

```java
@GetMapping("/shop-cart/listByCartItemIds")
@ApiOperation(value = "购物项表", notes = "")
public Result<List<NewBeeMallShoppingCartItem>>
cartItemListByIds(@RequestParam("cartItemIds") List<Long> cartItemIds) {
  if (CollectionUtils.isEmpty(cartItemIds)) {
    return ResultGenerator.genFailResult("error param");
  }
  return ResultGenerator.genSuccessResult(newBeeMallShoppingCartService.
getCartItemsByCartIds(cartItemIds));
}

@DeleteMapping("/shop-cart/deleteByCartItemIds")
@ApiOperation(value = "批量删除购物项", notes = "")
public Result<Boolean> deleteByCartItemIds(@RequestParam("cartItemIds")
List<Long> cartItemIds) {
  if (CollectionUtils.isEmpty(cartItemIds)) {
    return ResultGenerator.genFailResult("error param");
  }
  return ResultGenerator.genSuccessResult(newBeeMallShoppingCartService.
deleteCartItemsByCartIds(cartItemIds) > 0);
}
```

在 newbee-mall-cloud-shop-cart-api 目录下新建 ltd.shopcart.cloud.newbee.openfeign 包，之后在该包下新增 NewBeeCloudShopCartServiceFeign 类，用于创建对购物车模块中相关接口的 Feign 调用。

NewBeeCloudShopCartServiceFeign 类的代码如下：

```java
package ltd.shopcart.cloud.newbee.openfeign;

import ltd.common.cloud.newbee.dto.Result;
import ltd.shopcart.cloud.newbee.dto.NewBeeMallShoppingCartItemDTO;
```

```
import org.springframework.cloud.openfeign.FeignClient;
import org.springframework.web.bind.annotation.DeleteMapping;
import org.springframework.web.bind.annotation.GetMapping;
import org.springframework.web.bind.annotation.RequestParam;

import java.util.List;

@FeignClient(value = "newbee-mall-cloud-shop-cart-service", path =
"/shop-cart")
public interface NewBeeCloudShopCartServiceFeign {

    @GetMapping(value = "/listByCartItemIds")
    Result<List<NewBeeMallShoppingCartItemDTO>> listByCartItemIds
(@RequestParam(value = "cartItemIds") List<Long> cartItemIds);

    @DeleteMapping(value = "/deleteByCartItemIds")
    Result<Boolean> deleteByCartItemIds(@RequestParam(value = "cartItemIds")
List<Long> cartItemIds);
}
```

如果其他微服务需要与购物车微服务进行远程通信并获取相关数据，就可以引入 newbee-mall-cloud-shop-cart-api 模块作为依赖，并且直接调用 NewBeeCloudShopCartServiceFeign 类中的 listByCartItemIds()方法和 deleteByCartItemIds()方法完成对购物项数据的批量查询和删除。如果在后续的开发过程中，购物车微服务中有其他接口需要暴露以供其他微服务调用，就可以继续在 NewBeeCloudShopCartServiceFeign 类中编码。

本章主要讲解购物车模块在微服务架构下的编码改造，在购物车微服务开发完成后，本书实战项目已经完成了大部分功能模块的开发与测试，即将进入尾声。希望读者能够根据笔者提供的开发步骤顺利地完成本章的项目改造。

第 8 章

订单微服务编码实践及功能讲解

本章讲解微服务架构项目的最后一个模块——订单微服务,介绍其主要功能模块、功能设计思路及代码的改造过程。

8.1 订单微服务主要功能模块介绍

8.1.1 订单模块介绍

在把心仪的商品添加到购物车并确定需要购买的商品和对应的数量后,就可以执行提交订单的操作。此时就由购物车模块切换到另一个电商流程——订单流程。接下来主要介绍订单模块相关功能的开发。关于订单的生成和后续处理流程,不同的公司或不同的商城项目的需求与业务场景会有一些差异,但是从订单生成到订单完成大体的流程是类似的。

新蜂商城中从订单生成到订单完成,主要有以下几个步骤,如图 8-1 所示。

① 提交订单(由新蜂商城用户发起)。

② 订单入库(后台逻辑,用户无感知)。

③ 支付订单(由新蜂商城用户发起)。

④ 订单处理(包括确认订单、取消订单、修改订单信息等操作,新蜂商城用户和管理员用户都可以对订单进行处理)。

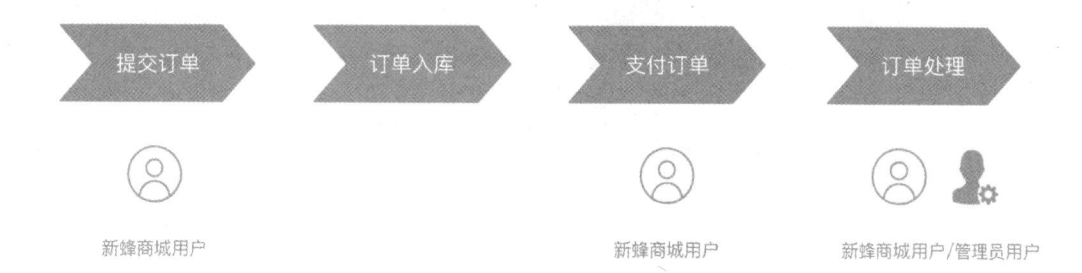

提交订单 → 订单入库 → 支付订单 → 订单处理

新蜂商城用户　　　　　　　　　新蜂商城用户　　　　新蜂商城用户/管理员用户

图 8-1　新蜂商城订单处理流程

8.1.2　订单模块的表结构设计

1. 订单主表及关联表设计

新蜂商城系统的订单模块主要涉及数据库中的三张表。一次下单行为可能购买一件商品，也可能购买多件商品，因此除订单主表 tb_newbee_mall_order 外，还有订单项表 tb_newbee_mall_order_item 和订单地址表 tb_newbee_mall_order_address。订单主表中存储订单的相关信息，而订单项表中主要存储关联的商品字段。

订单主表 tb_newbee_mall_order 表结构设计的主要字段如下。

① user_id：用户的 id。根据这个字段来确定是哪个用户下的订单。

② order_no：订单号。订单号是订单的唯一标识，在后续查询订单时会用到。这是每个电商系统都有的设计。

③ paystatus、paytype、pay_time：支付信息字段，包括支付状态、支付方式、支付时间。

④ order_status：订单状态字段。

⑤ create_time：订单生成时间。

订单主表的主要字段代码如下：

```
# 创建订单微服务所需数据
CREATE DATABASE /*!32312 IF NOT EXISTS*/'newbee_mall_cloud_order_db'
/*!40100 DEFAULT CHARACTER SET utf8 */;

USE 'newbee_mall_cloud_order_db';

DROP TABLE IF EXISTS 'tb_newbee_mall_order';
CREATE TABLE 'tb_newbee_mall_order' (
```

```
  'order_id' bigint(20) NOT NULL AUTO_INCREMENT COMMENT '订单主表主键id',
  'order_no' varchar(20) NOT NULL DEFAULT '' COMMENT '订单号',
  'user_id' bigint(20) NOT NULL DEFAULT '0' COMMENT '用户主键id',
  'total_price' int(11) NOT NULL DEFAULT '1' COMMENT '订单总价',
  'pay_status' tinyint(4) NOT NULL DEFAULT '0' COMMENT '支付状态 0:未支付 1:支付成功  -1:支付失败',
  'pay_type' tinyint(4) NOT NULL DEFAULT '0' COMMENT '0:无  1:支付宝支付 2:微信支付',
  'pay_time' datetime DEFAULT NULL COMMENT '支付时间',
  'order_status' tinyint(4) NOT NULL DEFAULT '0' COMMENT '订单状态 0:待支付 1:已支付  2:配货完成  3:出库成功  4:交易成功  -1:手动关闭  -2:超时关闭  -3:商家关闭',
  'extra_info' varchar(100) NOT NULL DEFAULT '' COMMENT '订单body',
  'is_deleted' tinyint(4) NOT NULL DEFAULT '0' COMMENT '删除标识字段(0:未删除 1:已删除)',
  'create_time' datetime NOT NULL DEFAULT CURRENT_TIMESTAMP COMMENT '创建时间',
  'update_time' datetime NOT NULL DEFAULT CURRENT_TIMESTAMP COMMENT '最新修改时间',
  PRIMARY KEY ('order_id')
) ENGINE=InnoDB DEFAULT CHARSET=utf8;
```

订单项表 tb_newbee_mall_order_item 表结构设计的主要字段如下。

① order_id：关联的订单主键id，标识该订单项是哪个订单中的数据。

② goodsid、goodsname、goodscoverimg、sellingprice、goodscount：订单中的商品信息，记录当时的商品信息。

③ create_time：记录生成时间。

订单项表的字段代码如下：

```
USE 'newbee_mall_cloud_order_db';

DROP TABLE IF EXISTS 'tb_newbee_mall_order_item';
CREATE TABLE 'tb_newbee_mall_order_item' (
  'order_item_id' bigint(20) NOT NULL AUTO_INCREMENT COMMENT '订单关联购物项主键id',
  'order_id' bigint(20) NOT NULL DEFAULT 0 COMMENT '订单主键id',
  'goods_id' bigint(20) NOT NULL DEFAULT 0 COMMENT '关联商品id',
  'goods_name' varchar(200) CHARACTER SET utf8 COLLATE utf8_general_ci NOT NULL DEFAULT '' COMMENT '下单时商品的名称(订单快照)',
  'goods_cover_img' varchar(200) CHARACTER SET utf8 COLLATE utf8_general_ci NOT NULL DEFAULT '' COMMENT '下单时商品的主图(订单快照)',
  'selling_price' int(11) NOT NULL DEFAULT 1 COMMENT '下单时商品的价格(订单快照)',
```

```sql
  'goods_count' int(11) NOT NULL DEFAULT 1 COMMENT '数量(订单快照)',
  'create_time' datetime(0) NOT NULL DEFAULT CURRENT_TIMESTAMP COMMENT '创建时间',
  PRIMARY KEY ('order_item_id') USING BTREE
) ENGINE = InnoDB AUTO_INCREMENT = 1 CHARACTER SET = utf8 COLLATE = utf8_general_ci ROW_FORMAT = Dynamic;
```

订单地址表 tb_newbee_mall_order_address 表结构设计的主要字段如下。

① order_id：关联的订单主键 id，标识该地址是哪个订单中的收货地址数据。

② username、userphone、user_address：收货信息字段，最好在订单主表或关联表中设置这几个字段，后端 API 项目中设计成 tb_newbee_mall_order 表和 tb_newbee_mall_order_address 表，二者为一对一的关系，将地址关联表中的字段全部设置在订单主表中也是可以的。有些商城的收货地址设计只关联一个地址 id 字段，这样做是不合理的，因为该 id 关联的地址表中的记录可以被修改和删除。也就是说，如果修改或删除，那么订单中的收件信息就不是下单时的数据了。因此，需要把这些字段放到订单主表中，并记录下单时的收货信息。

用户收货地址表和订单地址表的字段代码如下：

```sql
USE 'newbee_mall_cloud_order_db';

DROP TABLE IF EXISTS 'tb_newbee_mall_order_address';
CREATE TABLE 'tb_newbee_mall_order_address' (
  'order_id' bigint(20) NOT NULL,
  'user_name' varchar(30) NOT NULL DEFAULT '' COMMENT '收货人姓名',
  'user_phone' varchar(11) NOT NULL DEFAULT '' COMMENT '收货人手机号',
  'province_name' varchar(32) NOT NULL DEFAULT '' COMMENT '省',
  'city_name' varchar(32) NOT NULL DEFAULT '' COMMENT '城',
  'region_name' varchar(32) NOT NULL DEFAULT '' COMMENT '区',
  'detail_address' varchar(64) NOT NULL DEFAULT '' COMMENT '收货详细地址(街道/楼宇/单元)',
  PRIMARY KEY ('order_id')
) ENGINE=InnoDB DEFAULT CHARSET=utf8 COMMENT='订单地址表';

DROP TABLE IF EXISTS 'tb_newbee_mall_user_address';

CREATE TABLE 'tb_newbee_mall_user_address' (
                                            'address_id' bigint(20) NOT NULL AUTO_INCREMENT,
                                            'user_id' bigint(20) NOT NULL DEFAULT '0' COMMENT '用户主键id',
                                            'user_name' varchar(30) NOT NULL
```

```
                                                      DEFAULT '' COMMENT '收货人姓名',
  'user_phone' varchar(11) NOT NULL DEFAULT '' COMMENT '收货人手机号',
  'default_flag' tinyint(4) NOT NULL DEFAULT '0' COMMENT '是否为默认 0:非默认 1:默认',
  'province_name' varchar(32) NOT NULL DEFAULT '' COMMENT '省',
  'city_name' varchar(32) NOT NULL DEFAULT '' COMMENT '城',
  'region_name' varchar(32) NOT NULL DEFAULT '' COMMENT '区',
  'detail_address' varchar(64) NOT NULL DEFAULT '' COMMENT '收货详细地址(街道/楼宇/单元)',
  'is_deleted' tinyint(4) NOT NULL DEFAULT '0' COMMENT '删除标识字段(0:未删除  1:已删除)',
  'create_time' timestamp NOT NULL DEFAULT CURRENT_TIMESTAMP COMMENT '添加时间',
  'update_time' timestamp NOT NULL DEFAULT CURRENT_TIMESTAMP COMMENT '修改时间',
  PRIMARY KEY ('address_id')
) ENGINE=InnoDB DEFAULT CHARSET=utf8 COMMENT='用户收货地址表';
```

每个字段对应的含义在上面的 SQL 语句中都有介绍，读者可以对照理解，并正确地把建表语句导入数据库。关于两张表中的快照字段，包括收件信息字段和商品信息字段，读者可以参考淘宝网的订单快照来理解。这些信息都是可以更改的，因此不能只关联一个主键 id。比如，订单中存储的是下单时的数据，而商品信息是可以随时更改的，如果没有这几个字段，只用商品 id 关联，则商品信息一旦被更改，订单信息也随之被更改，就不再是下单时的数据了，这不符合逻辑。

2. 订单项表的设计思路

下面介绍订单项表 tb_newbee_mall_orderitem 和购物项表 tb_newbee_mall_shopping_cart_ite 的差异，以及单独设计一张订单项表的原因。

购物车模块的购物项和订单模块的订单项是很多商城项目都有的设计。只是有些商城项目为了简化开发，在实现的时候选择将两者等同。本来应该设计两张表，减少为只设计一张购物项表，订单生成后会在购物项表中增加与订单主键 id 的关联。

其实订单项和购物项二者的区别是很明显的，它们是相似却完全不同的两个对象。购物项是商品与购物车之间抽象出的一个对象，而订单项是商品与订单之间抽象出的一个对象。它们都与商品相关，并且在页面数据展示时也类似，所以有简化为一张表的实现方式。但是笔者并不赞同这种实现方式，虽然二者很相似，但是依然有很多不同的地方。购物项的相关操作在购物车中,而在生成订单后该购物项就不再存在了,即该对象已经被删除了,它的生命周期到此为止。既然生命周期已经终结,再与订单做关联就说不通了。

以订单快照为例,它需要记录下单时的商品内容和订单信息。比如,淘宝网的订单设计,下单时的订单快照数据就是下单时的商品数据,而不是最新的商品数据。购物车则不需要快照,直接读取最新的商品相关信息即可。

购物车模块的购物项和订单模块的订单项是两个不同的对象,因此笔者选择分别设计两张表。

3. 用户收货地址表

最终实战的微服务架构项目中有用户收货地址管理的模块,用户收货地址管理页面如图 8-2 所示。在确认订单页面中可以选择收货地址,也可以添加收货地址,再提交并生成一个订单信息。

图 8-2 用户收货地址管理页面

用户收货地址表 tb_newbee_mall_user_address 的字段代码如下:

```
USE 'newbee_mall_cloud_order_db';
```

```sql
DROP TABLE IF EXISTS 'tb_newbee_mall_user_address';

CREATE TABLE 'tb_newbee_mall_user_address' (
 'address_id' bigint(20) NOT NULL AUTO_INCREMENT,
 'user_id' bigint(20) NOT NULL DEFAULT '0' COMMENT '用户主键id',
 'user_name' varchar(30) NOT NULL DEFAULT '' COMMENT '收货人姓名',
 'user_phone' varchar(11) NOT NULL DEFAULT '' COMMENT '收货人手机号',
 'default_flag' tinyint(4) NOT NULL DEFAULT '0' COMMENT '是否为默认 0:非默认 1:默认',
 'province_name' varchar(32) NOT NULL DEFAULT '' COMMENT '省',
 'city_name' varchar(32) NOT NULL DEFAULT '' COMMENT '城',
 'region_name' varchar(32) NOT NULL DEFAULT '' COMMENT '区',
 'detail_address' varchar(64) NOT NULL DEFAULT '' COMMENT '收货详细地址(街道/楼宇/单元)',
 'is_deleted' tinyint(4) NOT NULL DEFAULT '0' COMMENT '删除标识字段(0:未删除 1:已删除)',
 'create_time' datetime NOT NULL DEFAULT CURRENT_TIMESTAMP COMMENT '添加时间',
 'update_time' datetime NOT NULL DEFAULT CURRENT_TIMESTAMP COMMENT '修改时间',
 PRIMARY KEY ('address_id')
) ENGINE=InnoDB DEFAULT CHARSET=utf8 COMMENT='收货地址表';
```

每个字段对应的含义在上面的 SQL 语句中都有介绍，读者可以对照理解，并正确地把建表语句导入数据库。

8.1.3 订单模块中的主要功能分析

1. 商城中的订单确认步骤

订单确认步骤是订单生成过程中一个很重要的功能，日常使用的商城项目基本上都有这个步骤。以淘宝网的订单确认页面为例，如图 8-3 所示。

这个页面中有用户在购物车中选择的商品信息，还有收货地址信息、运费信息、优惠信息等。在购物车中只有商品信息，而订单确认页面是多种信息的集合。只有信息齐全才能够生成订单数据。订单确认页面的设置可以理解为一个信息确认的过程，所有信息都确认无误后，才能进行提交订单的操作，之后生成一条订单记录。

订单确认步骤是在购物车页面发起的，如图 8-4 所示。单击购物车页面中的"结算"按钮即可进入订单确认页面。

根据购物车中的待结算商品数量来判断"结算"按钮是否显示正常。如果购物车中

第 8 章 订单微服务编码实践及功能讲解

无数据,则页面中无商品列表展示,也没有"结算"按钮。只有购物车列表数据正常,才会出现"结算"按钮,单击后才能进入订单确认页面。

图 8-3 淘宝网的订单确认页面

图 8-4 新蜂商城的购物车页面

订单确认页面显示的商品数据与购物车列表中显示的商品数据类似,除此之外,还有用户数据和支付数据,以及收件人的收货地址信息,其他的内容还有运费金额、优惠金额、实际支付金额等。新蜂商城的订单确认页面显示的信息如图 8-5 所示。

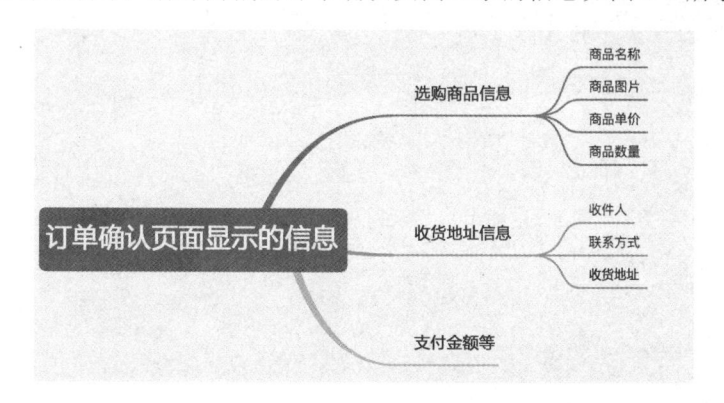

图 8-5 新蜂商城的订单确认页面显示的信息

因此，这里需要编写一个接口，用于查询订单确认页面的数据，根据用户所选择的购物项查询订单确认页面中需要显示的信息。

2. 订单生成功能

在订单确认页面处理完毕后，紧接着就是生成订单的环节。此时用户单击"提交订单"按钮，商城系统就会对应生成一笔订单数据并保存在数据库中。

用户单击"提交订单"按钮后，后台会进行一系列的操作，包括数据查询、数据判断、数据整合等。因此，这里需要编写一个接口，用于生成一条订单数据，并删除对应的购物项数据，修改商品的库存数据。

新蜂商城的订单生成流程如图 8-6 所示。

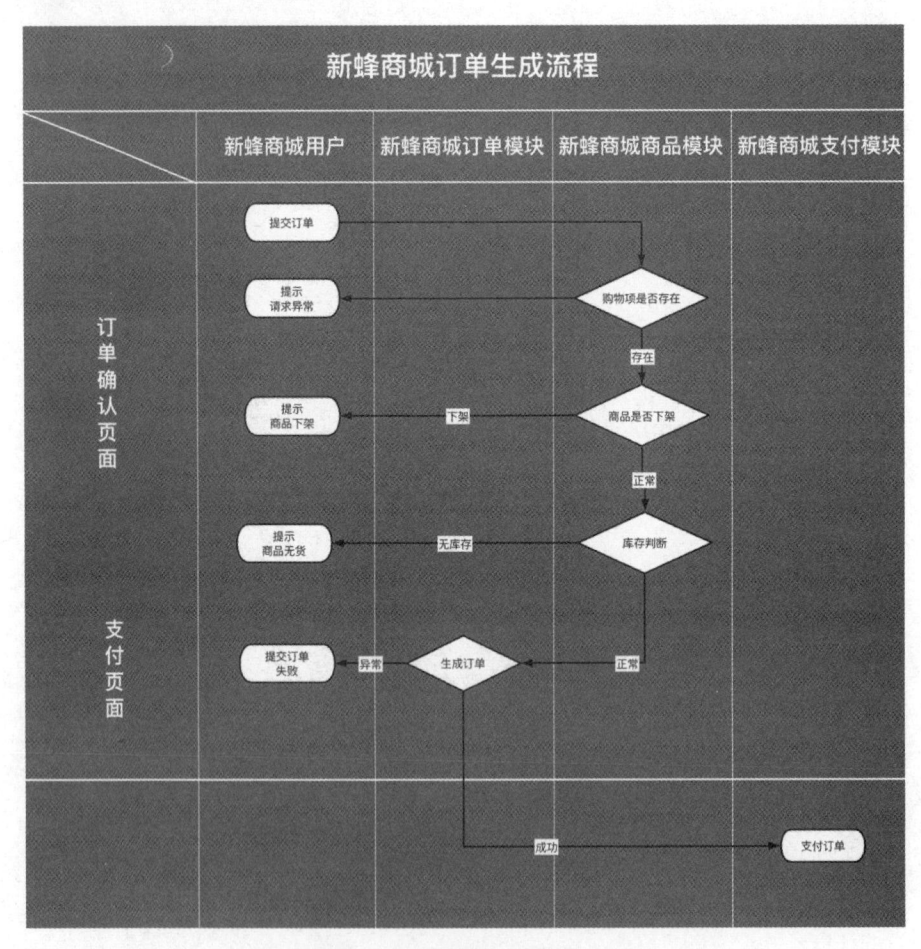

图 8-6 新蜂商城的订单生成流程

当然，订单生成只是订单模块中的第一步，后续还有一些步骤需要完成。

3. 订单支付模拟接口实现

前端在接收到生成订单接口的成功响应后，会在订单确认页面中显示一个支付的底部弹窗，即模拟支付功能也在订单确认页面，如图 8-7 所示。由于没有公司资质，无法申请微信和支付宝相关的接口接入权限，因此只能模拟订单支付的功能。

单击模拟支付窗口中的任意一个按钮，就会向支付回调地址发送请求，该按钮是模拟支付成功的接口回调，表示已经支付成功，可以修改订单的状态了。

这里需要编写一个接口，用于模拟订单支付成功的过程，根据订单号查询订单，并进行非空判断和订单状态的判断。如果订单已经不是"待支付"状态下的订单，则不进行后续操作。如果验证通过，则将该订单的相关状态和支付时间进行修改，之后调用数据层的方法进行实际的入库操作。

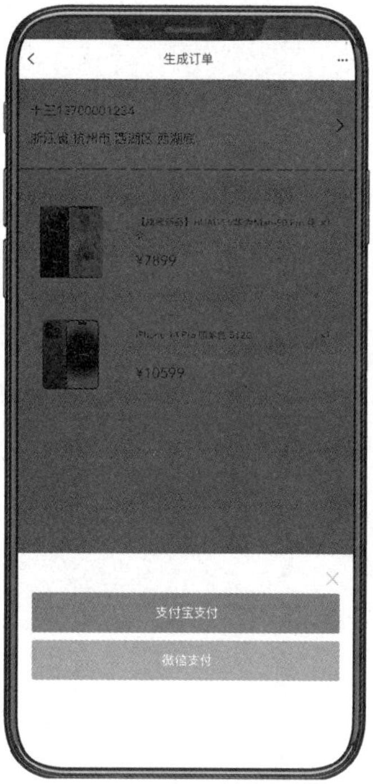

4. 订单详情功能

图 8-7　订单模拟支付窗口

制作订单详情页面是订单流程中非常重要的一个环节，该页面是商家与用户之间最直接的一个纽带。对于商家来说，体现了商家提供的销售服务，商品信息、订单信息、物流信息都在该页面实时展示给用户。对于用户来说，该页面会显示订单的重要信息，用户关心的所有信息都显示在这里，用户也可以在该页面实时关注订单的变化和新的动态。

订单详情页面主要有两个功能。

（1）显示基本的订单信息。

① 订单基本信息：订单号、订单状态、价格等信息。

② 配送信息：物流信息、收货地址信息。

③ 商品信息：商品名称、商品图、购买数量等。

④ 发票信息：发票信息和开票状态。

⑤ 客服：线上联系商家或拨打电话。

（2）为用户提供订单操作。

用户可以在订单详情页面进行支付订单、取消订单、确认订单等操作。

这里需要编写一个接口，用于查询订单详情页面所需的数据，包括订单的基本信息，如订单号、订单状态、下单时间等，以及订单中所关联的商品数据，如商品图片、名称、单价等，如图 8-8 所示。

5. "我的订单"列表功能

订单生成后就能够在个人中心的订单列表中看到相关数据了。各种状态的订单都会在这个列表中显示，并且商城端的订单列表支持分页功能。新蜂商城订单列表页面如图 8-9 所示。

图 8-8　订单详情页面　　　　　　图 8-9　订单列表页面

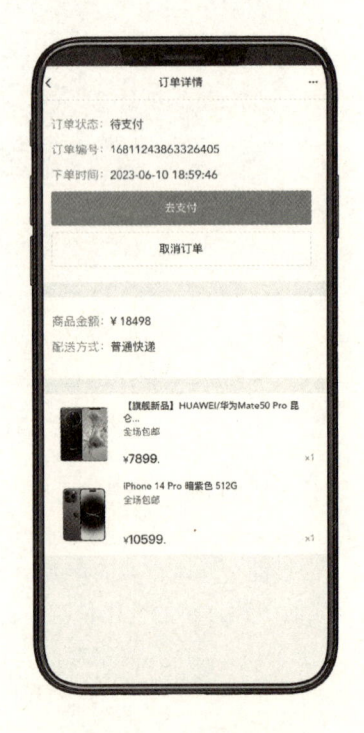

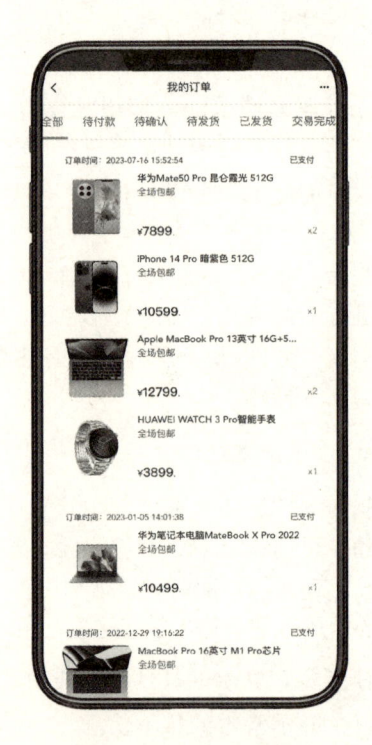

这里需要编写一个接口，用于查询当前登录用户的订单列表数据。订单列表是一个 List 对象，后端返回数据时需要一个订单列表对象。对象中的字段有订单主表中的字段，也有订单项表中的字段。在列表数据中有订单状态、订单交易时间、订单总价、商品标题字段、商品预览图字段、商品价格字段、商品购买数量字段。一个订单中可能有多个订单项，所以订单 VO 对象中也有一个订单项 VO 的列表对象。

订单模块除具有以上主要功能外，还有商城用户取消订单的功能及后台管理系统中的订单管理功能，主要包括如下几个接口：

① 管理员用户查询所有订单分页列表的接口；
② 管理员用户修改订单状态的接口；
③ 管理员用户/商家关闭订单的接口；
④ 管理员用户查看订单详情的接口。

8.1.4　订单处理流程及订单状态的介绍

1. 订单处理流程

订单模块是整个电商系统的重中之重，甚至可以说它是电商系统的"心脏"，贯穿了电商系统的大部分流程。各个环节都与它密不可分，从用户提交订单并成功生成订单开始，后续的流程都是围绕着订单模块进行的，包括从支付成功到确认收货的正常订单流程，也包括订单取消、订单退款等一系列的异常订单流程。

正常订单流程如图 8-10 所示。

订单支付成功 → 订单信息确认 → 订单出库 → 确认收货

新蜂商城用户　　商城管理员用户　　商城管理员用户　　新蜂商城用户

图 8-10　正常订单流程

在订单生成后，用户正常进行支付操作，商家正常进行订单确认和订单发货操作，由用户进行最后一个步骤——确认收货。这样，整个订单流程就正常走完了。

异常订单流程如图 8-11 所示。

在订单入库后，用户选择不支付而直接取消订单，或者用户正常支付但是在后续流程中选择取消订单，于是订单就不是正常状态的订单了，因为它的流程并没有如预想的那样。不只是用户可以关闭订单，如果流程中出现了意外事件，商城管理员用户也可以在后台管理系统中关闭订单。

2. 订单状态的介绍

订单流程完善的编码实践都是围绕着订单状态的改变来做的功能实现。理解订单状态及如何发生状态转变的逻辑，对读者理解代码、理解商城业务有很大的帮助。

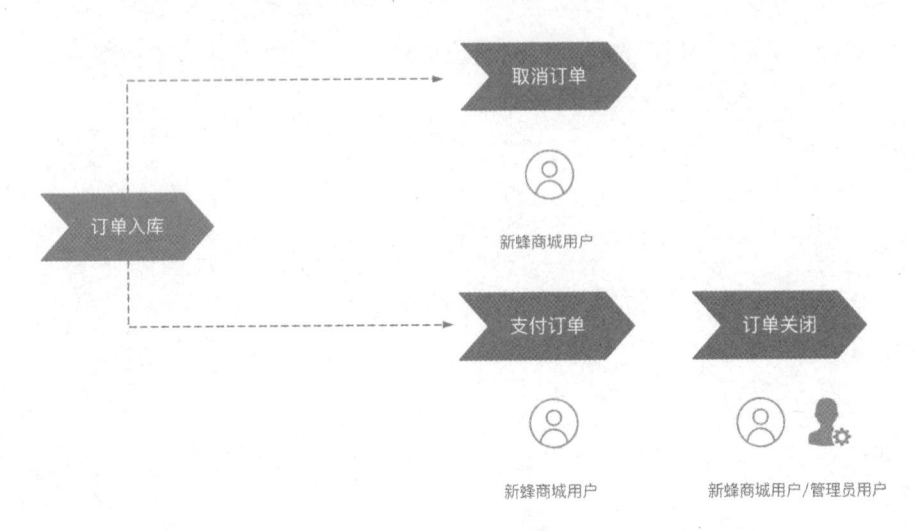

图 8-11 异常订单流程

订单主表中的 order_status 字段就是订单状态字段，新蜂商城订单状态的设计如下。

① 0：待支付。

② 1：已支付。

③ 2：配货完成。

④ 3：出库成功。

⑤ 4：交易成功。

⑥ -1：手动关闭。

⑦ -2：超时关闭。

⑧ -3：商家关闭。

以上是新蜂商城的订单状态存储的值及值对应的含义，与主流的商城设计类似，可能文案上有些小差别。比如，状态 0，新蜂商城用"待支付"表示，其他商城可能用"待付款"表示。数字的使用也可能有差异。比如，新蜂商城中订单的初始状态用数字 0 表示，其他的商城在实现时可能用数字 1 表示。

接下来详细介绍一下这些状态。

（1）待支付/待付款。

新蜂商城用数字 0 来表示这个状态。

用户提交订单后，会进行订单的入库、商品库存修改等操作，此时是订单的初始状态。目前主流的商城或常用的外卖平台，基本上在订单生成后就会唤起支付操作，所以订单的初始状态就被称为"待支付"或"待付款"。其实它的含义是订单成功入库，也就是初始状态。新蜂商城用"待支付"表示。

（2）已支付/已付款/待确认。

新蜂商城用数字 1 来表示这个状态。

用户完成订单支付，系统需要记录订单支付时间及支付方式等信息，等待商家进行订单确认以便进行后续操作。这个状态被称为"已支付"或"已付款"，也可以被称为"待商家确认"。这些称谓一般由产品经理或项目负责人来决定。新蜂商城用"已支付"表示。

（3）配货完成/已确认/待发货。

新蜂商城用数字 2 来表示这个状态。

商家确认订单正常，并且可以进行发货操作，就将订单状态修改为"已确认""待发货""配货完成"。新蜂商城用"配货完成"表示。

（4）出库成功/待收货/已发货。

新蜂商城用数字 3 来表示这个状态。

订单中的商品在出库并提交给物流系统后就进入了这个状态。对于仓库来说是"出库成功"，对于用户来说是"待收货"，而对于商家来说是"已发货"。新蜂商城用"出库成功"表示。

（5）交易成功/订单完成。

新蜂商城用数字 4 来表示这个状态。

用户收到购买的商品，单击商城订单系统中的"确认收货"按钮，表示订单已经完成了所有的正向步骤，此次交易成功。这个状态被称为"交易成功"或"订单完成"。新蜂商城用"交易成功"表示。

（6）订单关闭/已取消。

新蜂商城分别用数字-1、-2、-3 来表示"手动关闭""超时关闭""商家关闭"的状态。

它们属于订单异常的状态，在付款之前取消订单或在其他状态下选择主动取消订单都会进入这种状态，也可以统一称为"订单关闭"或"已取消"。

当然，现实中的订单流程还会涉及客服功能、订单售后、订单退款等逻辑，这些功能在本书的实战项目中并没有做具体的实现。

订单生成和各个状态的转换涉及多张表的数据更改，在测试时一定要注意数据库中商品、购物项、订单等数据是否被正确修改。

8.2 创建订单微服务模块

本节的源代码是在 newbee-mall-cloud-dev-step15 工程的基础上改造的,将工程命名为 newbee-mall-cloud-dev-step16。在工程中新增一个 newbee-mall-cloud-order-service 模块,并在 pom.xml 主文件中增加该模块的配置,代码如下:

```
<modules>
<!-- 新增订单微服务 -->
<module>newbee-mall-cloud-order-service</module>
<module>newbee-mall-cloud-shop-cart-service</module>
<module>newbee-mall-cloud-recommend-service</module>
<module>newbee-mall-cloud-goods-service</module>
<module>newbee-mall-cloud-user-service</module>
<module>newbee-mall-cloud-gateway-mall</module>
<module>newbee-mall-cloud-gateway-admin</module>
<module>newbee-mall-cloud-common</module>
</modules>
```

该模块的目录结构设置与其他功能模块的目录结构设置类似,如下所示:

```
newbee-mall-cloud-order-service          // 订单微服务
    ├── newbee-mall-cloud-order-api      // 存放订单模块中暴露的用于远程调用
    │                                       的 FeignClient 类
    └── newbee-mall-cloud-order-web      // 订单 API 的代码及逻辑
```

在新增订单微服务时,主要参考了新增商品微服务时的步骤。笔者直接将 newbee-mall-cloud-dev-step05 源代码下 newbee-mall-cloud-goods-service 模块中的代码复制过来,之后修改了模块名称和目录名称。

由于改造过程中直接复制了 newbee-mall-cloud-dev-step05 源代码下 newbee-mall-cloud-goods-service 模块中的代码,因此在修改完依赖配置后,需要修改包名,把 ltd.goods.cloud.×××改为 ltd.order.cloud.×××,之后修改 config 包中的代码,包括全局异常处理配置类、Swagger 配置类、自定义 MVC 配置类,主要修改了这些类的类名。这样就完成了一个订单微服务的初始构建工作。

订单微服务中的用户身份认证也需要处理,过程与之前微服务模块中的处理步骤类似。由于订单微服务中的接口涉及商城端和后台管理系统,因此与商城端用户和管理员用户两种身份相关的逻辑代码都要复制过来。接着,需要引入 LoadBalancer 依赖和 user-api 依赖,并修改商城端用户和管理员用户 token 处理类中的逻辑代码,并通过 OpenFeign 来调用用户微服务中的接口,在订单微服务中实现两种用户的鉴权和信息获取。

8.3 订单微服务编码

本节的内容是补充订单微服务中的业务代码，主要是把原单体 API 项目中的功能模块整合到订单微服务中。

打开订单微服务 newbee-mall-cloud-order-web 的工程目录，在 ltd.order.cloud.newbee 包下依次创建 config 包、dao 包、entity 包、service 包，在 resources 目录下新增 Mapper 文件夹用于存放 Mapper 文件。接着，将原单体 API 项目中与订单模块相关的业务代码和 Mapper 文件（如图 8-12 所示）依次复制过来。

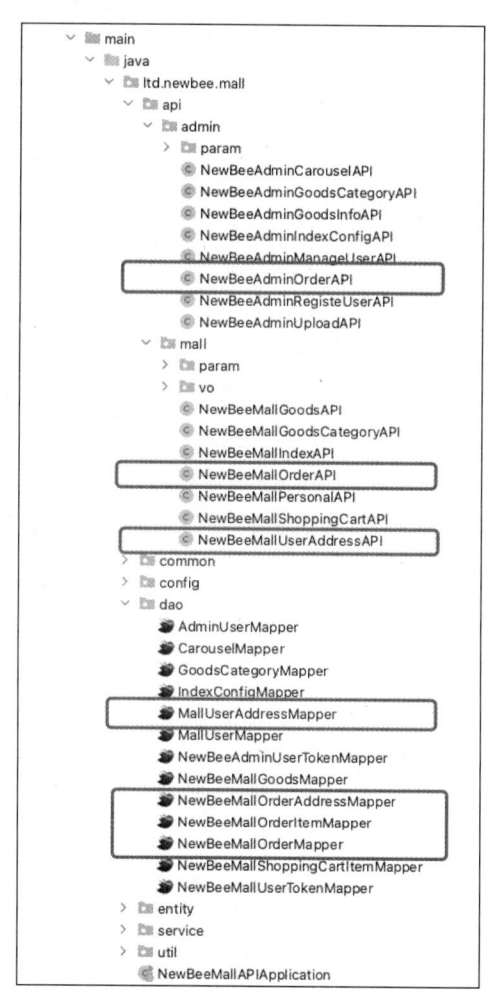

图 8-12 原单体 API 项目中与订单模块相关的业务代码和 Mapper 文件

上述步骤完成后，最终的目录结构如图 8-13 所示。

```
v newbee-mall-cloud-dev-step16 [newbee-mall-cloud]
  > .idea
  > newbee-mall-cloud-common
  > newbee-mall-cloud-gateway-admin
  > newbee-mall-cloud-gateway-mall
  > newbee-mall-cloud-goods-service
  v newbee-mall-cloud-order-service
    > newbee-mall-cloud-order-api
    v newbee-mall-cloud-order-web
      v src
        v main
          v java
            v ltd.order.cloud.newbee
              > config
              v controller
                > param
                > vo
                  NewBeeAdminOrderController
                  NewBeeMallOrderController
                  NewBeeMallUserAddressController
              v dao
                MallUserAddressMapper
                NewBeeMallOrderAddressMapper
                NewBeeMallOrderItemMapper
                NewBeeMallOrderMapper
              > entity
              v service
                > impl
                  NewBeeMallOrderService
                  NewBeeMallUserAddressService
              NewBeeMallCloudOrderServiceApplication
          > resources
      newbee-mall-cloud-order-web.iml
      pom.xml
```

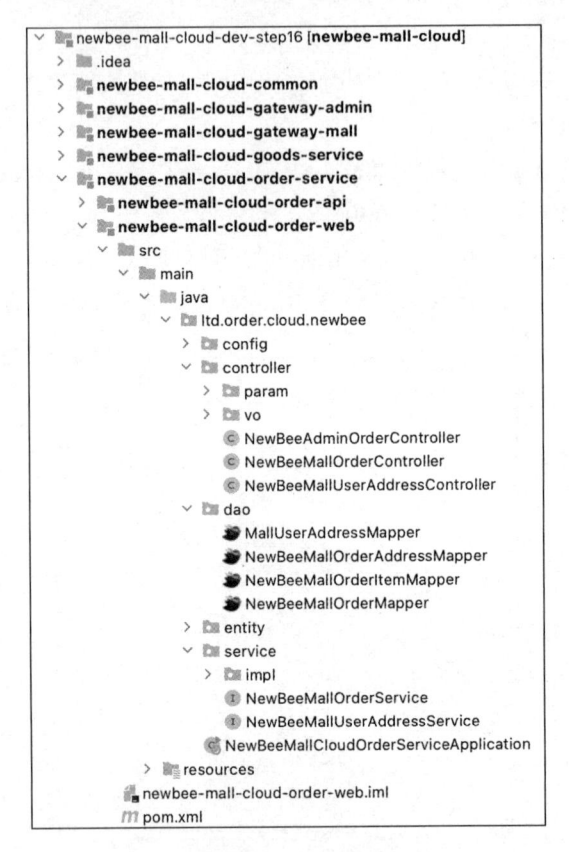

图 8-13　目录结构

修改 newbee-mall-cloud-order-web 工程中的 application.properties 配置文件，主要进行数据库连接参数及 MyBatis 扫描配置，代码如下：

```
# datasource config (MySQL)
spring.datasource.name=newbee-mall-cloud-order-datasource
spring.datasource.driverClassName=com.mysql.cj.jdbc.Driver
spring.datasource.url=jdbc:mysql://localhost:3306/newbee_mall_cloud_order_db?useUnicode=true&serverTimezone=Asia/Shanghai&characterEncoding=utf8&autoReconnect=true&useSSL=false
spring.datasource.username=root
spring.datasource.password=123456
spring.datasource.hikari.minimum-idle=5
spring.datasource.hikari.maximum-pool-size=15
spring.datasource.hikari.auto-commit=true
```

```
spring.datasource.hikari.idle-timeout=60000
spring.datasource.hikari.pool-name=hikariCP
spring.datasource.hikari.max-lifetime=600000
spring.datasource.hikari.connection-timeout=30000
spring.datasource.hikari.connection-test-query=SELECT 1

# mybatis config
mybatis.mapper-locations=classpath:mapper/*Mapper.xml
```

以上操作中的源代码涉及的数据库为 newbee_mall_cloud_order_db，数据库表为 tb_newbee_mall_order、tb_newbee_mall_order_item、tb_newbee_mall_user_address、tb_newbee_mall_order_address。除此之外，还会用到商品数据和用户数据，这部分数据就要使用远程调用技术了。

打开商城端网关 newbee-mall-cloud-gateway-mall 项目中的 application.properties 文件，新增关于订单微服务的路由信息，配置项为 spring.cloud.gateway.routes.*，新增内容如下：

```
# 订单接口的路由配置
spring.cloud.gateway.routes[5].id=order-service-route
spring.cloud.gateway.routes[5].uri=lb://newbee-mall-cloud-order-service
spring.cloud.gateway.routes[5].order=1
spring.cloud.gateway.routes[5].predicates[0]=Path=/orders/mall/**

# 收货地址接口的路由配置
spring.cloud.gateway.routes[6].id=order-service-route2
spring.cloud.gateway.routes[6].uri=lb://newbee-mall-cloud-order-service
spring.cloud.gateway.routes[6].order=1
spring.cloud.gateway.routes[6].predicates[0]=Path=/mall/address/**
```

如果访问网关项目的路径是以/orders/mall/和/mall/address/开头的，就路由到订单微服务实例。

打开后台管理系统网关 newbee-mall-cloud-gateway-admin 项目中的 application.properties 文件，新增关于订单微服务的路由信息，配置项为 spring.cloud.gateway.routes.*，新增内容如下：

```
# 订单接口的路由配置
spring.cloud.gateway.routes[5].id=order-service-route
spring.cloud.gateway.routes[5].uri=lb://newbee-mall-cloud-order-service
spring.cloud.gateway.routes[5].order=1
spring.cloud.gateway.routes[5].predicates[0]=Path=/orders/admin/**
```

如果访问网关项目的路径是以/orders/admin 开头的，就路由到订单微服务实例。

8.4 订单微服务远程调用商品微服务和购物车微服务编码实践

按照前文步骤对订单微服务进行编码后，代码中依然会被标红。被标红的代码及注释如下：

```
@PostMapping("/saveOrder")
@ApiOperation(value = "生成订单接口", notes = "传参为地址id和待结算的购物项id数组")
public Result<String> saveOrder(@ApiParam(value = "订单参数") @RequestBody SaveOrderParam saveOrderParam, @TokenToMallUser MallUser loginMallUser) {
  int priceTotal = 0;
  if (saveOrderParam == null || saveOrderParam.getCartItemIds() == null || saveOrderParam.getAddressId() == null) {
    NewBeeMallException.fail(ServiceResultEnum.PARAM_ERROR.getResult());
  }
  if (saveOrderParam.getCartItemIds().length < 1) {
    NewBeeMallException.fail(ServiceResultEnum.PARAM_ERROR.getResult());
  }
  // 根据购物项id列表查询购物项表并进行基本的逻辑验证
  List<NewBeeMallShoppingCartItemVO> itemsForSave = newBeeMallShoppingCartService.getCartItemsForSettle(Arrays.asList(saveOrderParam.getCartItemIds()), loginMallUser.getUserId());
  省略部分代码
  return ResultGenerator.genFailResult("生成订单失败");
}
```

在生成订单接口中，需要根据 cartItemIds 参数查询数据库中的一条或多条购物项数据进行基本的逻辑验证，之后才能执行后续的业务逻辑。而订单微服务未连接购物项表所在的数据库，无法直接查询对应的购物项数据，所以这部分代码会被标红，代码如下：

```
@Override
@Transactional
public String saveOrder(MallUser loginMallUser, MallUserAddress address, List<NewBeeMallShoppingCartItemVO> myShoppingCartItems) {
  List<Long> itemIdList = myShoppingCartItems.stream().map(NewBeeMallShoppingCartItemVO::getCartItemId).collect(Collectors.toList());
  List<Long> goodsIds = myShoppingCartItems.stream().map(NewBeeMallShoppingCartItemVO::getGoodsId).collect(Collectors.toList());
  // 查询商品数据
```

```java
    List<NewBeeMallGoods> newBeeMallGoods = newBeeMallGoodsMapper.selectBy
PrimaryKeys(goodsIds);
    // 检查是否包含已下架商品
    List<NewBeeMallGoods> goodsListNotSelling = newBeeMallGoods.stream()
        .filter(newBeeMallGoodsTemp ->
newBeeMallGoodsTemp.getGoodsSellStatus() != Constants.SELL_STATUS_UP)
        .collect(Collectors.toList());
    if (!CollectionUtils.isEmpty(goodsListNotSelling)) {
        // goodsListNotSelling 对象非空表示有下架商品
        NewBeeMallException.fail(goodsListNotSelling.get(0).getGoodsName() + "
已下架,无法生成订单");
    }
    Map<Long, NewBeeMallGoods> newBeeMallGoodsMap = newBeeMallGoods.stream().
collect(Collectors.toMap(NewBeeMallGoods::getGoodsId, Function.identity(),
(entity1, entity2) -> entity1));
    // 判断商品库存
    for (NewBeeMallShoppingCartItemVO shoppingCartItemVO :
myShoppingCartItems) {
        // 查询的商品中不存在购物车中的这条关联商品数据,直接返回错误提示
        if (!newBeeMallGoodsMap.containsKey(shoppingCartItemVO.getGoodsId())) {
            NewBeeMallException.fail(ServiceResultEnum.SHOPPING_ITEM_ERROR.
getResult());
        }
        // 存在数量大于库存的情况,直接返回错误提示
        if (shoppingCartItemVO.getGoodsCount() > newBeeMallGoodsMap.get
(shoppingCartItemVO.getGoodsId()).getStockNum()) {
            NewBeeMallException.fail(ServiceResultEnum.SHOPPING_ITEM_COUNT_
ERROR.getResult());
        }
    }
    // 删除购物项
    if (!CollectionUtils.isEmpty(itemIdList) && !CollectionUtils.isEmpty(goodsIds)
&& !CollectionUtils.isEmpty(newBeeMallGoods)) {
        if (newBeeMallShoppingCartItemMapper.deleteBatch(itemIdList) > 0) {
            List<StockNumDTO> stockNumDTOS = BeanUtil.copyList(myShoppingCartItems,
StockNumDTO.class);
            // 修改商品库存
            int updateStockNumResult = newBeeMallGoodsMapper.updateStockNum
(stockNumDTOS);
            if (updateStockNumResult < 1) {
                NewBeeMallException.fail(ServiceResultEnum.SHOPPING_ITEM_COUNT_
ERROR.getResult());
            }
```

```
    //生成订单号
    String orderNo = NumberUtil.genOrderNo();
    int priceTotal = 0;
    //保存订单
    NewBeeMallOrder newBeeMallOrder = new NewBeeMallOrder();
    newBeeMallOrder.setOrderNo(orderNo);
    newBeeMallOrder.setUserId(loginMallUser.getUserId());
    //总价
    for (NewBeeMallShoppingCartItemVO newBeeMallShoppingCartItemVO :
myShoppingCartItems) {
        priceTotal += newBeeMallShoppingCartItemVO.getGoodsCount() *
newBeeMallShoppingCartItemVO.getSellingPrice();
    }
    if (priceTotal < 1) {
        NewBeeMallException.fail(ServiceResultEnum.ORDER_PRICE_ERROR.getResult());
    }
    newBeeMallOrder.setTotalPrice(priceTotal);
    String extraInfo = "";
    省略部分代码
    NewBeeMallException.fail(ServiceResultEnum.DB_ERROR.getResult());
  }
  NewBeeMallException.fail(ServiceResultEnum.DB_ERROR.getResult());
 }
 NewBeeMallException.fail(ServiceResultEnum.SHOPPING_ITEM_ERROR.getResult());
 return ServiceResultEnum.SHOPPING_ITEM_ERROR.getResult();
}
```

同样，在生成订单的业务层代码中，也有针对购物项表和商品表的操作，如删除购物项数据、查询商品数据、修改商品库存，之后才能执行后续生成订单的业务逻辑。而订单微服务未连接购物项表和商品表所在的数据库，无法直接通过 NewBeeMall-ShoppingCartItemMapper 和 GoodsMapper 完成对应的功能，所以这部分代码会被标红。

因此，想要完成订单微服务的代码改造，必须进行购物项数据查询、购物项数据删除、商品数据查询和商品库存修改部分代码的改造，由原本直接操作购物项表和商品表改为远程调用购物车微服务和商品微服务中的接口来完成这个逻辑。订单微服务不仅要与用户微服务通信，还要与购物车微服务和商品微服务通信，代码改造步骤如下。

第一步，在商品微服务中增加修改库存的逻辑代码和对外暴露接口。

当前所需的查询购物项数据、删除购物项数据、查询商品数据和修改商品库存这 4 个远程调用操作，有 3 个功能都已经完成编码，只有修改商品库存还未完成编码。

打开 newbee-mall-cloud-goods-web 工程，在 NewBeeAdminGoodsInfoController 类中

新增接口，代码如下：

```java
/**
 * 修改商品库存
 */
@PutMapping("/updateStock")
@ApiOperation(value = "修改库存", notes = "")
public Result updateStock(@RequestBody UpdateStockNumDTO updateStockNumDTO) {
    return ResultGenerator.genSuccessResult(newBeeMallGoodsService.
updateStockNum(updateStockNumDTO.getStockNumDTOS()));
}
```

打开 newbee-mall-cloud-goods-api 工程，在 NewBeeCloudGoodsServiceFeign 类中定义该接口，使其可以被其他微服务调用，新增代码如下：

```java
@PutMapping(value = "/admin/updateStock")
Result<Boolean> updateStock(@RequestBody UpdateStockNumDTO updateStockNumDTO);
```

第二步，引入 goods-api 和 shop-cart-api 依赖。

打开 newbee-mall-cloud-order-web 工程下的 pom.xml 文件，增加与购物车微服务和商品微服务远程通信所需的 newbee-mall-cloud-shop-cart-api 模块和 newbee-mall-cloud-goods-api 模块，新增依赖配置如下：

```xml
<dependency>
  <groupId>ltd.goods.newbee.cloud</groupId>
  <artifactId>newbee-mall-cloud-goods-api</artifactId>
  <version>0.0.1-SNAPSHOT</version>
</dependency>

<dependency>
  <groupId>ltd.shopcart.newbee.cloud</groupId>
  <artifactId>newbee-mall-cloud-shop-cart-api</artifactId>
  <version>0.0.1-SNAPSHOT</version>
</dependency>
```

第三步，增加关于购物车微服务和商品微服务中 FeignClient 的配置。

打开 newbee-mall-cloud-goods-web 工程，对启动类 NewBeeMallCloudOrderService-Application 中的 @EnableFeignClients 注解进行修改，增加对 NewBeeCloudShopCart-ServiceFeign 类和 NewBeeCloudGoodsServiceFeign 类的声明，代码如下：

```java
@EnableFeignClients(basePackageClasses ={ltd.goods.cloud.newbee.openfeign.
NewBeeCloudGoodsServiceFeign.class, ltd.shopcart.cloud.newbee.openfeign.
NewBeeCloudShopCartServiceFeign.class, ltd.user.cloud.newbee.openfeign.
NewBeeCloudUserServiceFeign.class})
```

接下来就可以直接使用 NewBeeCloudShopCartServiceFeign 类和 NewBeeCloudGoods-ServiceFeign 类与购物车微服务、商品微服务进行远程通信了。

第四步，修改对商品表和购物项表操作的代码。

打开 newbee-mall-cloud-order-web 工程，修改 NewBeeMallOrderServiceImpl 类中 saveOrder()方法的逻辑代码。删除原本直接操作购物项表和商品表的代码，之后依次注入 NewBeeCloudShopCartServiceFeign 类和 NewBeeCloudGoodsServiceFeign 类，并通过调用购物车微服务和商品微服务来完成查询购物项数据、删除购物项数据、查询商品数据和修改商品库存 4 个逻辑操作。这里，笔者把购物项数据的查询和前置判断也转移到业务方法中了。修改后的代码及相应的注释如下：

```java
@Autowired
private NewBeeCloudGoodsServiceFeign goodsService;

@Autowired
private NewBeeCloudShopCartServiceFeign shopCartService;

@Override
@Transactional
public String saveOrder(Long mallUserId, MallUserAddress address, List<Long> cartItemIds) {
    //调用购物车微服务 Feign 获取数据
    Result<List<NewBeeMallShoppingCartItemDTO>> cartItemDTOListResult = shopCartService.listByCartItemIds(cartItemIds);
    if (cartItemDTOListResult == null || cartItemDTOListResult.getResultCode() != 200) {
        NewBeeMallException.fail("参数异常");
    }
    List<NewBeeMallShoppingCartItemDTO> itemsForSave = cartItemDTOListResult.getData();
    if (CollectionUtils.isEmpty(itemsForSave)) {
        //无数据
        NewBeeMallException.fail("参数异常");
    }
    List<Long> itemIdList = itemsForSave.stream().map(NewBeeMallShoppingCartItemDTO::getCartItemId).collect(Collectors.toList());
    List<Long> goodsIds = itemsForSave.stream().map(NewBeeMallShoppingCartItemDTO::getGoodsId).collect(Collectors.toList());
    //调用商品微服务 Feign 获取数据
    Result<List<NewBeeMallGoodsDTO>> goodsDTOListResult = goodsService.listBy
```

```java
GoodsIds(goodsIds);
    if (goodsDTOListResult == null || goodsDTOListResult.getResultCode() != 200)
{
        NewBeeMallException.fail("参数异常");
    }
    List<NewBeeMallGoodsDTO> newBeeMallGoods = goodsDTOListResult.getData();
    //检查是否包含已下架商品
    List<NewBeeMallGoodsDTO> goodsListNotSelling = newBeeMallGoods.stream()
        .filter(newBeeMallGoodsTemp ->
newBeeMallGoodsTemp.getGoodsSellStatus() != 0)
        .collect(Collectors.toList());
    if (!CollectionUtils.isEmpty(goodsListNotSelling)) {
        //goodsListNotSelling 对象非空表示有下架商品
        NewBeeMallException.fail(goodsListNotSelling.get(0).getGoodsName() + "
已下架,无法生成订单");
    }
    Map<Long, NewBeeMallGoodsDTO> newBeeMallGoodsMap = newBeeMallGoods.
stream().collect(Collectors.toMap(NewBeeMallGoodsDTO::getGoodsId,
Function.identity(), (entity1, entity2) -> entity1));
    //判断商品库存
    for (NewBeeMallShoppingCartItemDTO cartItemDTO : itemsForSave) {
        //查询的商品中不存在购物车中的这条关联商品数据,直接返回错误提示
        if (!newBeeMallGoodsMap.containsKey(cartItemDTO.getGoodsId())) {
            NewBeeMallException.fail(ServiceResultEnum.SHOPPING_ITEM_ERROR.
getResult());
        }
        //存在数量大于库存的情况,直接返回错误提示
        if (cartItemDTO.getGoodsCount() > newBeeMallGoodsMap.get(cartItem
DTO.getGoodsId()).getStockNum()) {
            NewBeeMallException.fail(ServiceResultEnum.SHOPPING_ITEM_COUNT_
ERROR.getResult());
        }
    }
    //删除购物项
    if (!CollectionUtils.isEmpty(itemIdList) && !CollectionUtils.isEmpty(goodsIds)
&& !CollectionUtils.isEmpty(newBeeMallGoods)) {

        //调用购物车微服务 Feign 删除数据
        Result<Boolean> deleteByCartItemIdsResult = shopCartService.
deleteByCartItemIds(itemIdList);
```

```
        if (deleteByCartItemIdsResult != null && deleteByCartItemIdsResult.
getResultCode() == 200) {


    List<StockNumDTO> stockNumDTOS = BeanUtil.copyList(itemsForSave,
StockNumDTO.class);
    UpdateStockNumDTO updateStockNumDTO = new UpdateStockNumDTO();
    updateStockNumDTO.setStockNumDTOS(stockNumDTOS);

    //调用商品微服务Feign修改库存数据
    Result<Boolean> updateStockResult = goodsService.updateStock(updateStockNumDTO);
    if (updateStockResult == null || updateStockResult.getResultCode() != 200) {
      NewBeeMallException.fail(ServiceResultEnum.PARAM_ERROR.getResult());
    }
    if (!updateStockResult.getData()) {
      NewBeeMallException.fail(ServiceResultEnum.SHOPPING_ITEM_COUNT_
ERROR.getResult());
    }
    省略部分代码
      NewBeeMallException.fail(ServiceResultEnum.DB_ERROR.getResult());
    }
  NewBeeMallException.fail(ServiceResultEnum.DB_ERROR.getResult());
 }
 NewBeeMallException.fail(ServiceResultEnum.SHOPPING_ITEM_ERROR. getResult());
 return ServiceResultEnum.SHOPPING_ITEM_ERROR.getResult();
}
```

8.5　订单微服务功能测试

代码修改完成后，测试步骤是不能漏掉的。一定要验证项目是否能正常启动、接口是否能正常调用，防止在代码移动过程中出现一些小问题，导致项目无法启动或代码报错。在项目启动前需要分别启动 Nacos Server 和 Redis Server，之后依次启动 newbee-mall-cloud-order-web 工程、newbee-mall-cloud-shop-cart-web 工程、newbee-mall-cloud-user-web 工程、newbee-mall-cloud-goods-web 工程下的主类（要测试网关功能，还需要再启动两个网关模块下的主类）。启动成功后，就可以进行本节的功能测试了。

打开用户微服务的 Swagger 页面，在浏览器中输入如下网址：http://localhost:29000/swagger-ui/index.html。

第 8 章 订单微服务编码实践及功能讲解

之后在该页面使用商城用户的登录接口获取一个 token 值，用于后续的功能测试。比如，笔者在测试时获取了一个值为 "c5b0a720e8068a186d9d3ff7ff3a28d8" 的 token 字段。

打开订单微服务的 Swagger 页面，在浏览器中输入如下网址：http://localhost:29040/swagger-ui/index.html。

接着，就可以在 Swagger 提供的 UI 页面进行订单微服务的接口测试了，接口文档显示内容如图 8-14 所示。

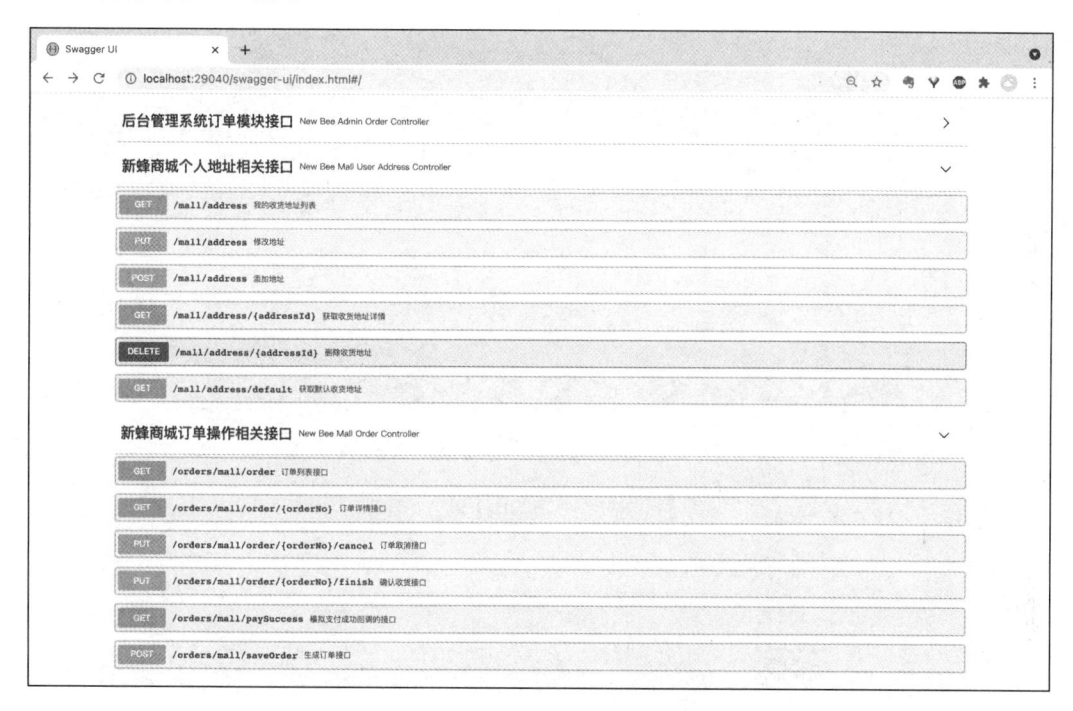

图 8-14 订单模块的接口文档显示内容

接口文档主要包括后台管理系统订单模块接口、新蜂商城个人地址相关接口和新蜂商城订单操作相关接口。由于篇幅有限，这里只对新蜂商城订单操作相关接口进行测试。

8.5.1 添加收货地址接口演示

下单时需要使用用户的收货地址信息，否则无法正确地生成订单数据。依次单击"添加地址""Try it out"按钮，在参数栏中输入收货地址的相关信息，在登录认证 token 的输入框中输入登录接口返回的 token 值，测试过程如图 8-15 所示。

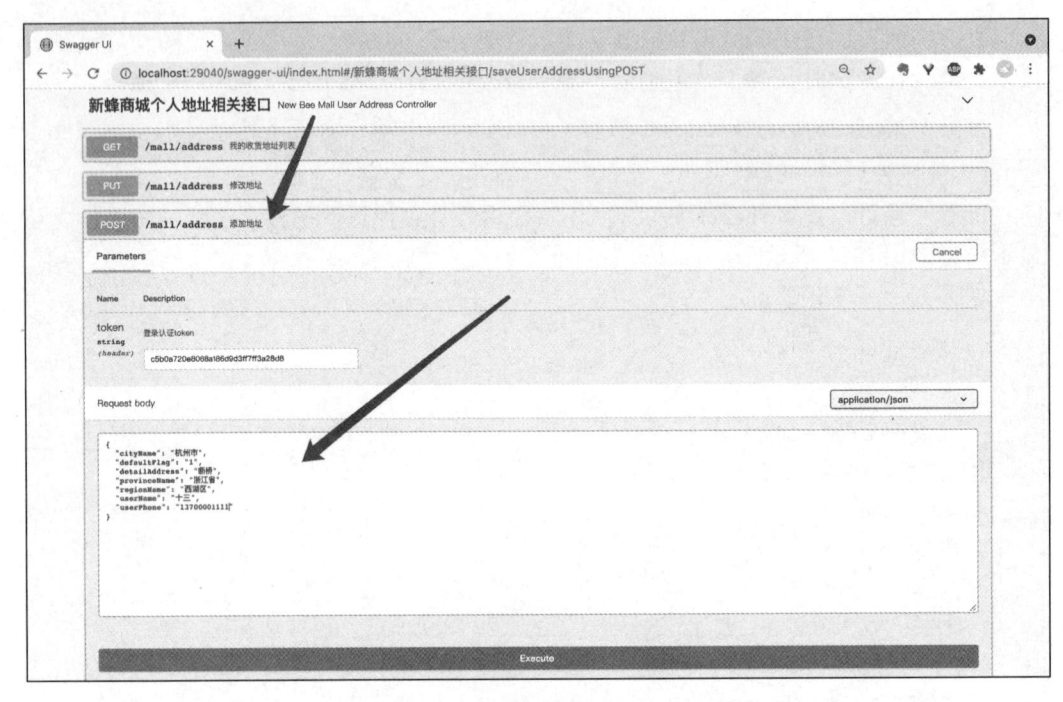

图 8-15　添加收货地址接口的测试过程

单击"Execute"按钮，接口的测试结果如图 8-16 所示。

图 8-16　添加收货地址接口的测试结果

若后端接口的测试结果中有"SUCCESS"，则表示收货地址信息添加成功。此时，再去订单微服务的数据库中查看收货地址表中的数据，可以看到已经新增了一条地址信息，该数据的主键 id 为 2164，后续生成订单时会用到。

8.5.2 生成订单接口演示

依次单击"生成订单接口""Try it out"按钮,在参数栏中输入当前用户的地址 id 和需要结算的购物项 id 列表,这里输入的数据都是刚刚演示时生成的数据(地址 id 为 2164,购物项 id 为购物车微服务实战章节生成的数据,分别是 7625 和 7626)。在登录认证 token 的输入框中输入登录接口返回的 token 值,测试过程如图 8-17 所示。

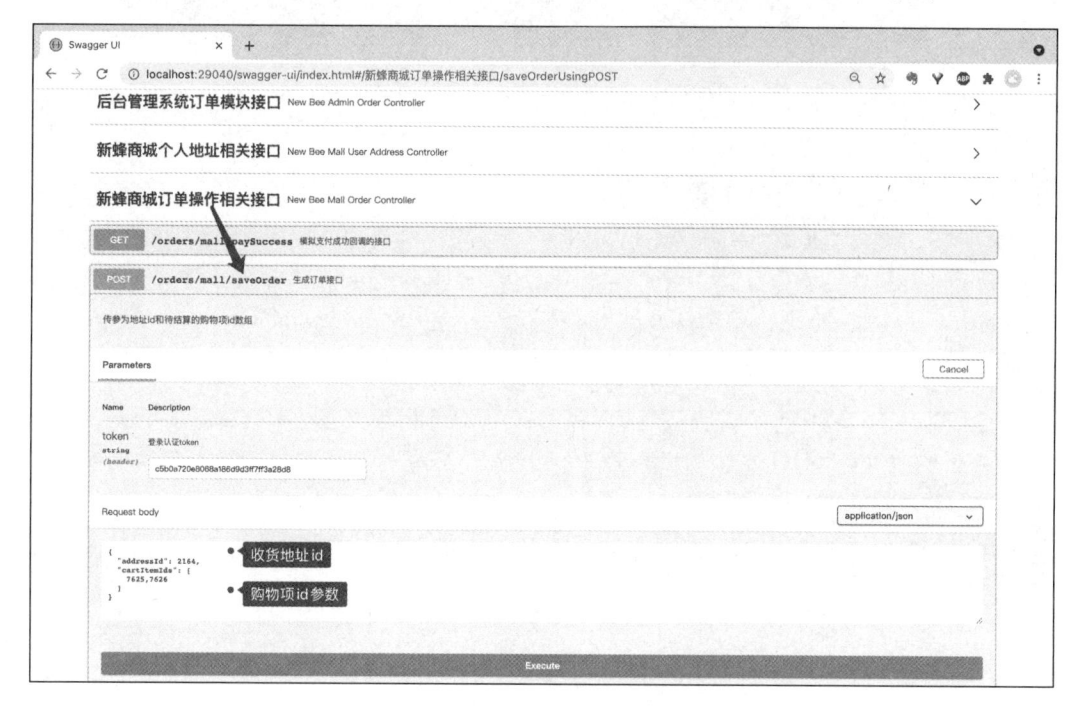

图 8-17 生成订单接口的测试过程

单击"Execute"按钮,接口的测试结果如图 8-18 所示。

如果结算时提交的数据都正确,就可以得到一个订单生成后的订单号字段,该字段的值在响应对象 Result 的 data 字段中。比如,当前接口的测试结果获取了值为"16629533442214515"的订单号,之后就能够使用该订单号来测试取消订单、模拟支付、查看订单详情的接口了。

至此,生成订单接口测试成功。

图 8-18　生成订单接口的测试结果

8.5.3　订单列表接口演示

依次单击"订单列表接口""Try it out"按钮，在参数栏中输入页码和订单状态字段，在登录认证 token 的输入框中输入登录接口返回的 token 值，就可以查询当前用户的订单列表数据了，测试过程如图 8-19 所示。

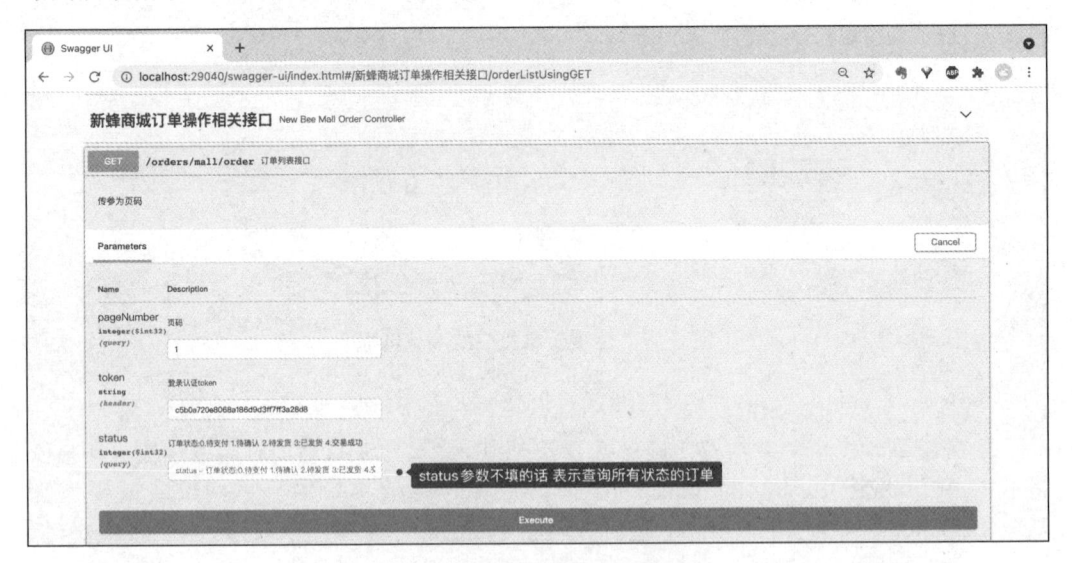

图 8-19　订单列表接口的测试过程

单击"Execute"按钮，接口的测试结果如图 8-20 所示。

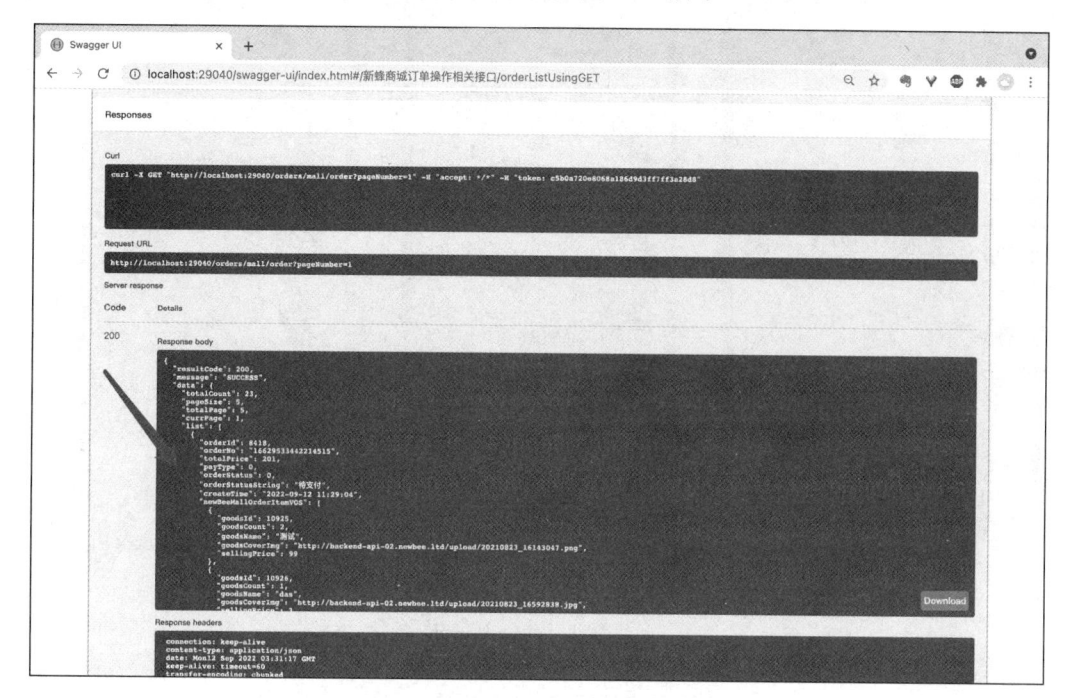

图 8-20　订单列表接口的测试结果

请求成功。订单列表中所需的数据在 Result 类的 data 属性中，其中有分页信息、订单列表数据，每条购物项中都包括订单号、订单状态、下单时间、订单中包含的商品等内容。

以上三个接口对应到实际的项目页面中，是新蜂商城项目的添加收货地址页面、订单确认页面和订单列表页面，显示效果如图 8-21 所示。

至此，下单流程中的部分功能就演示完成了，读者在测试时可以关注一下 MySQL 数据库中购物项表的变化。

功能测试完成且接口响应一切正常，表示订单微服务本身的功能编码完成，并且远程调用用户微服务、商品微服务、购物车微服务也一切正常。在测试时，读者也可以通过 debug 模式启动项目，打上几个断点来查看接口测试时的完整过程。由于篇幅有限，笔者这里只演示了部分接口的测试过程，读者在测试时可以查看其他接口。除在订单微服务架构项目的 Swagger 页面测试接口外，也可以通过商城端网关来访问这些接口进行功能测试。

图 8-21 添加收货地址页面、订单确认页面、订单列表页面

本章主要讲解订单模块在微服务架构下的编码改造,在订单微服务开发完成后,本书实战项目已经完成了原单体商城项目中所有功能模块的开发与测试。当然,这不是终点,后续依然会补充一些必要的知识点。希望读者能够根据笔者提供的开发步骤顺利地完成本章的项目改造。

第 9 章

Spring Cloud Gateway 聚合 Swagger 接口文档

本章讲解在网关层统一 Swagger 接口，将分散的 Swagger 接口文档聚合，方便后端接口的查看和调试。本章的源代码是在 newbee-mall-cloud-dev-step16 工程的基础上改造的，将工程命名为 newbee-mall-cloud-dev-step17。

9.1 为什么要聚合Swagger接口文档

在微服务架构的项目开发中，可能每个独立的微服务实例都有其独立的 Swagger 接口调试页面。比如，本书的微服务架构项目，有用户微服务、商品微服务、订单微服务等，这些微服务的工程源代码中都各自配置了 Swagger 接口文档工具的相关属性，实例启动后都能够通过访问特定的网址链接查看和调试相应的接口，各微服务实例的 Swagger 接口文档网址如表 9-1 所示。

表 9-1 各微服务实例的 Swagger 接口文档网址

服务名称	Swagger 接口文档网址
用户微服务	http://localhost:29000/swagger-ui/index.html
商品微服务	http://localhost:29010/swagger-ui/index.html
推荐微服务	http://localhost:29020/swagger-ui/index.html
购物车微服务	http://localhost:29030/swagger-ui/index.html
订单微服务	http://localhost:29040/swagger-ui/index.html

当前项目只有 5 个微服务实例，如果有更多的微服务实例，那么对真实开发场景中前后端开发人员联调接口会造成一些小困扰，毕竟在联调接口时不可能在浏览器中打开多个链接查看不同的 Swagger UI 页面。在微服务实例多的情况下，如果不聚合 Swagger 接口文档，那么访问每个服务的 API 文档都需要单独访问一个 Swagger UI 页面，很不方便。另外，一旦服务实例的 IP 地址或端口更换，就要重新沟通和修改，也比较麻烦。

在微服务架构项目中基本上都会整合网关服务，所有的请求都要经过网关层。既然请求有统一的入口，如果在网关层统一聚合这些 Swagger 接口文档，在网关层进行接口的查看和联调，就会更方便也更符合真实开发场景，毕竟在部署的时候不会对外暴露内部微服务实例工程的 IP 地址和端口。

当然，这个步骤并不是必需的，只是聚合接口文档会更方便、更规范一些。

9.2　网关层聚合Swagger接口文档的实现思路

在介绍实现思路前，要明确项目中的接口数据是怎样被渲染到 Swagger UI 页面中的。

打开项目中某个服务端 Swagger UI 页面。打开浏览器控制台中的"Network"面板，在页面加载过程中查看其中的请求，Swagger UI 页面中接口文档加载主要有 3 个请求，如图 9-1 所示。其中，/swagger-resources 和/v3/api-docs 两个请求的内容和作用对本次找出聚合 Swagger 接口文档的实现思路有一些帮助。

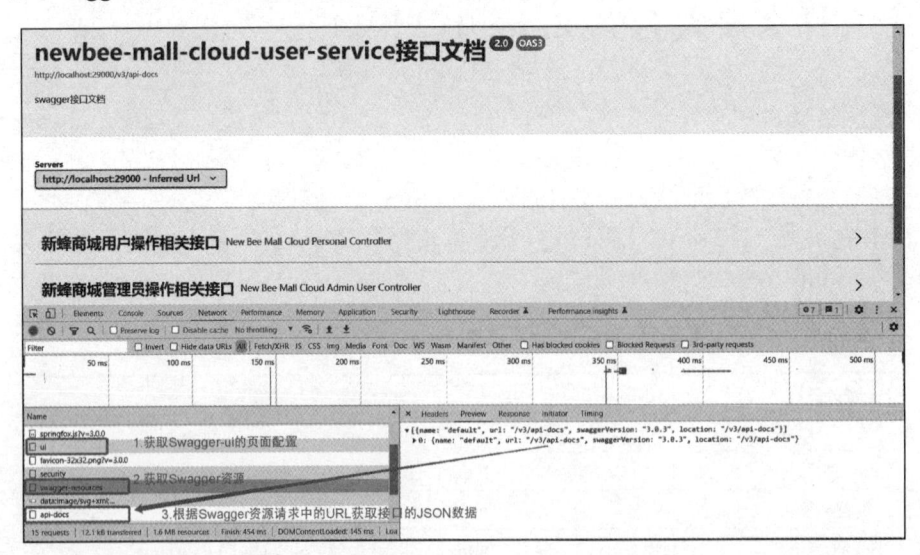

图 9-1　接口文档页面加载过程中的请求分析

/swagger-resources 请求用于获取 Swagger 资源,其响应的内容是 Swagger 资源数组。图 9-1 中返回的资源列表只有一个,资源名称为 default,资源网址为当前项目的基础路径加上 /v3/api-docs。获取 Swagger 资源后,直接请求这个资源网址获取接口文档的 JSON 数据,如图 9-2 所示。

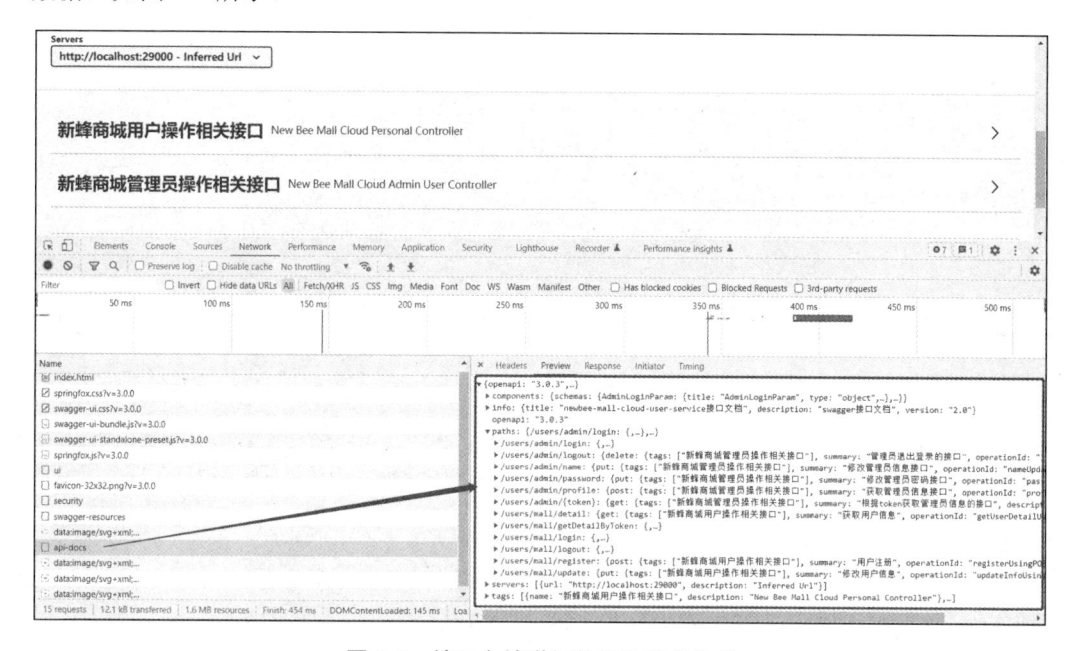

图 9-2 接口文档详细数据的请求分析

获取这些接口数据后,就可以渲染到 Swagger UI 页面中了。

好的,Swagger UI 页面中接口文档的加载过程清晰了,步骤如下:获取页面配置→获取 Swagger 资源→根据资源 URL 获取接口文档的 JSON 数据→将接口文档的 JSON 数据渲染到页面中。

聚合 Swagger 接口文档的实现思路就是在"获取 Swagger 资源"这个步骤中做一些修改,让当前的 Swagger UI 页面可以获取多个 Swagger 资源。具体做法是在网关层整合 Swagger 之后,从路由配置中获取底层多个微服务实例信息,根据这些实例信息组装对应的 Swagger 页面所需的资源信息,包括资源名称、资源 URL 等。

讲完了实现思路就要来聊一聊具体的编码实现,对于 Swagger 资源处理,Swagger 底层提供了一个 SwaggerResourcesProvider 接口,这是资源处理的关键接口,其唯一的默认实现类是 springfox.documentation.swagger.web.InMemorySwaggerResourcesProvider,在默认的情况下接口资源都会经由此类处理后返回给前端。其核心代码如下:

```
@Override
```

```java
public List<SwaggerResource> get() {
  List<SwaggerResource> resources = new ArrayList<>();

  for (Map.Entry<String, Documentation> entry : documentationCache.all().entrySet()) {
    String swaggerGroup = entry.getKey();
    if (swagger1Available) {
      SwaggerResource swaggerResource = resource(swaggerGroup, swagger1Url);
      swaggerResource.setSwaggerVersion("1.2");
      resources.add(swaggerResource);
    }

    if (swagger2Available) {
      SwaggerResource swaggerResource = resource(swaggerGroup, swagger2Url);
      swaggerResource.setSwaggerVersion("2.0");
      resources.add(swaggerResource);
    }

    if (oas3Available) {
      SwaggerResource swaggerResource = resource(swaggerGroup, oas3Url);
      swaggerResource.setSwaggerVersion("3.0.3");
      resources.add(swaggerResource);
    }
  }
  Collections.sort(resources);
  return resources;
}

private SwaggerResource resource(
    String swaggerGroup,
    String baseUrl) {
  SwaggerResource swaggerResource = new SwaggerResource();
  swaggerResource.setName(swaggerGroup);
  swaggerResource.setUrl(swaggerLocation(baseUrl, swaggerGroup));
  return swaggerResource;
}
```

至此，网关聚合的实现就清晰了。在网关层加入 Swagger 依赖并重写 Swagger-ResourcesProvider 接口中的 get()方法，将需要聚合路由的子模块 Swagge 资源组装成 SwaggerResource 对象并形成数据集合，这样就可以在 Swagger UI 页面中选择对应的服务模块，调用不同服务中的/v3/api-docs，获取具体的接口资源并展示到网关层的 Swagger UI 页面中。

9.3 网关层聚合Swagger接口文档编码

由于各个微服务实例都已经聚合了 Swagger 接口文档且一切功能正常,因此这里不再赘述。下面以商城端网关为例,直接讲解如何在网关层聚合对应的微服务实例 Swagger 接口文档。

第一步,添加 Swagger3 依赖。

打开商城端网关 newbee-mall-cloud-gateway-mall 项目下的 pom.xml 文件,增加 Swagger3 依赖,新增代码如下:

```xml
<dependency>
    <groupId>io.springfox</groupId>
    <artifactId>springfox-boot-starter</artifactId>
</dependency>
```

第二步,增加专门的路由配置来处理 Swagger 接口文档聚合。

网上关于"Spring Cloud Gateway 聚合 Swagger 接口文档"这个知识点的教程挺多,但是大部分是小案例,聚合时也是将配置文件中的所有路由配置都读取并做聚合。具体的业务实现中会有些不同,如当前项目中会对一个服务做两个路由配置,代码如下:

```
# 分类接口的路由配置
spring.cloud.gateway.routes[2].id=goods-service-route
spring.cloud.gateway.routes[2].uri=lb://newbee-mall-cloud-goods-service
spring.cloud.gateway.routes[2].order=1
spring.cloud.gateway.routes[2].predicates[0]=Path=/categories/mall/**

# 商品接口的路由配置
spring.cloud.gateway.routes[3].id=goods-service-route2
spring.cloud.gateway.routes[3].uri=lb://newbee-mall-cloud-goods-service
spring.cloud.gateway.routes[3].order=1
spring.cloud.gateway.routes[3].predicates[0]=Path=/goods/mall/**
```

其实这两个路由配置都指向商品微服务,如果读取所有路由,那么在网关层 Swagger UI 页面中就有重复的 Swagger 资源。因此,笔者在这里做了一些调整,增加了几条针对 Swagger 接口文档聚合的路由配置,在 newbee-mall-cloud-gateway-mall 项目下的 application.properties 文件中增加如下内容:

```
# 商品微服务 Swagger 接口文档的路由配置
spring.cloud.gateway.routes[7].id=goods-service-swagger-route
spring.cloud.gateway.routes[7].uri=lb://newbee-mall-cloud-goods-service
spring.cloud.gateway.routes[7].order=1
```

```
spring.cloud.gateway.routes[7].predicates[0]=Path=/goods/swagger/**
spring.cloud.gateway.routes[7].filters[0]=StripPrefix=2

# 推荐微服务 Swagger 接口文档的路由配置
spring.cloud.gateway.routes[8].id=recommend-service-swagger-route
spring.cloud.gateway.routes[8].uri=lb://newbee-mall-cloud-recommend-service
spring.cloud.gateway.routes[8].order=1
spring.cloud.gateway.routes[8].predicates[0]=Path=/indexConfigs/swagger/**
spring.cloud.gateway.routes[8].filters[0]=StripPrefix=2

# 订单微服务 Swagger 接口文档的路由配置
spring.cloud.gateway.routes[9].id=order-service-swagger-route
spring.cloud.gateway.routes[9].uri=lb://newbee-mall-cloud-order-service
spring.cloud.gateway.routes[9].order=1
spring.cloud.gateway.routes[9].predicates[0]=Path=/orders/swagger/**
spring.cloud.gateway.routes[9].filters[0]=StripPrefix=2

# 用户微服务 Swagger 接口文档的路由配置
spring.cloud.gateway.routes[10].id=user-service-swagger-route
spring.cloud.gateway.routes[10].uri=lb://newbee-mall-cloud-user-service
spring.cloud.gateway.routes[10].order=1
spring.cloud.gateway.routes[10].predicates[0]=Path=/users/swagger/**
spring.cloud.gateway.routes[10].filters[0]=StripPrefix=2

# 购物车微服务 Swagger 接口文档的路由配置
spring.cloud.gateway.routes[11].id=shop-cart-service-swagger-route
spring.cloud.gateway.routes[11].uri=lb://newbee-mall-cloud-shop-cart-service
spring.cloud.gateway.routes[11].order=1
spring.cloud.gateway.routes[11].predicates[0]=Path=/carts/swagger/**
spring.cloud.gateway.routes[11].filters[0]=StripPrefix=2
```

第三步，增加聚合配置类。

在 ltd.gateway.cloud.newbee.config 包下新建 PolymerizeSwaggerProvider 配置类，重写 get()方法，代码如下：

```java
package ltd.gateway.cloud.newbee.config;

import org.springframework.beans.factory.annotation.Autowired;
import org.springframework.cloud.gateway.config.GatewayProperties;
import org.springframework.cloud.gateway.support.NameUtils;
import org.springframework.context.annotation.Primary;
```

```java
import org.springframework.stereotype.Component;
import springfox.documentation.swagger.web.SwaggerResource;
import springfox.documentation.swagger.web.SwaggerResourcesProvider;

import java.util.ArrayList;
import java.util.List;

/**
 * 在网关层聚合底层微服务的Swagger资源
 *
 */
@Primary
@Component
public class PolymerizeSwaggerProvider implements SwaggerResourcesProvider {

    /**
     * Swagger Doc 的URL后缀
     */
    public static final String API_DOCS_URL = "/v3/api-docs";

    @Autowired
    private GatewayProperties gatewayProperties;

    @Override
    public List<SwaggerResource> get() {
        List<SwaggerResource> resources = new ArrayList<>();
        List<String> routes = new ArrayList<>();
        //需要聚合的路由配置
        routes.add("user-service-swagger-route");
        routes.add("recommend-service-swagger-route");
        routes.add("goods-service-swagger-route");
        routes.add("order-service-swagger-route");
        routes.add("shop-cart-service-swagger-route");
        gatewayProperties.getRoutes().stream().filter(routeDefinition ->
routes.contains(routeDefinition.getId()))
                .forEach(routeDefinition ->
routeDefinition.getPredicates(). stream()
                        .filter(predicateDefinition ->
("Path"). equalsIgnore Case(predicateDefinition.getName()))
```

```
                    .forEach(predicateDefinition ->
resources.add (swaggerResource(routeDefinition.getId(),
predicateDefinition.getArgs().get(NameUtils. GENERATED_NAME_PREFIX + "0")
                    .replace("/**", API_DOCS_URL)))));
        return resources;
    }

    private SwaggerResource swaggerResource(String name, String url) {
        SwaggerResource swaggerResource = new SwaggerResource();
        swaggerResource.setName(name);
        swaggerResource.setLocation(url);
        swaggerResource.setSwaggerVersion("3.0");
        return swaggerResource;
    }
}
```

这样，网关层 Swagger UI 页面在加载时请求的/swagger-resources 得到的就不再是一条数据，而是上述代码中根据 5 条路由信息整合后的 5 条 Swagger 资源了。更通俗一些理解，就是将 http://localhost:29000/v3/api-docs、http://localhost:29010/v3/api-docs、http://localhost:29020/v3/api-docs、http://localhost:29030/v3/api-docs、http://localhost:29040/v3/api-docs 这 5 个 URL 组装到 Swagger 资源列表中。

第四步，网关层过滤器对 Swagger 资源聚合请求放行。

由于网关层都做了全局拦截器，因此对获取具体的 Swagger 接口信息的请求要放行，否则这些请求无法正常读取到数据。

在 ValidMallUserTokenGlobalFilter 类下的 filter()方法中增加如下代码：

```
ignoreURLs.add("/indexConfigs/swagger/v3/api-docs");
ignoreURLs.add("/carts/swagger/v3/api-docs");
ignoreURLs.add("/orders/swagger/v3/api-docs");
ignoreURLs.add("/users/swagger/v3/api-docs");
ignoreURLs.add("/goods/swagger/v3/api-docs");
```

后台管理系统网关中聚合 Swagger 资源的编码步骤与此一致，这里不再赘述，读者可以直接下载本节源代码查看和学习。

最后来看一下实际效果。依次启动各个微服务架构项目及商城端网关，在浏览器中输入网关层 Swagger UI 页面的网址：http://localhost:29110/swagger-ui/index.html。

聚合后的接口文档页面如图 9-3 所示。

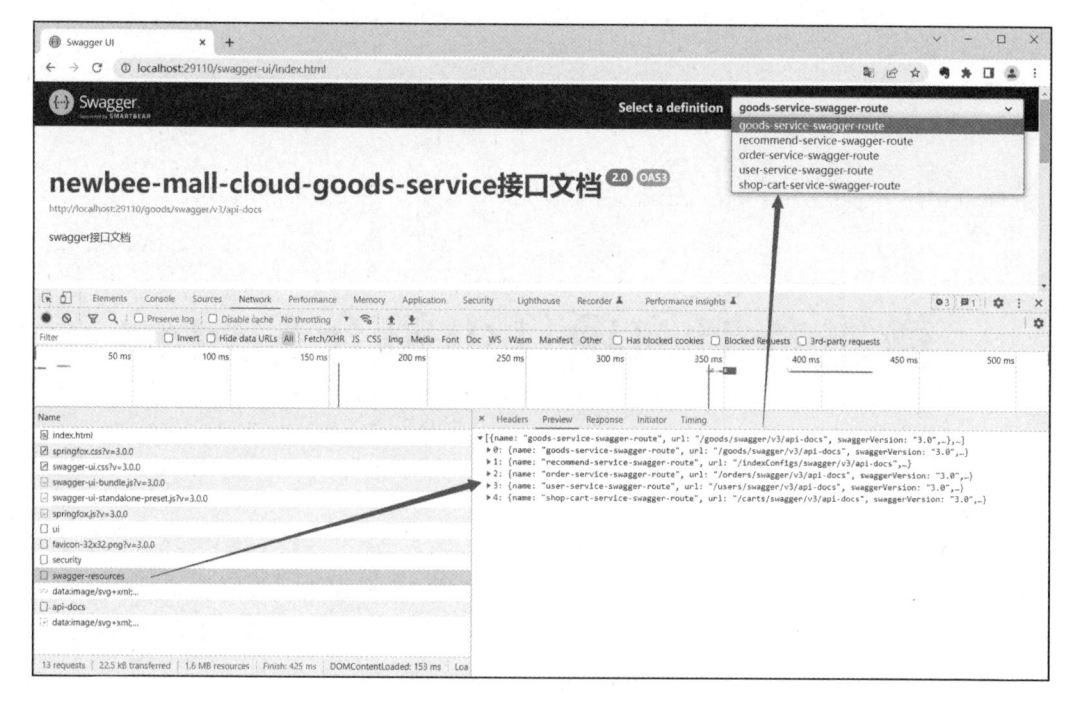

图 9-3 聚合后的接口文档页面

从图 9-3 中可以看出，Swagger UI 页面在加载时，对/swagger-resources 请求得到的不再是一条数据，而是上述代码中根据 5 条路由信息整合后的 5 条 Swagger 资源，与预期的效果一致，编码完成。

网关层聚合 Swagger 资源虽然不是一个特别复杂的知识点，却是企业开发中或真实的项目开发中不可缺少的一个步骤。在本章中，笔者介绍了聚合的原因、实现思路及具体的实现原理，目的是让读者更了解整个流程，而不是给几行代码让读者知道是怎么做的，却不知道为什么要这样编码。当然，本章还根据实战项目做了一些编码调整，并没有直接读取所有的路由配置来组装 Swagger 资源，毕竟不同的项目有不同的实现方式。

本章主要是在微服务架构项目中加入"网关层聚合 Swagger 资源"相关编码，对实战部分的讲解做补充和优化。希望读者能够根据笔者提供的开发步骤顺利地完成本章的项目改造。

第 10 章

微服务架构项目中整合 Seata

本章讲解在当前的微服务架构项目中整合 Seata，完成实战项目中所涉及的分布式事务处理。本章的源代码是在 newbee-mall-cloud-dev-step17 工程的基础上改造的，将工程命名为 newbee-mall-cloud-dev-step18。

10.1 实战项目中整合Seata编码实践

第一步，创建 undo_log 表。

在需要处理分布式事务的微服务实例下的数据库中创建 undo_log 表，实战项目中涉及分布式事务处理的微服务实例有订单微服务、购物车微服务和商品微服务，因此要在 newbee_mall_cloud_order_db 数据库、newbee_mall_cloud_cart_db 数据库和 newbee_mall_cloud_goods_db 数据库中依次创建 undo_log 表。

第二步，添加 Seata 依赖。

依次打开 newbee-mall-cloud-order-web、newbee-mall-cloud-shop-cart-web 和 newbee-mall-cloud-goods-web 3 个项目中的 pom.xml 文件，在 dependencies 节点下新增 Seata 的依赖项，配置代码如下：

```xml
<!-- Seata 依赖包 -->
<dependency>
    <groupId>com.alibaba.cloud</groupId>
    <artifactId>spring-cloud-starter-alibaba-seata</artifactId>
</dependency>
```

第三步,添加 Seata 配置项。

依次打开 newbee-mall-cloud-order-web、newbee-mall-cloud-shop-cart-web 和 newbee-mall-cloud-goods-web 3 个项目中的 application.properties 配置文件并进行修改,最终增加的配置项代码如下:

```
#seata config

seata.enabled=true
#3个不同的微服务被命名为不同的名称,如 goods-server、order-server、shopcart-server
seata.application-id=order-server
#事务分组配置
seata.tx-service-group=newbee_cloud_save_order_group
service.vgroupMapping.newbee_cloud_save_order_group=default

#连接Nacos服务中心的配置信息
seata.registry.type=nacos
seata.registry.nacos.application=seata-server
seata.registry.nacos.server-addr=127.0.0.1:8848
seata.registry.nacos.username=nacos
seata.registry.nacos.password=nacos
seata.registry.nacos.group=DEFAULT_GROUP
seata.registry.nacos.cluster=default
```

在 3 个项目的配置文件中依次添加上述配置代码即可,其他配置项可以不用配置,使用 Seata 的默认值即可。

第四步,添加 Seata 数据源代理。

依次打开 newbee-mall-cloud-order-web、newbee-mall-cloud-shop-cart-web 和 newbee-mall-cloud-goods-web 3 个项目中的 config 包,新增 SeataProxyConfiguration 类,代码如下:

```
import com.alibaba.druid.pool.DruidDataSource;
import io.seata.rm.datasource.DataSourceProxy;
import org.springframework.boot.context.properties.ConfigurationProperties;
import org.springframework.context.annotation.Bean;
import org.springframework.context.annotation.Configuration;
import org.springframework.context.annotation.Primary;

import javax.annotation.PostConstruct;
import javax.sql.DataSource;

@Configuration
public class SeataProxyConfiguration {
```

```java
//创建Druid数据源
@Bean
@ConfigurationProperties(prefix = "spring.datasource")
public DruidDataSource druidDataSource() {
    return new DruidDataSource();
}

//创建DataSource数据源代理
@Bean("dataSource")
@Primary
public DataSource dataSourceDelegation(DruidDataSource druidDataSource) {
    return new DataSourceProxy(druidDataSource);
}

/*
 * 解决druid日志报错：discard long time none received connection:xxx
 * */
@PostConstruct
public void setProperties(){
    System.setProperty("druid.mysql.usePingMethod","false");
}
}
```

第五步，添加@GlobalTransactional注解。

打开 newbee-mall-cloud-order-web 项目中的 ltd.order.cloud.newbee.service.impl.NewBeeMallOrderServiceImpl 类，在 saveOrder()方法上添加@GlobalTransactional 注解，代码修改如下：

```java
@Override
@Transactional
 //加上这个注解，开启Seata分布式事务
@GlobalTransactional
public String saveOrder(Long mallUserId, MallUserAddress address, List<Long> cartItemIds) {
省略部分代码
}
```

saveOrder()方法是一个涉及分布式事务的方法，在这个方法中会调用其他微服务来共同完成"下单"的流程，进而会操作 3 个独立的数据库。在这个方法上添加的@GlobalTransactional 注解是全局事务注解，作用是开启全局事务。当执行到 saveOrder()方法时，会自动开启全局事务。如果该方法中的代码逻辑都正常执行，则进行全局事务

的 Commit 操作；如果该方法中抛出异常，则进行 RollBack 操作。在购物车微服务和商品微服务的分支事务中不需要添加这个注解。

10.2 "分支事务不回滚"问题的复盘

10.2.1 发现问题

完成上述步骤后，依次启动 Nacos Server、Seata Server 及微服务实例，进行分布式事务处理的测试。然而验证的结果让笔者非常吃惊：分布式事务的处理并未成功。问题的具体表现为：在出现异常后，订单数据未生成，但是商品库存已扣减、购物车中的数据已删除。本来以为是自己看错了，但是经过笔者的多次测试，得到的结果都是如此，分支事务并未被正常处理。

笔者在工作笔记里查到了出现这个问题的确切时间，也想到了当天的情况。发现这个问题的时候，是 2022 年 4 月 29 日下午 4 点左右。

好的，发现问题后该怎么办呢？

10.2.2 尝试解决问题

在刚开始发现这个问题时，笔者并没有觉得这是一个大问题，认为自己可以轻轻松松地解决。当时并没有意识到严重性，也可以说是"轻敌"了。

当时主要觉得问题可能出现在配置和整合步骤上，于是做了如下 4 件事情。

① 检查整合步骤，看是不是漏掉了哪个环节，或者少了哪些配置，抑或哪个配置项因为粗心没配置好。

② 数据库中的表是否有问题。

③ 项目重启（遇事不决，重启试试）。

④ 查看日志，确认与 Seata 是否连接通信，是否正常注册 TM、RM 等。

因为觉得这是小问题，所以从 2022 年 4 月 29 日下午 4 点左右到 6 点，笔者一直在做上述的 4 件事情。结果就是，没发现步骤有问题，也没发现配置有问题，与 Seata Server 也正常通信，但是"分支事务不回滚"的问题依然在。笔者在这个时候其实已经隐隐觉得这可能不是一个小问题了，有点慌乱，但是检查了一段时间没头绪，脑子也乱了，索性就下班回家了。

本来想着第二天上班再处理这个问题，但是被这个问题搞得实在睡不着，那天晚上 11 点半打开电脑继续处理。处理的过程和下午一样，还是觉得可能是哪里不小心漏掉了步骤或配置不对，扩大了检查范围，除检查代码中的配置、数据库外，还检查了 Nacos Server 的配置、Seata Server 的配置，结果发现配置都没问题，也没有遗漏什么。

为什么笔者在发现这个问题时会觉得它是一个小问题，之后一直都在做检查配置项之类的事情？其原因就是之前已经整合过、配置过，而且验证通过，源代码也没问题，分布式事务的处理结果是正确的，就觉得只要整合步骤没有遗漏、配置项正确，分布式事务肯定会被正常处理。同样的代码、同样的配置、同样的测试环境，一个正常，另一个不正常，这的确有些出乎意料。然而，笔者尝试查询了几次问题后，就意识到问题的严重性了。确切地说，当时脑袋里一片空白，前一份整合代码中对分布式事务的处理完全正常，另一份整合代码则完全没反应，笔者有些慌乱了。

10.2.3 分析问题产生的原因

凌晨了，别犟了，换个思路吧！

冷静下来后，笔者觉得不能再骗自己了，这份代码肯定是有问题的，不然分支事务怎么会不回滚？但是确实不知道问题出在哪里，为什么同样的配置和步骤整合到 newbee-mall-cloud 实战项目里就不能用了呢？

然而，再去检查配置、对比代码已经没意义了。既然确定代码有问题，就根据 Seata 运行流程查一下哪里出了问题吧！具体做法是根据微服务实例的运行日志和 Seata Server 的运行日志来定位问题，最终的验证结果如下。

① 微服务实例与 Seata Server 通信正常。

② 3 个微服务实例都正常注册 TM、RM 等。

③ 全局事务能够正常开启。

④ 两个分支事务开启的日志一行都没出现。

不管是在微服务实例的运行日志中，还是在 Seata Server 的运行日志中，都没有看到两个分支事务的开启和处理。是的，没有任何信息和踪迹。再去数据库中确认，undo_log 表中也没有数据。虽然不知道哪里出了问题，但是至少有方向了。

全局事务能够正常开启和回滚，而两个分支事务不正常（与 Seata Server 正常通信，但是都没有生效）。到这里已经大致有了眉目，商品微服务和购物车微服务两个微服务实例和 Seata Server 的运行日志中都没有看到任何关于全局事务的信息，这也说明了两个分支事务可能根本就没有注册成功。全局事务正常开启和处理，而两个本应出现的分支事务没有出现，它们之间"失联"了。

10.2.4　查看源代码并确定问题所在

从代码层面来说，全局事务和分支事务的联系主要在一个变量上，这个变量就是全局事务的 ID——xid。现在它们"失联"了，只能通过这个变量的产生、传递、接收、处理等几个步骤来确认问题在哪里了。

此时，需要检查的内容就确定了下来，整理如下。

① 全局事务是否正常开启？xid 是否正确地生成了？

② xid 是否正确地传递给下游的调用实例？

③ 下游的调用实例是否正确地接收了 xid？

④ 接收 xid 后是否正确处理并开启了分支事务？

"问题不清晰，看源代码分析。"为了确认上述的几个检查内容，还是要用 debug 模式看一下 Seata 处理分布式事务过程中所涉及的源代码，由于涉及的源代码太多，因此笔者挑几个重要节点介绍一下。

对于全局事务是否正常开启、xid 是否正确地生成了，主要跟进了以下两个类的源代码：

① io.seata.spring.annotation.GlobalTransactionalInterceptor.java。

② io.seata.tm.api.TransactionalTemplate.java。

这两个类主要涉及全局事务的开启和处理，感兴趣的读者可以仔细地探索一下。当然，检查结果是这个步骤并没有问题，全局事务正常开启，xid 正确地生成。

难道是 xid 生成了却没有传递给下游？对于这个问题，笔者主要在 debug 模式下跟进了 com.alibaba.cloud.seata.feign.SeataFeignClient.java 这个类的源代码：

```
@Override
public Response execute(Request request, Request.Options options) throws
IOException {

    Request modifiedRequest = getModifyRequest(request);
    return this.delegate.execute(modifiedRequest, options);
}

private Request getModifyRequest(Request request) {
```

```
    String xid = RootContext.getXID();

    if (StringUtils.isEmpty(xid)) {
        return request;
    }

    Map<String, Collection<String>> headers = new HashMap<>(MAP_SIZE);
    headers.putAll(request.headers());

    List<String> seataXid = new ArrayList<>();
    seataXid.add(xid);
    // 把xid放入请求头中
    headers.put(RootContext.KEY_XID, seataXid);

    return Request.create(request.httpMethod(), request.url(), headers, request.body(),
                    request.charset(), null);
}
```

　　向下游微服务实例发送请求是由 SeataFeignClient 类来完成的，在这个类中会对 Request 对象进一步包装，把 xid 放进请求的 header 参数中并传递给下游方法，在 saverOrder()方法中使用 OpenFeign 调用商品微服务和购物车微服务中的方法前，会对 Request 对象做进一步的包装，再发起请求。当然，检查结果是这个步骤并没有问题，xid 被放入 header 参数中并传递给下游。

　　在查找问题的过程中，笔者还在购物车微服务的 deleteByCartItemIds()方法中添加了 request 参数，主要是为了查看该对象中是否有 xid 参数，代码如下：

```
@DeleteMapping("/shop-cart/deleteByCartItemIds")
@ApiOperation(value = "批量删除购物项", notes = "")
public Result<Boolean> deleteByCartItemIds(@RequestParam("cartItemIds")
List<Long> cartItemIds, HttpServletRequest request) {
  if (CollectionUtils.isEmpty(cartItemIds)) {
    return ResultGenerator.genFailResult("error param");
  }
  return
ResultGenerator.genSuccessResult(newBeeMallShoppingCartService.deleteCartItemsByCartIds(cartItemIds) > 0);
}
```

　　在 debug 模式下查看了 Request 对象中的内容，最终也确认了 header 参数中是有 xid 参数的，并进一步确认了上游微服务实例（订单微服务）正确地把 xid 传递给下游微服务，而且下游微服务实例也接收了 xid 参数，说明接收也没问题。

xid 的生产、传递、接收都没问题。到这里又卡住了，几个步骤好像都正常。笔者还是有些不敢相信这个结果，如果这些步骤都正常，全局事务和分支事务怎么会"失联"呢？

于是赶紧在代码中又加上了打印 RootContext.getXID()的语句，代码如下：

```
@DeleteMapping("/shop-cart/deleteByCartItemIds")
@ApiOperation(value = "批量删除购物项", notes = "")
public Result<Boolean> deleteByCartItemIds(@RequestParam("cartItemIds")
List<Long> cartItemIds, HttpServletRequest request) {
    // 通过 RootContext 获取 xid 字段
    System.out.println("RootContext.getXID()="+ RootContext.getXID());
    if (CollectionUtils.isEmpty(cartItemIds)) {
        return ResultGenerator.genFailResult("error param");
    }
    return ResultGenerator.genSuccessResult (newBeeMallShoppingCart
Service.deleteCartItemsByCartIds(cartItemIds) > 0);
}
```

如果正确接收到上游微服务实例传递的 xid，那么这个变量肯定不会有问题。重新启动项目并请求/saveOrder 验证整个分布式事务流程，打印 RootContext.getXID()的结果是 null，证明下游微服务实例确实没有正确地接收 xid。

为什么会这样呢？

此时，答案已经呼之欲出了。全局事务 ID——xid 正常地生成并正确地传递给下游微服务实例，看似成功地被下游微服务实例接收了，但是只是接收，并没有接收到。上游微服务实例传递了变量过去，下游微服务实例接收变量，但是没接收到。"没接收到"的意思就是到达下游微服务实例的请求中是有 xid 参数的，但是 xid 参数并没有被正常处理。xid 的传递在终点出现了问题，导致全局事务和分支事务"失联"了。

10.2.5 解决问题

下游微服务实例中 xid 参数接收和处理的类在哪里呢？在 com.alibaba.cloud.seata.web.SeataHandlerInterceptor 类中，源代码及注释如下：

```
public class SeataHandlerInterceptor implements HandlerInterceptor {
    private static final Logger log = LoggerFactory
            .getLogger(SeataHandlerInterceptor.class);
```

```java
    @Override
    public boolean preHandle(HttpServletRequest request, HttpServletResponse response,
            Object handler) {
    // 获取绑定后的 xid
        String xid = RootContext.getXID();
    // 获取请求头中的 xid
        String rpcXid = request.getHeader(RootContext.KEY_XID);
        if (log.isDebugEnabled()) {
            log.debug("xid in RootContext{}xid in RpcContext{}",xid,rpcXid);
        }
    // 如果未绑定
        if (StringUtils.isBlank(xid) && rpcXid != null) {
    // 则绑定 xid
            RootContext.bind(rpcXid);
            if (log.isDebugEnabled()) {
                log.debug("bind {} to RootContext", rpcXid);
            }
        }

        return true;
    }
省略部分代码
```

这个拦截器可以说是 xid 传递过程的终点，下游微服务实例会在这里接收请求头中的 xid 参数并进行绑定操作。如果这个拦截器中的方法正常运行，那么 xid 的传递就不会出问题，全局事务和分支事务也不会"失联"了。

在查找问题的过程中，笔者在这个拦截器的 preHandle() 方法中打了断点，发现在验证过程中根本没有进入这些断点，也就是说，这个拦截器根本没起作用。为什么这个拦截器没起作用呢？因为没有配置这个拦截器。newbee-mall-cloud-shop-cart-web 和 newbee-mall-cloud-goods-web 两个项目中分别定义 ShopCartServiceWebMvcConfigurer 和 GoodsServiceWebMvcConfigurer 两个类并继承了 WebMvcConfigurationSupport，如果一个拦截器要生效，就需要在这里进行配置。

解决办法就是在这两个项目中配置 SeataHandlerInterceptor 拦截器并使其生效，代码如下：

```java
public void addInterceptors(InterceptorRegistry registry) {
    registry.addInterceptor(new SeataHandlerInterceptor()).addPathPatterns
("/**");
}
```

好的，到这里，"分支事务不回滚"的问题就解决完了，一切都正常了。绕了那么一大圈，花费了那么多时间，分析了一堆源代码，结果仅仅是因为没有配置这个拦截器。

总体来说，这个问题围绕 Seata 分布式事务处理中"全局事务的开启与处理""xid 的生成与传递"两个知识点，如果不熟悉，建议读者看看这部分知识的文章和分析。

其实在真实的业务开发中，也有可能遇到这种情况。比如，写个简单的案例或小功能一切都正常，但是真正拿到企业开发的项目中，就出现了问题。毕竟写案例不会考虑太多，涉及的代码也少，能执行就行，而真实项目中有些是被忽略的或开发者不熟悉的配置，这都是需要注意的地方。

另外，扩展一下这个知识点。除"未配置 SeataHandlerInterceptor"会导致"分支事务不回滚"的问题外，全局事务失败的原因还有如下几种情形。

① 代码中的配置错误或配置项有遗漏，导致报错。处理办法：检查配置，修改正确即可。

② 数据源未被 Seata 代理，即未正确配置 io.seata.rm.datasource.DataSourceProxy 类。处理办法：修改代码，手动或自动配置 DataSourceProxy。

③ 依赖版本升级导致全局事务失效，笔者之前遇到过，在从 seata-spring-boot-starter 1.3.0 版本升级到 1.4.2 版本时，Seata 数据源自动配置逻辑的调整导致全局事务失败。处理办法：手动配置一下数据源代理。

本章主要讲解微服务架构项目中整合 Seata 完成分布式事务处理的相关编码过程，对实战部分的讲解做补充和优化，还对处理分布式事务代码时实际遇到的一个"坑"做了复盘，详细地记录了遇到问题后笔者的处理过程和思考过程，希望对读者有一些启发。

第 11 章

微服务架构项目中整合 Sentinel

本章讲解在微服务架构项目中整合 Sentinel 的过程,以及整合过程中遇到的问题和解决思路。本章的源代码是在 newbee-mall-cloud-dev-step18 工程的基础上改造的,将工程命名为 newbee-mall-cloud-dev-step20。

11.1 实战项目中整合Sentinel编码实践

在微服务架构项目中整合 Sentinel 的步骤如下。

第一步,引入 Sentinel 依赖。

依次打开 newbee-mall-cloud-user-web、newbee-mall-cloud-recommend-web、newbee-mall-cloud-order-web、newbee-mall-cloud-shop-cart-web 和 newbee-mall-cloud-goods-web 5 个微服务实例工程中的 pom.xml 文件,在 dependencies 节点下新增 Sentinel 的依赖项,配置代码如下:

```xml
<dependency>
    <groupId>com.alibaba.cloud</groupId>
    <artifactId>spring-cloud-starter-alibaba-sentinel</artifactId>
</dependency>
```

第二步,新增 Sentinel 配置项。

依次打开 newbee-mall-cloud-user-web、newbee-mall-cloud-recommend-web、newbee-mall-cloud-order-web、newbee-mall-cloud-shop-cart-web 和 newbee-mall-cloud-goods-web 5 个微服务实例工程中的 application.properties 配置文件，新增 Sentinel 配置项。

newbee-mall-cloud-user-web 工程：

```
# sentinel config
spring.cloud.sentinel.transport.port=8900
spring.cloud.sentinel.transport.clientIp=127.0.0.1
# 指定 Sentinel 控制台 IP 地址
spring.cloud.sentinel.transport.dashboard=127.0.0.1:9113
```

newbee-mall-cloud-recommend-web 工程：

```
# sentinel config
spring.cloud.sentinel.transport.port=8920
spring.cloud.sentinel.transport.clientIp=127.0.0.1
# 指定 Sentinel 控制台 IP 地址
spring.cloud.sentinel.transport.dashboard=127.0.0.1:9113
```

newbee-mall-cloud-order-web 工程：

```
# sentinel config
spring.cloud.sentinel.transport.port=8940
spring.cloud.sentinel.transport.clientIp=127.0.0.1
# 指定 Sentinel 控制台 IP 地址
spring.cloud.sentinel.transport.dashboard=127.0.0.1:9113
```

newbee-mall-cloud-shop-cart-web 工程：

```
# sentinel config
spring.cloud.sentinel.transport.port=8930
spring.cloud.sentinel.transport.clientIp=127.0.0.1
# 指定 Sentinel 控制台 IP 地址
spring.cloud.sentinel.transport.dashboard=127.0.0.1:9113
```

newbee-mall-cloud-goods-web 工程：

```
# sentinel config
spring.cloud.sentinel.transport.port=8910
spring.cloud.sentinel.transport.clientIp=127.0.0.1
# 指定 Sentinel 控制台 IP 地址
spring.cloud.sentinel.transport.dashboard=127.0.0.1:9113
```

到这里就完成了在实战项目中对 Sentinel 初步的整合和配置，非常简单。接下来，依次启动 5 个微服务实例（需保证 Nacos Server 和 Sentinel 控制台已经正常运行）。

启动后，登录 Sentinel 控制台，页面的导航栏依然是一片空白，无法看到这些微服务实例。本以为是懒加载机制导致的，只需要对微服务发起几次调用，触发服务信息等数据的上报，在客户端收集数据并传输给 Sentinel 控制台后，就会看到数据，但是访问了 5 个微服务实例中的一些请求后，页面中依然是空白的，如图 11-1 所示。

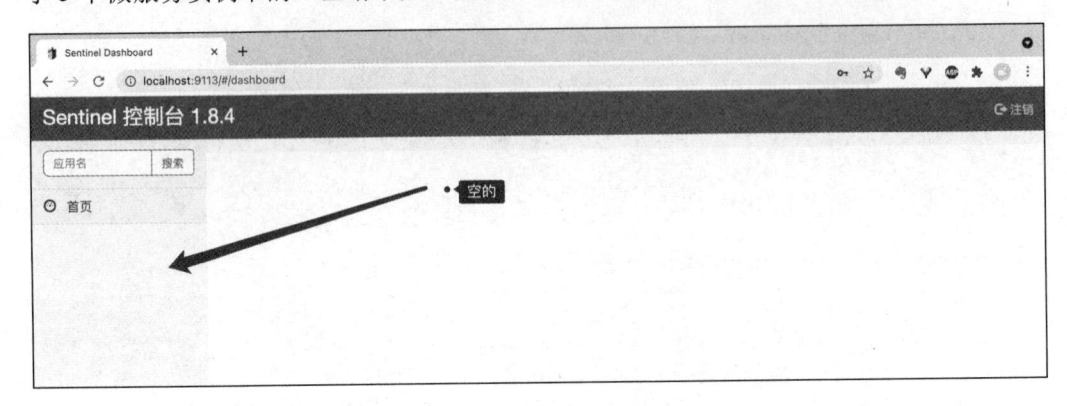

图 11-1 Sentinel 控制台页面空白

11.2 "Sentinel控制台页面中的微服务数据空白"问题的处理

整合步骤完成后，发现 Sentinel 控制台页面中的微服务数据空白，意外的问题出现了。写案例的时候正常，一放到具体的项目里就不能正常使用了。

紧接着，开始检查代码，看是不是配置有问题，结果一切正常。之后又刷新了 Maven 依赖，重启项目和 Sentinel Server，问题依然存在。

11.2.1 错误的解决思路

这个时候，笔者尝试了修改配置项，不使用懒加载的方式，在微服务实例启动时就将信息上报给 Sentinel 控制台，于是在配置文件中增加了如下配置项：

`spring.cloud.sentinel.eager=true`

再次重启项目，进入 Sentinel 控制台，页面左侧的微服务名称出现了，如图 11-2 所示。

第 11 章 微服务架构项目中整合 Sentinel

图 11-2　Sentinel 控制台页面左侧的微服务名称正常显示

微服务数据虽然出现了，但是每个微服务中"实时监控""簇点数据"页面中的数据都是空的，如图 11-3 所示。

图 11-3　微服务的监控数据显示为空

不管发起多少次请求，这些页面都空空如也。因此，微服务架构项目中整合 Sentinel 还是有问题，根本没有监控到任何数据。此时，笔者推测出两个问题：一是实例中的 Sentinel Client 与 Sentinel Server 间的通信有问题，二是实例中的 Sentinel Client 根本没有向 Sentinel Server 上报实例的监控信息。

在 Sentinel 控制台左侧的微服务列表中，单击"机器列表"选项，可以看到微服务实例的信息，而且微服务实例与 Sentinel Server 都部署在本地，也就是同一个网络环境中，通信肯定不存在问题。那么很大的可能就是 Sentinel Client 向 Sentinel Server 上报实例的监控信息步骤时出现了问题。

笔者查看实例的启动日志，发现了一条警告级别的日志（不是报错日志）：

```
Bean 'com.alibaba.cloud.sentinel.custom.SentinelAutoConfiguration' of type
[com.alibaba.cloud.sentinel.custom.SentinelAutoConfiguration] is not
eligible for getting processed by all BeanPostProcessors (for example: not
eligible for auto-proxying)
```

当时，笔者觉得这可能是个突破口，于是按照这条日志查询了相关的问题，并尝试着解决"Sentinel 控制台页面中微服务数据空白"的问题，但花了不少时间，尝试了好几种解决方案，都无功而返。

好了，以上就是笔者对这个问题的错误思路，花了不少时间，但是都没能处理好，这里就不再赘述了。

11.2.2 正确的解决思路

开发人员应该都有过"发现了问题却不知道如何下手才能解决"的无力感，整个人都会变得焦躁起来。

当感觉可能解决不了问题的时候，笔者忽然想到是不是自己的思路错了？犹记得之前整合 Seata 时也出现了类似的情况，最后发现问题就是没有配置拦截器而已。这次不会又如此吧！于是，笔者赶紧查了一下 spring-cloud-starter-alibaba-sentinel 的源代码，如图 11-4 所示，最终发现依然是拦截器失效的问题。

因为 newbee-mall-cloud-user-web、newbee-mall-cloud-recommend-web、newbee-mall-cloud-order-web、newbee-mall-cloud-shop-cart-web 和 newbee-mall-cloud-goods-web 5 个微服务实例工程中都自定义了 xxxWebMvcConfigurer，如果想要 Sentinel 中的这个拦截器生效，就需要分别在各个微服务实例中定义，代码修改如下：

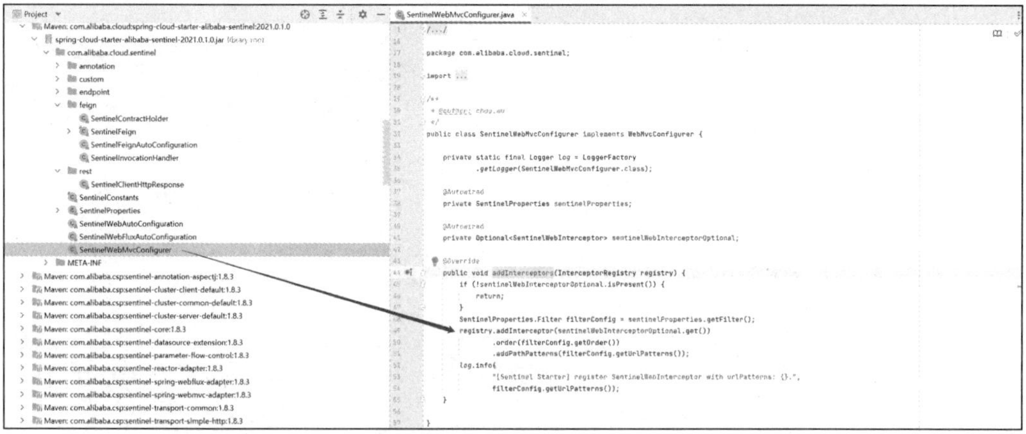

图 11-4 SentinelWebMvcConfigurer 代码截图

UserServiceWebMvcConfigurer 代码如下：

```
@Configuration
public class UserServiceWebMvcConfigurer extends WebMvcConfigurationSupport {

    private static final Logger log = LoggerFactory.getLogger
(UserServiceWebMvcConfigurer.class);

    @Autowired
    private SentinelProperties sentinelProperties;
    @Autowired
    private Optional<SentinelWebInterceptor> sentinelWebInterceptorOptional;

    @Autowired
    private TokenToAdminUserMethodArgumentResolver
tokenToAdminUserMethodArgumentResolver;

    @Autowired
    private TokenToMallUserMethodArgumentResolver tokenToMallUserMethod
ArgumentResolver;

    public void addArgumentResolvers(List<HandlerMethodArgumentResolver>
argumentResolvers) {
        argumentResolvers.add(tokenToAdminUserMethodArgumentResolver);
        argumentResolvers.add(tokenToMallUserMethodArgumentResolver);
    }

    @Override
```

```java
    public void addResourceHandlers(ResourceHandlerRegistry registry) {
        registry.
            addResourceHandler("/swagger-ui/**")
            .addResourceLocations("classpath:/META-INF/resources/webjars/springfox-swagger-ui/")
            .resourceChain(false);
    }

    public void addInterceptors(InterceptorRegistry registry) {
        // 配置Sentinel拦截器
        if (this.sentinelWebInterceptorOptional.isPresent()) {
            SentinelProperties.Filter filterConfig = this.sentinelProperties.getFilter();
            registry.addInterceptor((HandlerInterceptor) this.sentinelWebInterceptorOptional.get()).order(filterConfig.getOrder()).addPathPatterns(filterConfig.getUrlPatterns());
            log.info("[Sentinel Starter] register SentinelWebInterceptor with urlPatterns: {}.", filterConfig.getUrlPatterns());
        }
    }
}
```

RecommendServiceWebMvcConfigurer 代码如下：

```java
@Configuration
public class RecommendServiceWebMvcConfigurer extends WebMvcConfigurationSupport {

    private static final Logger log = LoggerFactory.getLogger(RecommendServiceWebMvcConfigurer.class);

    @Autowired
    private SentinelProperties sentinelProperties;
    @Autowired
    private Optional<SentinelWebInterceptor> sentinelWebInterceptorOptional;
    @Autowired
    private TokenToAdminUserMethodArgumentResolver tokenToAdminUserMethodArgumentResolver;

    public void addArgumentResolvers(List<HandlerMethodArgumentResolver> argumentResolvers) {
        argumentResolvers.add(tokenToAdminUserMethodArgumentResolver);
    }
```

```java
    @Override
    public void addResourceHandlers(ResourceHandlerRegistry registry) {
        registry.
                addResourceHandler("/swagger-ui/**")
                .addResourceLocations("classpath:/META-INF/resources/webjars/springfox-swagger-ui/")
                .resourceChain(false);
    }

    public void addInterceptors(InterceptorRegistry registry) {
        // 配置Sentinel拦截器
        if (this.sentinelWebInterceptorOptional.isPresent()) {
            SentinelProperties.Filter filterConfig = this.sentinelProperties.getFilter();
            registry.addInterceptor((HandlerInterceptor) this.sentinelWebInterceptorOptional.get()).order(filterConfig.getOrder()).addPathPatterns(filterConfig.getUrlPatterns());
            log.info("[Sentinel Starter] register SentinelWebInterceptor with urlPatterns: {}.", filterConfig.getUrlPatterns());
        }
    }
}
```

OrderServiceWebMvcConfigurer代码如下:

```java
@Configuration
public class OrderServiceWebMvcConfigurer extends WebMvcConfigurationSupport {

    private static final Logger log = LoggerFactory.getLogger(OrderServiceWebMvcConfigurer.class);

    @Autowired
    private SentinelProperties sentinelProperties;
    @Autowired
    private Optional<SentinelWebInterceptor> sentinelWebInterceptorOptional;

    @Autowired
    private TokenToAdminUserMethodArgumentResolver tokenToAdminUserMethodArgumentResolver;

    @Autowired
    private TokenToMallUserMethodArgumentResolver tokenToMallUserMethodArgumentResolver;
```

```java
    public void addArgumentResolvers(List<HandlerMethodArgumentResolver> argumentResolvers) {
        argumentResolvers.add(tokenToAdminUserMethodArgumentResolver);
        argumentResolvers.add(tokenToMallUserMethodArgumentResolver);
    }

    @Override
    public void addResourceHandlers(ResourceHandlerRegistry registry) {
        registry.
                addResourceHandler("/swagger-ui/**")
                .addResourceLocations("classpath:/META-INF/resources/webjars/springfox-swagger-ui/")
                .resourceChain(false);
    }

    public void addInterceptors(InterceptorRegistry registry) {
        // 配置Sentinel拦截器
        if (this.sentinelWebInterceptorOptional.isPresent()) {
            SentinelProperties.Filter filterConfig = this.sentinelProperties.getFilter();
            registry.addInterceptor((HandlerInterceptor) this.sentinelWebInterceptorOptional.get()).order(filterConfig.getOrder()).addPathPatterns(filterConfig.getUrlPatterns());
            log.info("[Sentinel Starter] register SentinelWebInterceptor with urlPatterns: {}.", filterConfig.getUrlPatterns());
        }
    }
}
```

ShopCartServiceWebMvcConfigurer 代码如下：

```java
@Configuration
public class ShopCartServiceWebMvcConfigurer extends WebMvcConfigurationSupport {

    private static final Logger log = LoggerFactory.getLogger(ShopCartServiceWebMvcConfigurer.class);

    @Autowired
    private SentinelProperties sentinelProperties;
    @Autowired
    private Optional<SentinelWebInterceptor> sentinelWebInterceptorOptional;
    @Autowired
    private TokenToMallUserMethodArgumentResolver tokenToMallUserMethodArgumentResolver;
```

```java
    public void addArgumentResolvers(List<HandlerMethodArgumentResolver> argumentResolvers) {
        argumentResolvers.add(tokenToMallUserMethodArgumentResolver);
    }

    @Override
    public void addResourceHandlers(ResourceHandlerRegistry registry) {
        registry.
                addResourceHandler("/swagger-ui/**")
                .addResourceLocations("classpath:/META-INF/resources/webjars/springfox-swagger-ui/")
                .resourceChain(false);
    }

    public void addInterceptors(InterceptorRegistry registry) {
        registry.addInterceptor(new SeataHandlerInterceptor()).addPathPatterns("/**");

        // 增加对Sentinel拦截器的配置
        if (this.sentinelWebInterceptorOptional.isPresent()) {
            SentinelProperties.Filter filterConfig = this.sentinelProperties.getFilter();
            registry.addInterceptor((HandlerInterceptor) this.sentinelWebInterceptorOptional.get()).order(filterConfig.getOrder()).addPathPatterns(filterConfig.getUrlPatterns());
            log.info("[Sentinel Starter] register SentinelWebInterceptor with urlPatterns: {}.", filterConfig.getUrlPatterns());
        }
    }
}
```

GoodsServiceWebMvcConfigurer 代码如下：

```java
@Configuration
public class GoodsServiceWebMvcConfigurer extends WebMvcConfigurationSupport {

    private static final Logger log = LoggerFactory.getLogger(GoodsServiceWebMvcConfigurer.class);

    @Autowired
    private SentinelProperties sentinelProperties;
    @Autowired
```

```java
    private Optional<SentinelWebInterceptor> sentinelWebInterceptorOptional;

    @Autowired
    private TokenToAdminUserMethodArgumentResolver tokenToAdminUserMethodArgumentResolver;

    @Autowired
    private TokenToMallUserMethodArgumentResolver tokenToMallUserMethodArgumentResolver;

    public void addArgumentResolvers(List<HandlerMethodArgumentResolver> argumentResolvers) {
        argumentResolvers.add(tokenToAdminUserMethodArgumentResolver);
        argumentResolvers.add(tokenToMallUserMethodArgumentResolver);
    }

    @Override
    public void addResourceHandlers(ResourceHandlerRegistry registry) {
        registry.
                addResourceHandler("/swagger-ui/**")
                .addResourceLocations("classpath:/META-INF/resources/webjars/springfox-swagger-ui/")
                .resourceChain(false);
    }

    public void addInterceptors(InterceptorRegistry registry) {
        registry.addInterceptor(new SeataHandlerInterceptor()).addPathPatterns("/**");

        // 增加对Sentinel拦截器的配置
        if (this.sentinelWebInterceptorOptional.isPresent()) {
            SentinelProperties.Filter filterConfig = this.sentinelProperties.getFilter();
            registry.addInterceptor((HandlerInterceptor) this.sentinelWebInterceptorOptional.get()).order(filterConfig.getOrder()).addPathPatterns(filterConfig.getUrlPatterns());
            log.info("[Sentinel Starter] register SentinelWebInterceptor with urlPatterns: {}.", filterConfig.getUrlPatterns());
        }
    }

}
```

修改完成后，重启所有的微服务实例。访问后，所有的服务信息都出现在 Sentinel Dashboard 左侧的列表中，并且"实例监控""簇点链路"等信息都正常显示，如图 11-5 所示。

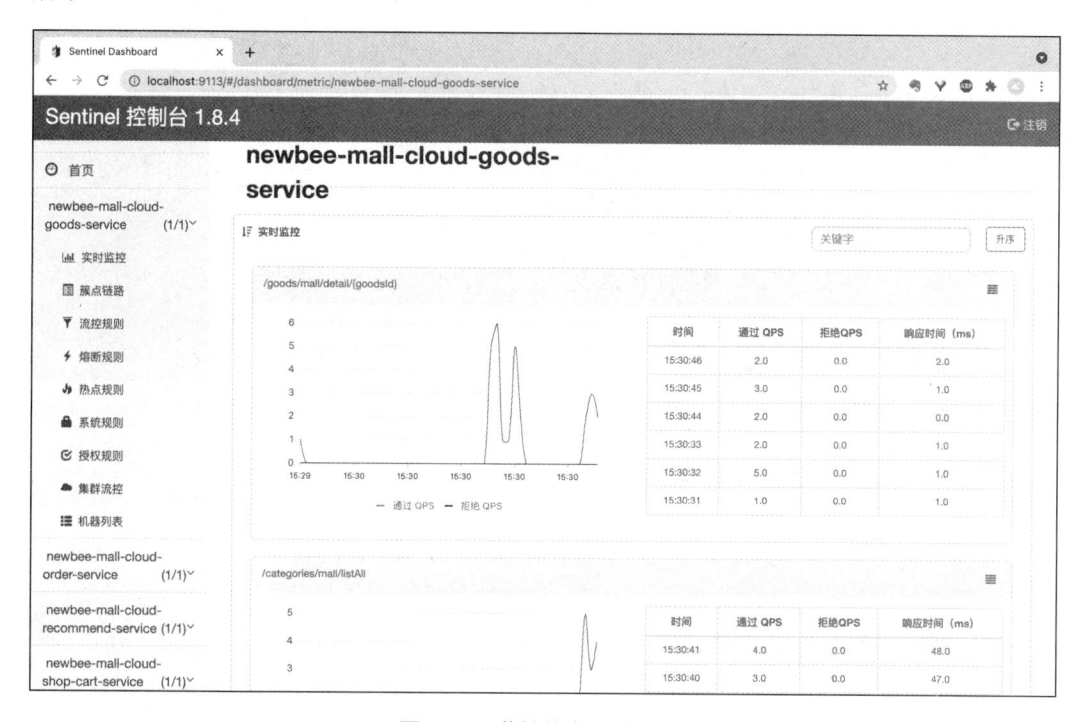

图 11-5　监控信息正常显示

"Sentinel Dashboard 页面中微服务数据空白"的问题解决了。到这里，才算真正地把 Sentinel 整合到微服务架构项目中。之后，就可以对一些资源进行限流配置及降级熔断配置了。

扩展一下这个知识点，除未配置 Sentinel 拦截器会出现"Sentinel Dashboard 页面中微服务数据空白"的问题外，还有如下几种情形也会出现该问题。

① 代码配置错误或配置项有遗漏，导致页面空白。解决办法：检查配置，主要是 IP 地址和端口号，若因为粗心漏掉了一些，修改正确即可。

② 默认的懒加载原因，导致页面空白。解决办法：发起几次请求就可以了。

③ 微服务实例与 Sentinel Dashboard 间的网络不通。比如，一个在内网，另一个在公网，或者由于防火墙原因导致二者不能正常通信。解决办法：保证网络畅通，都部署到内网或都部署到公网，抑或使用内网穿透技术。

④ 在使用 Docker 或 Kubernetes 等容器化技术部署时导致的网络不通，和问题③类似。

网上有很多这个问题的解决办法，不过大部分都是关于上述 4 种情形的。笔者并未搜索到未配置 Sentinel 拦截器导致的页面空白问题，因此导致在错误的思路上花费了很多的时间。

本章主要讲解微服务架构项目中整合 Sentinel 的相关编码过程，对实战部分的讲解做补充和优化。之后对整合 Sentinel 时遇到的一个"坑"做了复盘，详细地记录了遇到问题后笔者的处理过程和思考过程，希望对读者有一些启发。

第 12 章

微服务架构项目中整合 Seluth、Zipkin

本章讲解在微服务架构项目中整合 Sleuth、Zipkin，本章的源代码是在 newbee-mall-cloud-dev-step20 工程的基础上改造的，将工程命名为 newbee-mall-cloud-dev-step21。

12.1 整合Sleuth编码实践

在微服务架构项目中整合 Sleuth 的具体步骤如下。

第一步，开启 OpenFeign 的日志输出功能。

为了增强后续功能演示的效果，需要开启 OpenFeign 的日志输出功能，这样就能够把 OpenFeign 远程调用接口的日志内容打印出来。

在 newbee-mall-cloud-recommend-web、newbee-mall-cloud-order-web、newbee-mall-cloud-shop-cart-web 和 newbee-mall-cloud-goods-web 4 个微服务实例工程中的 config 包下新增 OpenFeignConfiguration 类，用于设置 OpenFeign 的日志级别，代码如下：

```
@Configuration
public class OpenFeignConfiguration {
    @Bean
    public Logger.Level openFeignLogLevel() {
        // 设置OpenFeign的日志级别
        return Logger.Level.FULL;
```

```
    }
}
```

因为用户微服务并不会通过 OpenFeign 调用其他微服务，所以这里不需要在 newbee-mall-cloud-user-web 工程中添加这个配置类。

第二步，在配置文件中增加关于日志输出的配置。

修改 newbee-mall-cloud-user-web、newbee-mall-cloud-recommend-web、newbee-mall-cloud-order-web、newbee-mall-cloud-shop-cart-web 和 newbee-mall-cloud-goods-web 5 个微服务实例工程中的 application.properties 文件，设置日志输出的相关配置，分别新增如下配置项。

newbee-mall-cloud-user-web：

```
# 演示需要，开启当前项目中的 debug 级别日志
logging.level.ltd.user.cloud.newbee=debug
```

日志输出级别为 debug，即 debug、info 和 error 级别的日志都会被输出。

newbee-mall-cloud-recommend-web：

```
# 演示需要，开启 OpenFeign 和当前项目中的 debug 级别日志
logging.level.ltd.user.cloud.newbee.openfeign=debug
logging.level.ltd.goods.cloud.newbee.openfeign=debug
logging.level.ltd.recommend.cloud.newbee=debug
```

由于推荐微服务会通过 OpenFeign 调用用户微服务和商品微服务，因此不仅设置了推荐微服务工程包目录下的日志输出级别，还设置了 OpenFeign 的日志输出级别，与 OpenFeign 相关的日志也会被输出。

另外，读者需要注意这里的包名配置。logging.level.ltd.recommend.cloud.newbee=debug 配置的是推荐微服务工程下的包，其他两行配置项分别是用户微服务和商品微服务 OpenFeign 包下的配置，因为推荐微服务在 pom.xml 文件中引入了相关的 api 依赖，所以要想输出推荐微服务通过 OpenFeign 调用商品微服务和用户微服务时的日志，这两行日志就是必须加上的，注意包名设置分别是用户微服务和商品微服务两个工程下的包名配置 ltd.user.cloud.newbee.openfeign 和 ltd.goods.cloud.newbee.openfeign，而不是推荐微服务工程下的包名配置 ltd.recommend.cloud.newbee。

newbee-mall-cloud-order-web：

```
# 演示需要，开启 OpenFeign 和当前项目中的 debug 级别日志
logging.level.ltd.user.cloud.newbee.openfeign=debug
logging.level.ltd.goods.cloud.newbee.openfeign=debug
logging.level.ltd.shopcart.cloud.newbee.openfeign=debug
logging.level.ltd.order.cloud.newbee=debug
```

由于订单微服务会通过 OpenFeign 调用购物车微服务、用户微服务和商品微服务，因此不仅设置了订单微服务工程包目录下的日志输出级别，还设置了 OpenFeign 的日志输出级别，与 OpenFeign 相关的日志也会被输出。

另外，读者需要注意这里的包名配置。logging.level.ltd.order.cloud.newbee=debug 配置的是订单微服务工程下的包，其他 3 行配置项分别是用户微服务、商品微服务和购物车微服务 OpenFeign 包下的配置，因为订单微服务在 pom.xml 文件中引入了相关的 api 依赖，所以要想输出订单微服务通过 OpenFeign 调用购物车微服务、商品微服务和用户微服务时的日志，这 3 行日志就是必须加上的，注意包名设置分别是购物车微服务、用户微服务和商品微服务 3 个工程下的包名配置，即 logging.level.ltd.shopcart.cloud.newbee.openfeign、ltd.user.cloud.newbee.openfeign 和 ltd.goods.cloud.newbee.openfeign，而不是订单微服务工程下的包名配置 ltd.order.cloud.newbee。

newbee-mall-cloud-goods-web：

```
# 演示需要，开启 OpenFeign 和当前项目中的 debug 级别日志
logging.level.ltd.user.cloud.newbee.openfeign=debug
logging.level.ltd.goods.cloud.newbee=debug
```

由于推荐微服务会通过 OpenFeign 调用用户微服务，因此不仅设置了商品微服务工程包目录下的日志输出级别，还设置了 OpenFeign 的日志输出级别，与 OpenFeign 相关的日志也会被输出，读者需要注意这里的包名设置。

newbee-mall-cloud-shop-cart-web：

```
# 演示需要，开启 OpenFeign 和当前项目中的 debug 级别日志
logging.level.ltd.user.cloud.newbee.openfeign=debug
logging.level.ltd.goods.cloud.newbee.openfeign=debug
logging.level.ltd.shopcart.cloud.newbee=debug
```

由于购物车微服务会通过 OpenFeign 调用用户微服务和商品微服务，因此不仅设置了购物车微服务工程包目录下的日志输出级别，还设置了 OpenFeign 的日志输出级别，与 OpenFeign 相关的日志也会被输出，读者需要注意这里的包名设置。

第三步，引入 Sleuth 依赖。

依次打开 newbee-mall-cloud-user-web、newbee-mall-cloud-recommend-web、newbee-mall-cloud-order-web、newbee-mall-cloud-shop-cart-web 和 newbee-mall-cloud-goods-web 5 个微服务实例工程中的 pom.xml 文件，在 dependencies 节点下新增 Spring Cloud Sleuth 的依赖项，配置如下：

```
<!-- Sleuth 依赖项 -->
<dependency>
  <groupId>org.springframework.cloud</groupId>
```

```xml
<artifactId>spring-cloud-starter-sleuth</artifactId>
</dependency>
```

之后，启动 5 个微服务实例工程，此时控制台上输出的日志信息就变了。在这里，笔者通过推荐微服务的 Swagger UI 页面调用了首页接口，控制台中输出的日志内容如下，包括基础的日志信息、MyBatis 执行的日志信息及 OpenFeign 调用时的日志信息。

推荐微服务实例中输出的日志信息：

```
2023-07-16 01:46:06.308  INFO [newbee-mall-cloud-recommend-service,d402ccf7f2e520b5,d402ccf7f2e520b5] 12688 --- [io-29020-exec-5] com.zaxxer.hikari.HikariDataSource       : hikariCP - Starting...
2023-07-16 01:46:06.612  INFO [newbee-mall-cloud-recommend-service,d402ccf7f2e520b5,d402ccf7f2e520b5] 12688 --- [io-29020-exec-5] com.zaxxer.hikari.HikariDataSource       : hikariCP - Start completed.
2023-07-16 01:46:06.617 DEBUG [newbee-mall-cloud-recommend-service,d402ccf7f2e520b5,d402ccf7f2e520b5] 12688 --- [io-29020-exec-5] l.r.c.n.d.C.findCarouselsByNum           : ==>  Preparing: select carousel_id, carousel_url, redirect_url, carousel_rank, is_deleted, create_time, create_user, update_time, update_user from tb_newbee_mall_carousel where is_deleted = 0 order by carousel_rank desc limit ?
2023-07-16 01:46:06.633 DEBUG [newbee-mall-cloud-recommend-service,d402ccf7f2e520b5,d402ccf7f2e520b5] 12688 --- [io-29020-exec-5] l.r.c.n.d.C.findCarouselsByNum           : ==> Parameters: 5(Integer)
2023-07-16 01:46:06.661 DEBUG [newbee-mall-cloud-recommend-service,d402ccf7f2e520b5,d402ccf7f2e520b5] 12688 --- [io-29020-exec-5] l.r.c.n.d.C.findCarouselsByNum           : <==      Total: 2
2023-07-16 01:46:06.672 DEBUG [newbee-mall-cloud-recommend-service,d402ccf7f2e520b5,d402ccf7f2e520b5] 12688 --- [io-29020-exec-5] l.r.c.n.d.I.findIndexConfigsByTypeAndNum : ==>  Preparing: select config_id, config_name, config_type, goods_id, redirect_url, config_rank, is_deleted, create_time, create_user, update_time, update_user from tb_newbee_mall_index_config where config_type = ? and is_deleted = 0 order by config_rank desc limit ?
2023-07-16 01:46:06.673 DEBUG [newbee-mall-cloud-recommend-service,d402ccf7f2e520b5,d402ccf7f2e520b5] 12688 --- [io-29020-exec-5] l.r.c.n.d.I.findIndexConfigsByTypeAndNum : ==> Parameters: 3(Integer), 4(Integer)
2023-07-16 01:46:06.691 DEBUG [newbee-mall-cloud-recommend-service,d402ccf7f2e520b5,d402ccf7f2e520b5] 12688 --- [io-29020-exec-5] l.r.c.n.d.I.findIndexConfigsByTypeAndNum : <==      Total: 4
2023-07-16 01:46:06.694 DEBUG [newbee-mall-cloud-recommend-service,d402ccf7f2e520b5,d402ccf7f2e520b5] 12688 --- [io-29020-exec-5] l.g.c.n.o.NewBeeCloudGoodsServiceFeign   : [NewBeeCloudGoodsServiceFeign#listByGoodsIds] ---> GET
```

```
http://newbee-ma**-cloud-goods-service/goods/admin/listByGoodsIds?goodsI
ds=10918&goodsIds=10908&goodsIds=10906&goodsIds=10902 HTTP/1.1
2023-07-16 01:46:06.694 DEBUG [newbee-mall-cloud-recommend-service,
d402ccf7f2e520b5,d402ccf7f2e520b5] 12688 --- [io-29020-exec-5]
l.g.c.n.o.NewBeeCloudGoodsServiceFeign   :
[NewBeeCloudGoodsServiceFeign#listByGoodsIds] Accept-Encoding: gzip
2023-07-16 01:46:06.694 DEBUG [newbee-mall-cloud-recommend-service,
d402ccf7f2e520b5,d402ccf7f2e520b5] 12688 --- [io-29020-exec-5]
l.g.c.n.o.NewBeeCloudGoodsServiceFeign   :
[NewBeeCloudGoodsServiceFeign#listByGoodsIds] Accept-Encoding: deflate
2023-07-16 01:46:06.694 DEBUG [newbee-mall-cloud-recommend-service,
d402ccf7f2e520b5,d402ccf7f2e520b5] 12688 --- [io-29020-exec-5]
l.g.c.n.o.NewBeeCloudGoodsServiceFeign   :
[NewBeeCloudGoodsServiceFeign#listByGoodsIds] ---> END HTTP (0-byte body)
2023-07-16 01:46:07.116 DEBUG [newbee-mall-cloud-recommend-service,
d402ccf7f2e520b5,d402ccf7f2e520b5] 12688 --- [io-29020-exec-5]
l.g.c.n.o.NewBeeCloudGoodsServiceFeign   :
[NewBeeCloudGoodsServiceFeign#listByGoodsIds] <--- HTTP/1.1 200 (421ms)
2023-07-16 01:46:07.117 DEBUG [newbee-mall-cloud-recommend-service,
d402ccf7f2e520b5,d402ccf7f2e520b5] 12688 --- [io-29020-exec-5]
l.g.c.n.o.NewBeeCloudGoodsServiceFeign   :
[NewBeeCloudGoodsServiceFeign#listByGoodsIds] connection: keep-alive
2023-07-16 01:46:07.117 DEBUG [newbee-mall-cloud-recommend-service,
d402ccf7f2e520b5,d402ccf7f2e520b5] 12688 --- [io-29020-exec-5]
l.g.c.n.o.NewBeeCloudGoodsServiceFeign   :
[NewBeeCloudGoodsServiceFeign#listByGoodsIds] content-type:
application/json
2023-07-16 01:46:07.117 DEBUG [newbee-mall-cloud-recommend-service,
d402ccf7f2e520b5,d402ccf7f2e520b5] 12688 --- [io-29020-exec-5]
l.g.c.n.o.NewBeeCloudGoodsServiceFeign   :
[NewBeeCloudGoodsServiceFeign#listByGoodsIds] date: Thu, 01 Dec 2022
14:46:07 GMT
2023-07-16 01:46:07.117 DEBUG [newbee-mall-cloud-recommend-service,
d402ccf7f2e520b5,d402ccf7f2e520b5] 12688 --- [io-29020-exec-5]
l.g.c.n.o.NewBeeCloudGoodsServiceFeign   :
[NewBeeCloudGoodsServiceFeign#listByGoodsIds] keep-alive: timeout=60
2023-07-16 01:46:07.117 DEBUG [newbee-mall-cloud-recommend-service,
d402ccf7f2e520b5,d402ccf7f2e520b5] 12688 --- [io-29020-exec-5]
l.g.c.n.o.NewBeeCloudGoodsServiceFeign   :
[NewBeeCloudGoodsServiceFeign#listByGoodsIds] transfer-encoding: chunked
2023-07-16 01:46:07.117 DEBUG [newbee-mall-cloud-recommend-service,
d402ccf7f2e520b5,d402ccf7f2e520b5] 12688 --- [io-29020-exec-5]
l.g.c.n.o.NewBeeCloudGoodsServiceFeign   :
```

[NewBeeCloudGoodsServiceFeign#listByGoodsIds]
2023-07-16 01:46:07.125 DEBUG [newbee-mall-cloud-recommend-service,
d402ccf7f2e520b5,d402ccf7f2e520b5] 12688 --- [io-29020-exec-5]
l.g.c.n.o.NewBeeCloudGoodsServiceFeign   :
[NewBeeCloudGoodsServiceFeign#listByGoodsIds]
{"resultCode":200,"message":"SUCCESS","data":[{"goodsId":10918,"goodsNam
e":"Apple AirPods（第三代）","goodsIntro":"AirPods 第三代搭载空间音频和自适应均衡
全新登场！更长续航\\无线充电\\抗汗抗水等更多功能提供美妙体验！","goodsCategoryId":33,
"goodsCoverImg":"https://newbee-ma**.oss-cn-beijing.aliyuncs.com/images/
MME73_AV4_GEO_CN.jpeg","goodsCarousel":"https://newbee-ma**.oss-cn-beiji
ng.aliyuncs.com/images/MME73_AV4_GEO_CN.jpeg","originalPrice":1599,"sell
ingPrice":1399,"stockNum":10000,"tag":"美妙新声","goodsSellStatus":0,
"createUser":0, "createTime":"2021-12-28 14:04:07","updateUser":0,
"updateTime":"2021-12-28 18:58:35","goodsDetailContent":null},{"goodsId":
10908,"goodsName":"HUAWEI Mate 40 Pro 全网通5G手机 8GB+512GB（黄色）",
"goodsIntro":"5nm 麒麟9000旗舰芯片 | 超感光徕卡电影影象","goodsCategoryId":
46,"goodsCoverImg": "https://newbee-ma**.oss-cn-beijing.aliyuncs.com/
images/mate40-white.png","goodsCarousel":"https://newbee-ma**.oss-cn-bei
jing.aliyuncs.com/images/mate40-white.png","originalPrice":6488,"selling
Price":6488,"stockNum":9999,"tag":"跃见非凡","goodsSellStatus":0,
"createUser":0,"createTime": "2020-10-22 22:08:42","updateUser":0,
"updateTime":"2021-04-20 21:08:32", "goodsDetailContent":null},
{"goodsId":10906,"goodsName":"Apple iPhone12 Pro (A2408) 128GB 海蓝色 支持移
动联通电信5G 双卡双待手机","goodsIntro":"A14 仿生芯片，6.1英寸超视网膜XDR显示屏，
激光雷达扫描仪，超瓷晶面板，现实力登场！","goodsCategoryId":47,
"goodsCoverImg":"https://newbee-ma**.oss-cn-beijing.aliyuncs.com/images/
iphone-12-pro-blue-hero.png","goodsCarousel":"https://newbee-ma**.oss-cn
-beijing.aliyuncs.com/images/iphone-12-pro-blue-hero.png","originalPrice
":8499,"sellingPrice":8499,"stockNum":9999,"tag":"自我再飞跃",
"goodsSellStatus":0,"createUser":0,"createTime":"2020-10-14
10:32:55","updateUser":0,"updateTime":"2020-10-16 17:13:43",
"goodsDetailContent":null},{"goodsId":10902,"goodsName":"华为 HUAWEI P40
冰霜银 全网通5G手机","goodsIntro":"麒麟990 5G SoC芯片 5000万超感知徕卡三摄 30
倍数字变焦 6GB+128GB","goodsCategoryId":46,"goodsCoverImg":
"https://newbee-ma**.oss-cn-beijing.aliyuncs.com/images/p40-silver.png",
"goodsCarousel":"https://newbee-ma**.oss-cn-beijing.aliyuncs.com/images/
p40-silver.png","originalPrice":4399,"sellingPrice":4299,"stockNum":10000,
"tag":"超感知影像","goodsSellStatus":0,"createUser":0,"createTime":
"2020-03-27 10:07:37","updateUser":0,"updateTime":"2020-05-15
17:18:30","goodsDetailContent":null}]}
2023-07-16 01:46:07.125 DEBUG [newbee-mall-cloud-recommend-service,
d402ccf7f2e520b5,d402ccf7f2e520b5] 12688 --- [io-29020-exec-5]
l.g.c.n.o.NewBeeCloudGoodsServiceFeign   :

```
[NewBeeCloudGoodsServiceFeign#listByGoodsIds] <--- END HTTP (2582-byte body)
2023-07-16 01:46:07.140 DEBUG [newbee-mall-cloud-recommend-service,
d402ccf7f2e520b5,d402ccf7f2e520b5] 12688 --- [io-29020-exec-5]
l.r.c.n.d.I.findIndexConfigsByTypeAndNum  : ==>  Preparing: select config_id,
config_name, config_type, goods_id, redirect_url, config_rank, is_deleted,
create_time, create_user, update_time, update_user from
tb_newbee_mall_index_config where config_type = ? and is_deleted = 0 order
by config_rank desc limit ?
2023-07-16 01:46:07.140 DEBUG [newbee-mall-cloud-recommend-service,
d402ccf7f2e520b5,d402ccf7f2e520b5] 12688 --- [io-29020-exec-5]
l.r.c.n.d.I.findIndexConfigsByTypeAndNum  : ==> Parameters: 4(Integer),
6(Integer)
2023-07-16 01:46:07.159 DEBUG [newbee-mall-cloud-recommend-service,
d402ccf7f2e520b5,d402ccf7f2e520b5] 12688 --- [io-29020-exec-5]
l.r.c.n.d.I.findIndexConfigsByTypeAndNum  : <==      Total: 6
2023-07-16 01:46:07.160 DEBUG [newbee-mall-cloud-recommend-service,
d402ccf7f2e520b5,d402ccf7f2e520b5] 12688 --- [io-29020-exec-5]
l.g.c.n.o.NewBeeCloudGoodsServiceFeign   :
[NewBeeCloudGoodsServiceFeign#listByGoodsIds] ---> GET
http://newbee-ma**-cloud-goods-service/goods/admin/listByGoodsIds?goodsI
ds=10925&goodsIds=10926&goodsIds=10915&goodsIds=10920&goodsIds=10921&goo
dsIds=10919 HTTP/1.1
2023-07-16 01:46:07.160 DEBUG [newbee-mall-cloud-recommend-service,
d402ccf7f2e520b5,d402ccf7f2e520b5] 12688 --- [io-29020-exec-5]
l.g.c.n.o.NewBeeCloudGoodsServiceFeign   :
[NewBeeCloudGoodsServiceFeign#listByGoodsIds] Accept-Encoding: gzip
2023-07-16 01:46:07.160 DEBUG [newbee-mall-cloud-recommend-service,
d402ccf7f2e520b5,d402ccf7f2e520b5] 12688 --- [io-29020-exec-5]
l.g.c.n.o.NewBeeCloudGoodsServiceFeign   :
[NewBeeCloudGoodsServiceFeign#listByGoodsIds] Accept-Encoding: deflate
2023-07-16 01:46:07.160 DEBUG [newbee-mall-cloud-recommend-service,
d402ccf7f2e520b5,d402ccf7f2e520b5] 12688 --- [io-29020-exec-5]
l.g.c.n.o.NewBeeCloudGoodsServiceFeign   :
[NewBeeCloudGoodsServiceFeign#listByGoodsIds] ---> END HTTP (0-byte body)
2023-07-16 01:46:07.183 DEBUG [newbee-mall-cloud-recommend-service,
d402ccf7f2e520b5,d402ccf7f2e520b5] 12688 --- [io-29020-exec-5]
l.g.c.n.o.NewBeeCloudGoodsServiceFeign   :
[NewBeeCloudGoodsServiceFeign#listByGoodsIds] <--- HTTP/1.1 200 (22ms)
2023-07-16 01:46:07.183 DEBUG [newbee-mall-cloud-recommend-service,
d402ccf7f2e520b5,d402ccf7f2e520b5] 12688 --- [io-29020-exec-5]
l.g.c.n.o.NewBeeCloudGoodsServiceFeign   :
[NewBeeCloudGoodsServiceFeign#listByGoodsIds] connection: keep-alive
2023-07-16 01:46:07.183 DEBUG [newbee-mall-cloud-recommend-service,
```

```
d402ccf7f2e520b5,d402ccf7f2e520b5] 12688 --- [io-29020-exec-5]
l.g.c.n.o.NewBeeCloudGoodsServiceFeign      :
[NewBeeCloudGoodsServiceFeign#listByGoodsIds] content-type:
application/json
2023-07-16 01:46:07.183 DEBUG [newbee-mall-cloud-recommend-service,
d402ccf7f2e520b5,d402ccf7f2e520b5] 12688 --- [io-29020-exec-5]
l.g.c.n.o.NewBeeCloudGoodsServiceFeign      :
[NewBeeCloudGoodsServiceFeign#listByGoodsIds] date: Thu, 01 Dec 2022
14:46:07 GMT
2023-07-16 01:46:07.183 DEBUG [newbee-mall-cloud-recommend-service,
d402ccf7f2e520b5,d402ccf7f2e520b5] 12688 --- [io-29020-exec-5]
l.g.c.n.o.NewBeeCloudGoodsServiceFeign      :
[NewBeeCloudGoodsServiceFeign#listByGoodsIds] keep-alive: timeout=60
2023-07-16 01:46:07.183 DEBUG [newbee-mall-cloud-recommend-service,
d402ccf7f2e520b5,d402ccf7f2e520b5] 12688 --- [io-29020-exec-5]
l.g.c.n.o.NewBeeCloudGoodsServiceFeign      :
[NewBeeCloudGoodsServiceFeign#listByGoodsIds] transfer-encoding: chunked
2023-07-16 01:46:07.183 DEBUG [newbee-mall-cloud-recommend-service,
d402ccf7f2e520b5,d402ccf7f2e520b5] 12688 --- [io-29020-exec-5]
l.g.c.n.o.NewBeeCloudGoodsServiceFeign      :
[NewBeeCloudGoodsServiceFeign#listByGoodsIds]
2023-07-16 01:46:07.183 DEBUG [newbee-mall-cloud-recommend-service,
d402ccf7f2e520b5,d402ccf7f2e520b5] 12688 --- [io-29020-exec-5]
l.g.c.n.o.NewBeeCloudGoodsServiceFeign      :
[NewBeeCloudGoodsServiceFeign#listByGoodsIds] {"resultCode":200,"message":
"SUCCESS","data":[{"goodsId":10925,"goodsName":"HUAWEI P50 Pocket 4G 全网通
超光谱影像系统 创新双屏操作体验 P50 宝盒 12GB+512GB 鎏光金华为折叠屏手机",
"goodsIntro":"华为 P50Pocket 新品, 华为年度旗舰折叠屏手机",
"goodsCategoryId":46,"goodsCoverImg":"https://newbee-ma**.oss-cn-beijing
.aliyuncs.com/images/p50-pocket-gold.png","goodsCarousel":"https://newbe
e-ma**.oss-cn-beijing.aliyuncs.com/images/p50-pocket-gold.png","original
Price":12988,"sellingPrice":10988,"stockNum":9998,"tag":"折叠万象",
"goodsSellStatus":0,"createUser":0,"createTime":"2021-12-28 16:02:50",
"updateUser":0,"updateTime":"2021-12-28 16:02:50",
"goodsDetailContent":null},{"goodsId":10926,"goodsName":"华为笔记本电脑
MateBook X Pro 2022","goodsIntro":"原色全面屏, 轻薄高能, 超级终端",
"goodsCategoryId":33,"goodsCoverImg":"https://newbee-ma**.oss-cn-beijing
.aliyuncs.com/images/MateBook%20X%20Pro.png","goodsCarousel":"https://ne
wbee-ma**.oss-cn-beijing.aliyuncs.com/images/MateBook%20X%20Pro.png","or
iginalPrice":11699,"sellingPrice":10499,"stockNum":9997,"tag":"入目惊鸿",
"goodsSellStatus":0,"createUser":0,"createTime":"2021-12-28 17:31:30",
"updateUser":0,"updateTime":"2021-12-28 17:46:17",
"goodsDetailContent":null},{"goodsId":10915,"goodsName":"Apple iPhone 13
```

(A2634) 256GB 粉色","goodsIntro":"解锁超能力！超先进双摄系统，超强耐用性，超劲续航大提升！","goodsCategoryId":47,"goodsCoverImg":"https://newbee-mall.oss-cn-beijing.aliyuncs.com/images/iphone-13-pink-select-2021.png","goodsCarousel":"https://newbee-mall.oss-cn-beijing.aliyuncs.com/images/iphone-13-pink-select-2021.png","originalPrice":6999,"sellingPrice":6799,"stockNum":9999,"tag":"解锁超能力","goodsSellStatus":0,"createUser":0,"createTime":"2021-12-28 13:32:06","updateUser":0,"updateTime":"2021-12-28 13:32:06","goodsDetailContent":null},{"goodsId":10920,"goodsName":"MacBook Pro 16 英寸 M1 Pro 芯片","goodsIntro":"16G 512G 银色。M1Pro 和 M1Max 芯片，霸气不封顶，120Hz 高刷 Mini-LED 屏幕，更长续航。","goodsCategoryId":35,"goodsCoverImg":"https://newbee-ma**.oss-cn-beijing.aliyuncs.com/images/mbp16-silver-select-202110_GEO_CN.jpeg","goodsCarousel":"https://newbee-ma**.oss-cn-beijing.aliyuncs.com/images/mbp16-silver-select-202110_GEO_CN.jpeg","originalPrice":20999,"sellingPrice":18999,"stockNum":9999,"tag":"强者的强","goodsSellStatus":0,"createUser":0,"createTime":"2021-12-28 15:13:27","updateUser":0,"updateTime":"2021-12-28 17:48:16","goodsDetailContent":null},{"goodsId":10921,"goodsName":"HUAWEI P50 Pro 4G 全网通 8GB+512GB 可可茶金","goodsIntro":"麒麟 9000 芯片,万象双环设计","goodsCategoryId":46,"goodsCoverImg":"https://newbee-ma**.oss-cn-beijing.aliyuncs.com/images/p50-gold.png","goodsCarousel":"https://newbee-ma**.oss-cn-beijing.aliyuncs.com/images/p50-gold.png","originalPrice":7888,"sellingPrice":7488,"stockNum":9999,"tag":"万象新生","goodsSellStatus":0,"createUser":0,"createTime":"2021-12-28 15:23:01","updateUser":0,"updateTime":"2021-12-28 17:45:10","goodsDetailContent":null},{"goodsId":10919,"goodsName":"Apple Watch Series 7 智能手表","goodsIntro":"全盘坚固表里如一！显示屏更大，耐用性提升，充电速度更快！","goodsCategoryId":33,"goodsCoverImg":"https://newbee-ma**.oss-cn-beijing.aliyuncs.com/images/MKUU3_VW_34FR%2Bwatch-45-alum-midnight-cell-7s_VW_34FR_WF_CO.jpeg","goodsCarousel":"https://newbee-ma**.oss-cn-beijing.aliyuncs.com/images/MKUU3_VW_34FR%2Bwatch-45-alum-midnight-cell-7s_VW_34FR_WF_CO.jpeg","originalPrice":2999,"sellingPrice":3399,"stockNum":10000,"tag":"全屏先手","goodsSellStatus":0,"createUser":0,"createTime":"2021-12-28 14:10:45","updateUser":0,"updateTime":"2021-12-28 17:45:38","goodsDetailContent":null}]}
2023-07-16 01:46:07.183 DEBUG [newbee-mall-cloud-recommend-service,d402ccf7f2e520b5,d402ccf7f2e520b5] 12688 --- [io-29020-exec-5] l.g.c.n.o.NewBeeCloudGoodsServiceFeign   : [NewBeeCloudGoodsServiceFeign#listByGoodsIds] <--- END HTTP (3804-byte body)
2023-07-16 01:46:07.185 DEBUG [newbee-mall-cloud-recommend-service,d402ccf7f2e520b5,d402ccf7f2e520b5] 12688 --- [io-29020-exec-5] l.r.c.n.d.I.findIndexConfigsByTypeAndNum : ==> Preparing: select config_id, config_name, config_type, goods_id, redirect_url, config_rank, is_deleted, create_time, create_user, update_time, update_user from tb_newbee_mall

```
index_config where config_type = ? and is_deleted = 0 order by config_rank
desc limit ?
2023-07-16 01:46:07.185 DEBUG [newbee-mall-cloud-recommend-service,
d402ccf7f2e520b5,d402ccf7f2e520b5] 12688 --- [io-29020-exec-5]
l.r.c.n.d.I.findIndexConfigsByTypeAndNum  : ==> Parameters: 5(Integer),
10(Integer)
2023-07-16 01:46:07.205 DEBUG [newbee-mall-cloud-recommend-service,
d402ccf7f2e520b5,d402ccf7f2e520b5] 12688 --- [io-29020-exec-5]
l.r.c.n.d.I.findIndexConfigsByTypeAndNum  : <==      Total: 10
2023-07-16 01:46:07.206 DEBUG [newbee-mall-cloud-recommend-service,
d402ccf7f2e520b5,d402ccf7f2e520b5] 12688 --- [io-29020-exec-5]
l.g.c.n.o.NewBeeCloudGoodsServiceFeign    :
[NewBeeCloudGoodsServiceFeign#listByGoodsIds] ---> GET
http://newbee-ma**-cloud-goods-service/goods/admin/listByGoodsIds?goodsI
ds=10922&goodsIds=10930&goodsIds=10916&goodsIds=10927&goodsIds=10920&goo
dsIds=10929&goodsIds=10928&goodsIds=10233&goodsIds=10907&goodsIds=10154
HTTP/1.1
2023-07-16 01:46:07.206 DEBUG [newbee-mall-cloud-recommend-service,
d402ccf7f2e520b5,d402ccf7f2e520b5] 12688 --- [io-29020-exec-5]
l.g.c.n.o.NewBeeCloudGoodsServiceFeign    :
[NewBeeCloudGoodsServiceFeign#listByGoodsIds] Accept-Encoding: gzip
2023-07-16 01:46:07.206 DEBUG [newbee-mall-cloud-recommend-service,
d402ccf7f2e520b5,d402ccf7f2e520b5] 12688 --- [io-29020-exec-5]
l.g.c.n.o.NewBeeCloudGoodsServiceFeign    :
[NewBeeCloudGoodsServiceFeign#listByGoodsIds] Accept-Encoding: deflate
2023-07-16 01:46:07.206 DEBUG [newbee-mall-cloud-recommend-service,
d402ccf7f2e520b5,d402ccf7f2e520b5] 12688 --- [io-29020-exec-5]
l.g.c.n.o.NewBeeCloudGoodsServiceFeign    :
[NewBeeCloudGoodsServiceFeign#listByGoodsIds] ---> END HTTP (0-byte body)
2023-07-16 01:46:07.231 DEBUG
[newbee-mall-cloud-recommend-service,d402ccf7f2e520b5,d402ccf7f2e520b5]
12688 --- [io-29020-exec-5] l.g.c.n.o.NewBeeCloudGoodsServiceFeign    :
[NewBeeCloudGoodsServiceFeign#listByGoodsIds] <--- HTTP/1.1 200 (24ms)
2023-07-16 01:46:07.231 DEBUG [newbee-mall-cloud-recommend-service,
d402ccf7f2e520b5,d402ccf7f2e520b5] 12688 --- [io-29020-exec-5]
l.g.c.n.o.NewBeeCloudGoodsServiceFeign    :
[NewBeeCloudGoodsServiceFeign#listByGoodsIds] connection: keep-alive
2023-07-16 01:46:07.231 DEBUG [newbee-mall-cloud-recommend-service,
d402ccf7f2e520b5,d402ccf7f2e520b5] 12688 --- [io-29020-exec-5]
l.g.c.n.o.NewBeeCloudGoodsServiceFeign    :
[NewBeeCloudGoodsServiceFeign#listByGoodsIds] content-type:
application/json
2023-07-16 01:46:07.231 DEBUG [newbee-mall-cloud-recommend-service,
```

d402ccf7f2e520b5,d402ccf7f2e520b5] 12688 --- [io-29020-exec-5]
l.g.c.n.o.NewBeeCloudGoodsServiceFeign   :
[NewBeeCloudGoodsServiceFeign#listByGoodsIds] date: Thu, 01 Dec 2022
14:46:07 GMT
2023-07-16 01:46:07.231 DEBUG [newbee-mall-cloud-recommend-service,
d402ccf7f2e520b5,d402ccf7f2e520b5] 12688 --- [io-29020-exec-5]
l.g.c.n.o.NewBeeCloudGoodsServiceFeign   :
[NewBeeCloudGoodsServiceFeign#listByGoodsIds] keep-alive: timeout=60
2023-07-16 01:46:07.231 DEBUG [newbee-mall-cloud-recommend-service,
d402ccf7f2e520b5,d402ccf7f2e520b5] 12688 --- [io-29020-exec-5]
l.g.c.n.o.NewBeeCloudGoodsServiceFeign   :
[NewBeeCloudGoodsServiceFeign#listByGoodsIds] transfer-encoding: chunked
2023-07-16 01:46:07.231 DEBUG [newbee-mall-cloud-recommend-service,
d402ccf7f2e520b5,d402ccf7f2e520b5] 12688 --- [io-29020-exec-5]
l.g.c.n.o.NewBeeCloudGoodsServiceFeign   :
[NewBeeCloudGoodsServiceFeign#listByGoodsIds]
2023-07-16 01:46:07.231 DEBUG
[newbee-mall-cloud-recommend-service,d402ccf7f2e520b5,d402ccf7f2e520b5]
12688 --- [io-29020-exec-5] l.g.c.n.o.NewBeeCloudGoodsServiceFeign   :
[NewBeeCloudGoodsServiceFeign#listByGoodsIds]
{"resultCode":200,"message":"SUCCESS","data":[{"goodsId":10922,"goodsNam
e":"HUAWEI P50 雪域白","goodsIntro":"骁龙888,4G全网通,原色双影像单元,万象双环设计",
"goodsCategoryId":46,"goodsCoverImg":"https://newbee-ma**.oss-cn-beijing
.aliyuncs.com/images/p50-white.png","goodsCarousel":"https://newbee-ma**
.oss-cn-beijing.aliyuncs.com/images/p50-white.png","originalPrice":4888,
"sellingPrice":4488,"stockNum":10000,"tag":"万象新生",
"goodsSellStatus":0,"createUser":0,"createTime":"2021-12-28
15:45:32","updateUser":0,"updateTime":"2021-12-28
18:57:21","goodsDetailContent":null},{"goodsId":10930,"goodsName":"华为
FreeBuds Pro 无线耳机","goodsIntro":"主动降噪真无线蓝牙耳机/入耳式耳机/环境音/人声
透传/双连接","goodsCategoryId":33,"goodsCoverImg":
"https://newbee-ma**.oss-cn-beijing.aliyuncs.com/images/freebuds-pro.png",
"goodsCarousel":"https://newbee-ma**.oss-cn-beijing.aliyuncs.com/images/
freebuds-pro.png","originalPrice":899,"sellingPrice":749,"stockNum":10000,
"tag":"动态降噪 听我想听","goodsSellStatus":0,"createUser":0,
"createTime":"2021-12-28 18:47:15","updateUser":0,"updateTime":
"2021-12-28 18:47:15","goodsDetailContent":null},
{"goodsId":10916,"goodsName":"Apple iPhone 13 Pro 远峰蓝色","goodsIntro":"
自适应高刷新率、画面更流畅、响应更灵敏,电影效果模式随手拍大片!",
"goodsCategoryId":47,"goodsCoverImg":
"https://newbee-ma**.oss-cn-beijing.aliyuncs.com/images/iphone-13-pro-
family-hero.png","goodsCarousel":"https://newbee-ma**.oss-cn-beijing.
aliyuncs.com/images/iphone-13-pro-family-hero.png","originalPrice":8199,

"sellingPrice":7999,"stockNum":10000,"tag":"强得很","goodsSellStatus":0,
"createUser":0,"createTime":"2021-12-28 13:39:01","updateUser": 0,
"updateTime":"2021-12-28 19:00:56","goodsDetailContent":null},
{"goodsId":10927,"goodsName":"HUAWEI MatePad Pro 12.6英寸",
"goodsIntro":"8+256GB Wi-Fi 曜石灰 麒麟9000E OLED全面屏平板电脑",
"goodsCategoryId":34,"goodsCoverImg":"https://newbee-ma**.oss-cn-beijing
.aliyuncs.com/images/matepad-pro.png","goodsCarousel":"https://newbee-ma
**.oss-cn-beijing.aliyuncs.com/images/matepad-pro.png","originalPrice":
4899,"sellingPrice":4699,"stockNum":10000,"tag":"创造无界",
"goodsSellStatus":0,"createUser":0,"createTime":"2021-12-28 17:50:31",
"updateUser":0,"updateTime":"2021-12-28
18:56:13","goodsDetailContent":null},{"goodsId":10920,"goodsName":"MacBo
ok Pro 16英寸 M1 Pro芯片","goodsIntro":"16G 512G 银色。M1Pro和M1Max芯片，霸
气不封顶，120Hz高刷Mini-LED屏幕，更长续航。","goodsCategoryId":35,
"goodsCoverImg":"https://newbee-ma**.oss-cn-beijing.aliyuncs.com/images/
mbp16-silver-select-202110_GEO_CN.jpeg","goodsCarousel":"https://newbee-
ma**.oss-cn-beijing.aliyuncs.com/images/mbp16-silver-select-202110_GEO_C
N.jpeg","originalPrice":20999,"sellingPrice":18999,"stockNum":9999,"tag":
"强者的强","goodsSellStatus":0,"createUser":0,"createTime":"2021-12-28
15:13:27","updateUser":0,"updateTime":"2021-12-28 17:48:16",
"goodsDetailContent":null},{"goodsId":10929,"goodsName":"华为 HUAWEI Sound
X 2021","goodsIntro":"智能音箱幻彩光随声动","goodsCategoryId":33,
"goodsCoverImg":"https://newbee-ma**.oss-cn-beijing.aliyuncs.com/images/
soundx.png","goodsCarousel":"https://newbee-ma**.oss-cn-beijing.aliyuncs
.com/images/soundx.png","originalPrice":2399,"sellingPrice":2199,"stockN
um":10000,"tag":"声声出色","goodsSellStatus":0,"createUser":0,
"createTime":"2021-12-28 18:41:39","updateUser":0,"updateTime":
"2021-12-28 18:41:39","goodsDetailContent":null},
{"goodsId":10928,"goodsName":"HUAWEI WATCH 3 Pro智能手表","goodsIntro":"运
动智能手表 尊享款 eSIM独立通话强劲续航心脏与呼吸健康","goodsCategoryId":33,
"goodsCoverImg":"https://newbee-ma**.oss-cn-beijing.aliyuncs.com/images/
watch-3-pro.png","goodsCarousel":"https://newbee-ma**.oss-cn-beijing.ali
yuncs.com/images/watch-3-pro.png","originalPrice":3999,"sellingPrice":38
99,"stockNum":10000,"tag":"一表万象","goodsSellStatus":0,"createUser": 0,
"createTime":"2021-12-28 18:39:05","updateUser":0,"updateTime":
"2021-12-28 18:39:05","goodsDetailContent":null},
{"goodsId":10233,"goodsName":"纪梵希高定香榭天鹅绒唇膏306#","goodsIntro":"(小
羊皮口红 法式红 雾面哑光 持久锁色) 新老包装随机发货","goodsCategoryId":
86,"goodsCoverImg":"/goods-img/04949c0e-87df-445b-96dd-29e7fc69f734.jpg",
"goodsCarousel":"/goods-img/04949c0e-87df-445b-96dd-29e7fc69f734.jpg",
"originalPrice":355,"sellingPrice":355,"stockNum":10000,"tag":"雾面哑光 持
久锁色","goodsSellStatus":0,"createUser":0,"createTime":"2019-09-18

```
13:25:08","updateUser":0,"updateTime":"2019-09-18 17:40:58",
"goodsDetailContent":null},{"goodsId":10907,"goodsName":"HUAWEI Mate 40
Pro 全网通5G手机 8GB+512GB(秘银色)","goodsIntro":"5nm麒麟9000旗舰芯片 | 超感
光徕卡电影影象","goodsCategoryId":46,"goodsCoverImg":
"https://newbee-ma**.oss-cn-beijing.aliyuncs.com/images/mate40-silver.pn
g","goodsCarousel":"https://newbee-ma**.oss-cn-beijing.aliyuncs.com/imag
es/mate40-silver.png","originalPrice":6488,"sellingPrice":6488,"stockNum
":10000,"tag":"跃见非凡","goodsSellStatus":0,"createUser":0,"createTime":
"2020-10-22 22:07:32","updateUser":0,"updateTime":"2020-10-22
22:12:37","goodsDetailContent":null},{"goodsId":10154,"goodsName":"无印良
品 MUJI 塑料浴室座椅","goodsIntro":"原色","goodsCategoryId":0,
"goodsCoverImg":"/goods-img/15395057-94e9-4545-a8ee-8aee025f40c5.jpg","g
oodsCarousel":"/goods-img/15395057-94e9-4545-a8ee-8aee025f40c5.jpg","ori
ginalPrice":85,"sellingPrice":85,"stockNum":10000,"tag":"无印良品
","goodsSellStatus":0,"createUser":0,"createTime":"2019-09-18
13:19:35","updateUser":0,"updateTime":"2019-09-18 13:19:35",
"goodsDetailContent":null}]]
2023-07-16 01:46:07.232 DEBUG [newbee-mall-cloud-recommend-service,
d402ccf7f2e520b5,d402ccf7f2e520b5] 12688 --- [io-29020-exec-5]
l.g.c.n.o.NewBeeCloudGoodsServiceFeign    :
[NewBeeCloudGoodsServiceFeign#listByGoodsIds] <--- END HTTP (5767-byte
body)
```

商品微服务实例中输出的日志信息：

```
2023-07-16 01:46:07.053 DEBUG [newbee-mall-cloud-goods-service,
d402ccf7f2e520b5,7ca7b3749f418cec] 1980 --- [io-29010-exec-1]
l.g.c.n.d.N.selectByPrimaryKeys           : ==> Preparing: select goods_id,
goods_name, goods_intro,goods_category_id, goods_cover_img, goods_carousel,
original_price, selling_price, stock_num, tag, goods_sell_status,
create_user, create_time, update_user, update_time from
tb_newbee_mall_goods_info where goods_id in ( ? , ? , ? , ? ) order by
field(goods_id, ? , ? , ? , ? );
2023-07-16 01:46:07.068 DEBUG
[newbee-mall-cloud-goods-service,d402ccf7f2e520b5,7ca7b3749f418cec] 1980
--- [io-29010-exec-1] l.g.c.n.d.N.selectByPrimaryKeys           : ==>
Parameters: 10918(Long), 10908(Long), 10906(Long), 10902(Long), 10918(Long),
10908(Long), 10906(Long), 10902(Long)
2023-07-16 01:46:07.094 DEBUG
[newbee-mall-cloud-goods-service,d402ccf7f2e520b5,7ca7b3749f418cec] 1980
--- [io-29010-exec-1] l.g.c.n.d.N.selectByPrimaryKeys           : <==
Total: 4
2023-07-16 01:46:07.164 DEBUG
[newbee-mall-cloud-goods-service,d402ccf7f2e520b5,2d023392884aa421] 1980
```

```
--- [io-29010-exec-2] l.g.c.n.d.N.selectByPrimaryKeys        : ==>
Preparing: select goods_id, goods_name, goods_intro,goods_category_id,
goods_cover_img, goods_carousel, original_price, selling_price, stock_num,
tag, goods_sell_status, create_user, create_time, update_user, update_time
from tb_newbee_mall_goods_info where goods_id in ( ? , ? , ? , ? , ? , ? )
order by field(goods_id, ? , ? , ? , ? , ? , ? );
2023-07-16 01:46:07.164 DEBUG
[newbee-mall-cloud-goods-service,d402ccf7f2e520b5,2d023392884aa421] 1980
--- [io-29010-exec-2] l.g.c.n.d.N.selectByPrimaryKeys        : ==>
Parameters: 10925(Long), 10926(Long), 10915(Long), 10920(Long), 10921(Long),
10919(Long), 10925(Long), 10926(Long), 10915(Long), 10920(Long),
10921(Long), 10919(Long)
2023-07-16 01:46:07.181 DEBUG
[newbee-mall-cloud-goods-service,d402ccf7f2e520b5,2d023392884aa421] 1980
--- [io-29010-exec-2] l.g.c.n.d.N.selectByPrimaryKeys        : <==
Total: 6
2023-07-16 01:46:07.208 DEBUG
[newbee-mall-cloud-goods-service,d402ccf7f2e520b5,b1862faf3cc65f33] 1980
--- [io-29010-exec-3] l.g.c.n.d.N.selectByPrimaryKeys        : ==>
Preparing: select goods_id, goods_name, goods_intro,goods_category_id,
goods_cover_img, goods_carousel, original_price, selling_price, stock_num,
tag, goods_sell_status, create_user, create_time, update_user, update_time
from tb_newbee_mall_goods_info where goods_id in
( ? , ? , ? , ? , ? , ? , ? , ? , ? , ? ) order by
field(goods_id, ? , ? , ? , ? , ? , ? , ? , ? , ? , ? );
2023-07-16 01:46:07.209 DEBUG
[newbee-mall-cloud-goods-service,d402ccf7f2e520b5,b1862faf3cc65f33] 1980
--- [io-29010-exec-3] l.g.c.n.d.N.selectByPrimaryKeys        : ==>
Parameters: 10922(Long), 10930(Long), 10916(Long), 10927(Long), 10920(Long),
10929(Long), 10928(Long), 10233(Long), 10907(Long), 10154(Long),
10922(Long), 10930(Long), 10916(Long), 10927(Long), 10920(Long),
10929(Long), 10928(Long), 10233(Long), 10907(Long), 10154(Long)
2023-07-16 01:46:07.228 DEBUG
[newbee-mall-cloud-goods-service,d402ccf7f2e520b5,b1862faf3cc65f33] 1980
--- [io-29010-exec-3] l.g.c.n.d.N.selectByPrimaryKeys        : <==
Total: 10
```

因篇幅有限，这里只给出了部分日志。读者可自行测试在调用其他功能时会输出哪些日志。微服务架构项目中整合 Spring Cloud Seluth 完成。

另外，当前的日志配置级别为 debug，级别很低，因此输出的日志是非常丰富的，在开发环境、测试环境或 beta 环境中可以使用，在生产环境中一般不建议输出如此丰富的日志信息，可以适当调高级别。

12.2　在全局异常处理类中增加日志

Sleuth 及后续的 ELK 日志中心都与日志信息相关，因此除基础的整合外，在项目的一些类中也要加上日志输出的代码。在这里，笔者主要在全局异常处理类中增加报错信息的日志输出，代码如下。

GoodsServiceExceptionHandler 类中增加的报错信息的日志输出：

```
@RestControllerAdvice
public class GoodsServiceExceptionHandler {

    // 声明日志对象
    private static final Logger log = LoggerFactory.getLogger(GoodsServiceExceptionHandler.class);

    @ExceptionHandler(BindException.class)
    public Object bindException(BindException e) {
        // 输出异常信息，日志级别为 error
        log.error("GoodsServiceExceptionHandler:",e);
        Result result = new Result();
        result.setResultCode(510);
        BindingResult bindingResult = e.getBindingResult();
        result.setMessage(Objects.requireNonNull (bindingResult.getFieldError()).getDefaultMessage());
        return result;
    }

    @ExceptionHandler(MethodArgumentNotValidException.class)
    public Object bindException(MethodArgumentNotValidException e) {
        // 输出异常信息，日志级别为 error
        log.error("GoodsServiceExceptionHandler:",e);
        Result result = new Result();
        result.setResultCode(510);
        BindingResult bindingResult = e.getBindingResult();
        result.setMessage(Objects.requireNonNull (bindingResult.getFieldError()).getDefaultMessage());
        return result;
    }

    @ExceptionHandler(Exception.class)
    public Object handleException(Exception e, HttpServletRequest req) {
        // 输出异常信息，日志级别为 error
```

```
        log.error("GoodsServiceExceptionHandler:",e);
        Result result = new Result();
        result.setResultCode(500);
        // 区分是否为自定义异常
        if (e instanceof NewBeeMallException) {
            result.setMessage(e.getMessage());
            if (e.getMessage().equals(ServiceResultEnum.ADMIN_NOT_LOGIN_
ERROR.getResult()) || e.getMessage().equals(ServiceResultEnum.ADMIN_TOKEN_
EXPIRE_ERROR.getResult())) {
                result.setResultCode(419);
            }
            if (e.getMessage().equals(ServiceResultEnum.NOT_LOGIN_
ERROR.getResult()) || e.getMessage().equals(ServiceResultEnum.TOKEN_
EXPIRE_ERROR.getResult())) {
                result.setResultCode(416);
            }
        } else {
            e.printStackTrace();
            result.setMessage("未知异常，请查看控制台日志并检查配置文件。");
        }
        return result;
    }
}
```

ShopCartServiceExceptionHandler 类中增加的报错信息的日志输出：

```
@RestControllerAdvice
public class ShopCartServiceExceptionHandler {

    // 声明日志对象
    private static final Logger log = LoggerFactory.getLogger
(ShopCartServiceExceptionHandler.class);

    @ExceptionHandler(BindException.class)
    public Object bindException(BindException e) {
        // 输出异常信息，日志级别为 error
        log.error("ShopCartServiceExceptionHandler:",e);
        Result result = new Result();
        result.setResultCode(510);
        BindingResult bindingResult = e.getBindingResult();
        result.setMessage(Objects.requireNonNull
(bindingResult.getFieldError()).getDefaultMessage());
        return result;
    }
```

```java
    @ExceptionHandler(MethodArgumentNotValidException.class)
    public Object bindException(MethodArgumentNotValidException e) {
        // 输出异常信息，日志级别为error
        log.error("ShopCartServiceExceptionHandler:",e);
        Result result = new Result();
        result.setResultCode(510);
        BindingResult bindingResult = e.getBindingResult();
        result.setMessage(Objects.requireNonNull (bindingResult.
getFieldError()).getDefaultMessage());
        return result;
    }

    @ExceptionHandler(Exception.class)
    public Object handleException(Exception e, HttpServletRequest req) {
        // 输出异常信息，日志级别为error
        log.error("ShopCartServiceExceptionHandler:",e);
        Result result = new Result();
        result.setResultCode(500);
        // 区分是否为自定义异常
        if (e instanceof NewBeeMallException) {
            result.setMessage(e.getMessage());
            if (e.getMessage().equals(ServiceResultEnum.NOT_LOGIN_ERROR.
getResult()) || e.getMessage().equals(ServiceResultEnum.TOKEN_EXPIRE_ERROR.
getResult())) {
                result.setResultCode(416);
            }
        } else {
            e.printStackTrace();
            result.setMessage("未知异常，请查看控制台日志并检查配置文件。");
        }
        return result;
    }
}
```

在代码中主要修改了声明的日志对象，并且输出 error 级别的日志。其余 3 个微服务工程中的异常处理类的代码修改与此类似，就不再给出代码了。

12.3　整合Zipkin实践

首先，需要在每个微服务模块的 pom.xml 文件中添加 Zipkin 依赖。依次打开 newbee-mall-cloud-user-web、newbee-mall-cloud-recommend-web、newbee-mall-cloud- order-web、 newbee-mall-cloud-shop-cart-web 和 newbee-mall-cloud-goods-web 5 个微服务实例工程中的 pom.xml 文件，在 dependencies 标签下引入 Zipkin 的依赖文件，新增代码如下：

```xml
<!--Zipkin 依赖-->
<dependency>
  <groupId>org.springframework.cloud</groupId>
  <artifactId>spring-cloud-sleuth-zipkin</artifactId>
</dependency>
```

其次，在微服务实例中配置 Zipkin 的通信地址及采样率。依次打开 newbee-mall-cloud-user-web、newbee-mall-cloud-recommend-web、newbee-mall-cloud-order-web、newbee-mall-cloud-shop-cart-web 和 newbee-mall-cloud-goods-web 5 个微服务实例工程中的 application.properties 配置文件，分别新增如下配置项：

```
# Sleuth 采样率，取值范围为[0.1,1.0]，值越大，收集信息越及时，但对性能影响越大
spring.sleuth.sampler.probability=1.0
# 每秒数据采集量，最多 n 条/秒 Trace
spring.sleuth.sampler.rate=500

spring.zipkin.base-url=http://localhost:9411
```

最后，依次启动 5 个微服务实例，并打开浏览器访问一些接口。在这里，笔者通过各微服务的 Swagger UI 页面进行测试，访问了几个网址。之后进入 Zipkin 的 UI 页面，单击 "RUN QUERY" 按钮，便会出现调用的链路信息，如图 12-1 所示。

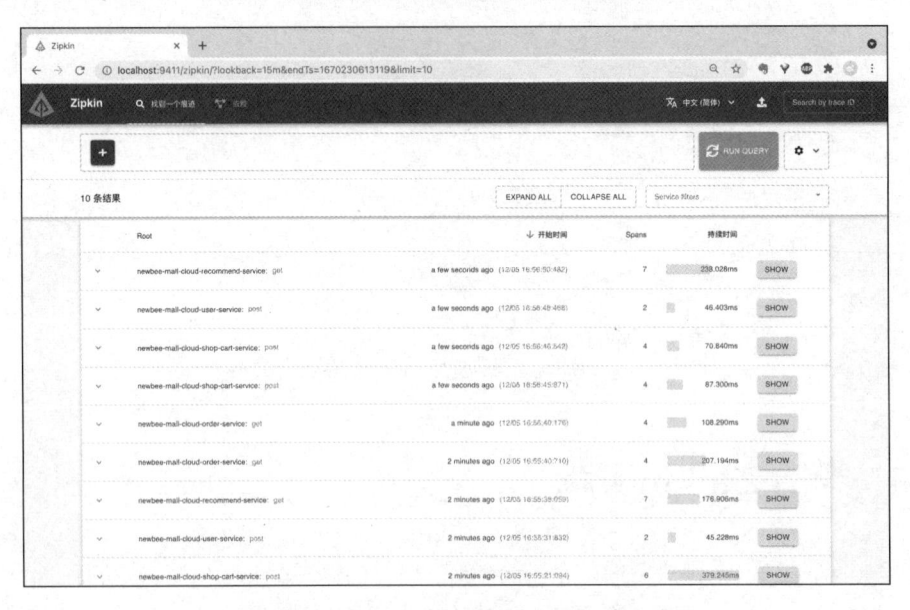

图 12-1 Zipkin 页面中显示的链路信息截图

可以看到请求信息出现在 Zipkin 页面中，包括微服务名称、请求方法、请求时间等基本信息。如果想看更详细的信息，可以单击旁边的 "SHOW" 按钮。

比如，首页接口调用链路详情页面，如图 12-2 所示。

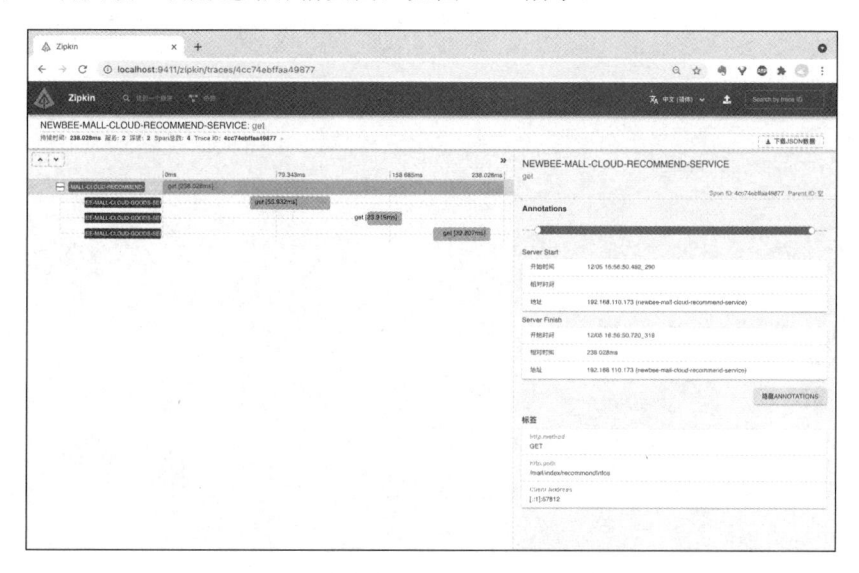

图 12-2　首页接口调用链路详情页面

添加商品至购物车接口调用链路详情页面，如图 12-3 所示。

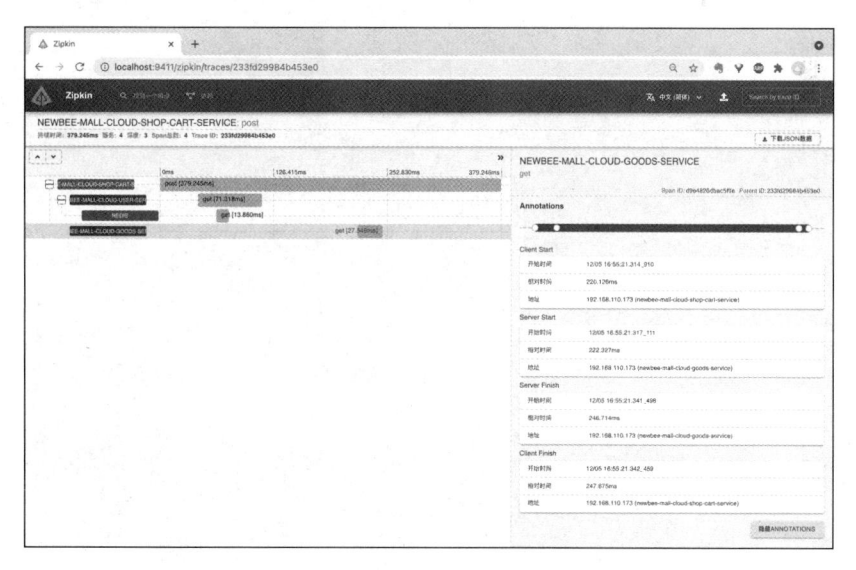

图 12-3　添加商品至购物车接口调用链路详情页面

本章主要讲解在微服务架构项目中整合 Seluth 和 Zipkin 的相关编码过程，对实战部分的讲解做补充和优化。编码改造比较简单，只是有些小细节要注意一下，毕竟在实战项目里整合与写一个小案例整合有一些差别。虽然简单，但是笔者整理完成后，可以让读者学习起来更顺畅、更简单。希望读者能够根据笔者提供的开发步骤顺利地完成本章的项目改造。

第 13 章

微服务架构项目中整合 ELK 日志中心

本章讲解在最终的微服务架构项目中进行配置的过程,最终把微服务架构项目中各个微服务实例的日志信息输出到 ELK 日志中心,也就是在微服务实例和 ELK 日志中间搭上"管子"。

13.1 微服务架构项目中的日志输出配置

详细的配置步骤如下。

第一步,引入 logstash-logback-encoder 依赖。

依次打开 newbee-mall-cloud-user-web、newbee-mall-cloud-recommend-web、newbee-mall-cloud-order-web、newbee-mall-cloud-shop-cart-web 和 newbee-mall-cloud-goods-web 5 个微服务实例工程中的 pom.xml 文件,在 dependencies 节点下新增 logstash-logback-encoder 的依赖项,配置代码如下:

```xml
<dependency>
  <groupId>net.logstash.logback</groupId>
  <artifactId>logstash-logback-encoder</artifactId>
  <version>7.0.1</version>
</dependency>
```

第二步，添加 Logback 日志配置文件。

依次打开 newbee-mall-cloud-user-web、newbee-mall-cloud-recommend-web、newbee-mall-cloud-order-web、newbee-mall-cloud-shop-cart-web 和 newbee-mall-cloud-goods-web 5 个微服务实例工程，在 src/main/resources 目录下创建 logback.xml 配置文件，代码如下：

```xml
<?xml version="1.0" encoding="UTF-8"?>
<!DOCTYPE configuration>
<configuration>
    <include resource="org/springframework/boot/logging/logback/defaults.xml"/>
    <include resource="org/springframework/boot/logging/logback/console-appender.xml"/>
    <!--应用名称-->
    <property name="APP_NAME" value="newbee-mall-cloud-order-service-log"/>
    <contextName>${APP_NAME}</contextName>

    <!-- 控制台的日志输出样式 -->
    <property name="CONSOLE_LOG_PATTERN" value="%clr(%d{yyyy-MM-dd HH:mm:ss.SSS}){faint} %clr(${LOG_LEVEL_PATTERN:-%5p}) %clr(${PID:- }){magenta} %clr(---){faint} %clr([%15.15t]){faint} %m%n${LOG_EXCEPTION_CONVERSION_WORD:-%wEx}}"/>

    <!-- 控制台输出 -->
    <appender name="CONSOLE" class="ch.qos.logback.core.ConsoleAppender">
        <filter class="ch.qos.logback.classic.filter.ThresholdFilter">
            <level>INFO</level>
        </filter>
        <!-- 日志输出编码 -->
        <encoder>
            <pattern>${CONSOLE_LOG_PATTERN}</pattern>
            <charset>utf8</charset>
        </encoder>
    </appender>

    <!-- 输出到 Logstash 开启的 TCP 端口 -->
    <appender name="LOGSTASH" class="net.logstash.logback.appender.LogstashTcpSocketAppender">
        <!--可以访问的 logstash 日志收集端口-->
        <destination>192.168.110.57:4560</destination>
        <filter class="ch.qos.logback.classic.filter.ThresholdFilter">
            <level>INFO</level>
        </filter>
```

```
        <encoder charset="UTF-8" class="net.logstash.logback.encoder.
LogstashEncoder"/>
    </appender>

    <root level="INFO">
        <appender-ref ref="CONSOLE"/>
        <appender-ref ref="LOGSTASH"/>
    </root>
</configuration>
```

配置信息基本一致,只是 APP_NAME 参数有差异。好的,"管子"搭好了,接下来启动项目测试一下,结果如图 13-1 所示。

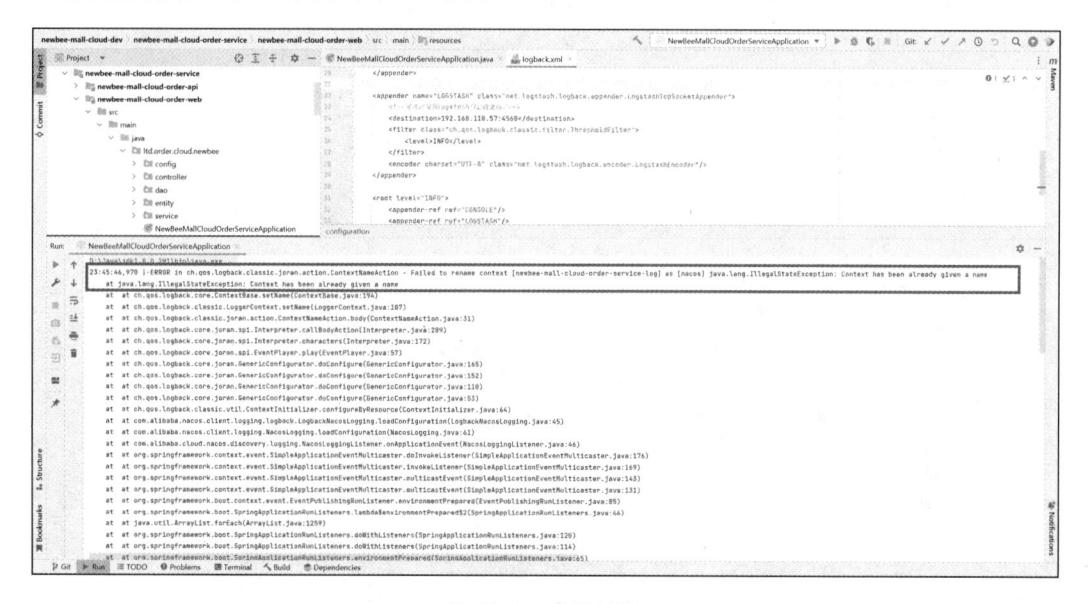

图 13-1 启动结果

代码运行日志中出现错误,错误信息如下:

```
23:45:46,970 |-ERROR in ch.qos.logback.classic.joran.action.
ContextNameAction - Failed to rename context
[newbee-mall-cloud-order-service-log] as [nacos]
java.lang.IllegalStateException: Context has been already given a name
    at java.lang.IllegalStateException: Context has been already given a name
```

报错的原因不复杂,是依赖冲突导致的。

Spring Boot 框架中已经集成了日志框架 Logback,而项目依赖 nacos-client 中也配置了 Logback(nacos-client 中的 Logback 加载要优先项目自身的 Logback 框架加载),在一个项目中，context_name 只能定义一次。所以，在项目启动时，nacos-client 中的 Logback 先加载完成后，再加载项目本身的 Logback 就出现了冲突。

这个报错并不影响使用，但是最好处理一下，需要在启动类中增加如下代码：

```
System.setProperty("nacos.logging.default.config.enabled","false");
```

只加载自定义的 Logback 配置，不使用 nacos-client 依赖中的配置，这样就不会造成冲突了。因此，需要在 newbee-mall-cloud-user-web、newbee-mall-cloud-recommend-web、newbee-mall-cloud-order-web、newbee-mall-cloud-shop-cart-web 和 newbee-mall-cloud-goods-web 5 个微服务实例工程下的启动类中添加上面这行代码。

第三步，新增日志输出的测试代码。

这里主要为了模拟平时 error 日志的输出和测试在 Kibana 中查询日志，在 newbee-mall-cloud-goods-web 工程的 NewBeeMallGoodsController 类中新增如下代码：

```
@GetMapping("/test1")
public Result<String> test1() throws BindException {
  throw new BindException(1,"BindException");
}

@GetMapping("/test2")
public Result<String> test2() throws NewBeeMallException {
  NewBeeMallException.fail("NewBeeMallException");
  return ResultGenerator.genSuccessResult("test2");
}

@GetMapping("/test3")
public Result<String> test3() throws Exception {
  int i=1/0;
  return ResultGenerator.genSuccessResult("test2");
}
```

因为已经在全局异常处理类中配置了异常的拦截和日志输出，所以在浏览器的地址栏中访问上述代码中的网址，就会直接输出 3 条 error 级别的测试日志。

13.2 通过Kibana查询日志

13.2.1 查看日志

微服务工程中的实例启动后，日志都可以在 Kibana 中查看。如果想再仔细验证，则可以在搜索框中输入关键字进行更精准的匹配。比如，笔者分别搜索了订单微服务和用户微服务的启动类名称，搜索结果如图 13-2 和图 13-3 所示。

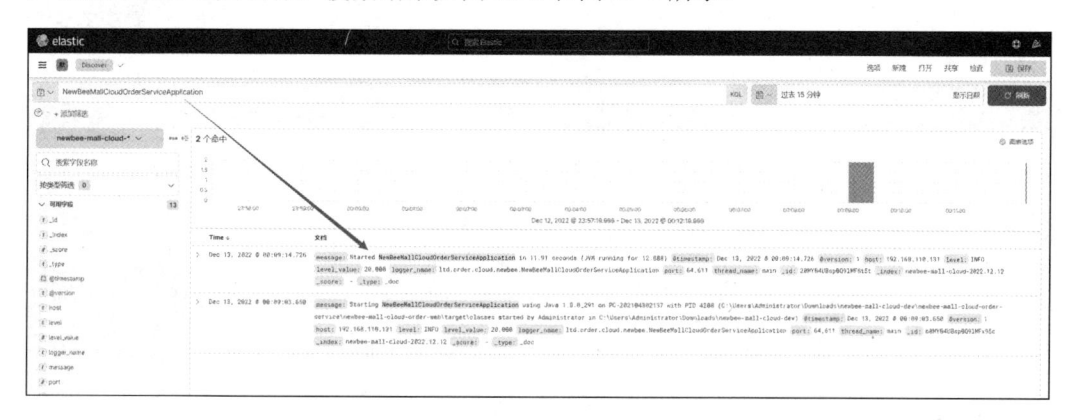

图 13-2　根据 NewBeeMallCloudOrderServiceApplication 关键字搜索日志

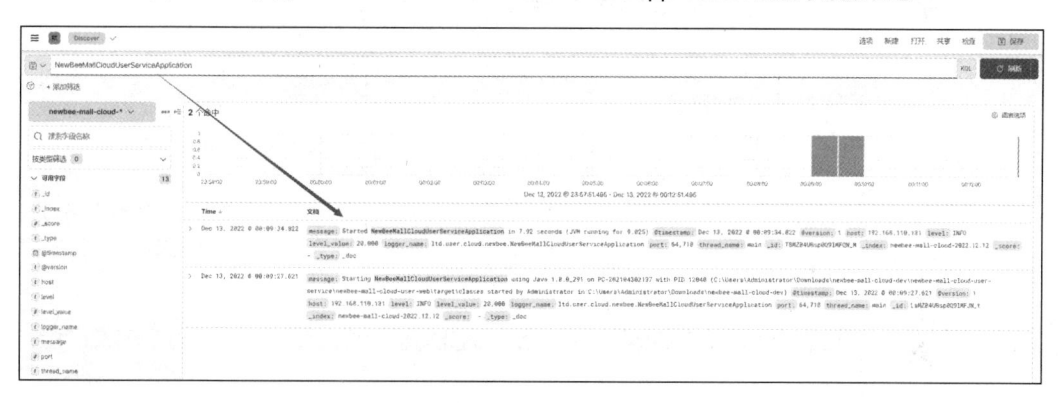

图 13-3　根据 NewBeeMallCloudUserServiceApplication 关键字搜索日志

另外，搜索的时候一定要注意时间和区间，有时搜索不到结果可能是因为时间选择不对，页面右上角有时间选择器，单击"切换"按钮即可。

13.2.2　日志定时刷新

当然，有些读者会有疑问，页面里的日志怎么不刷新？明明打印了日志且时间选择也是正确的，但是 Kibana 页面中就是没有显示。因为在默认情况下，Kibana 中的 Discover 页面不会定时刷新，需要手动单击右上角的"刷新"按钮，如果想设置自动刷新，则可以按照图 13-4 中的示意进行配置。

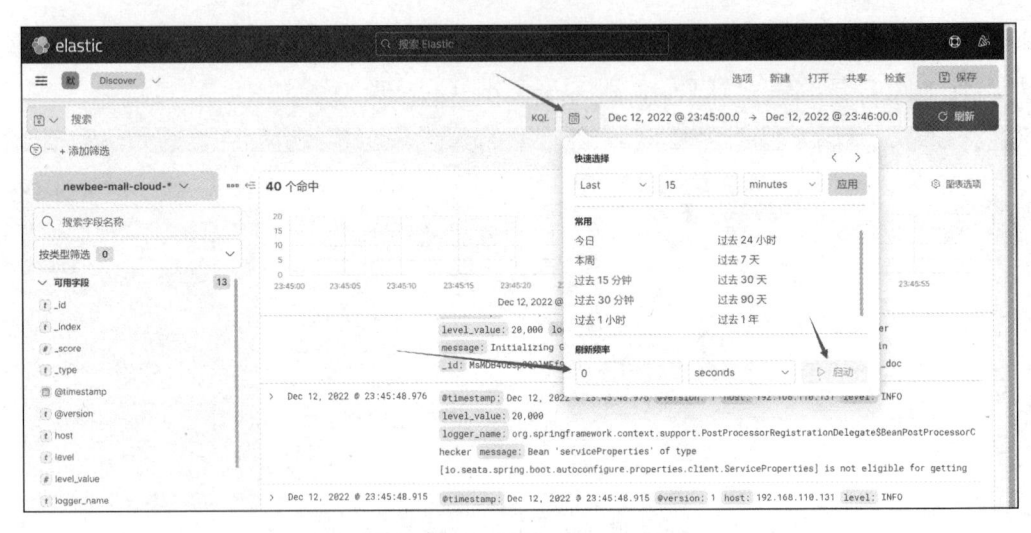

图 13-4 设置日志自动刷新

13.2.3 常用的日志搜索条件

除输入一些关键字外，还可以根据一些关键字段搜索日志。比如，直接搜索最近 15 分钟 error 级别的日志，就可以在输入框中输入"level:error"来搜索，相关的信息就会出现，如图 13-5 所示。

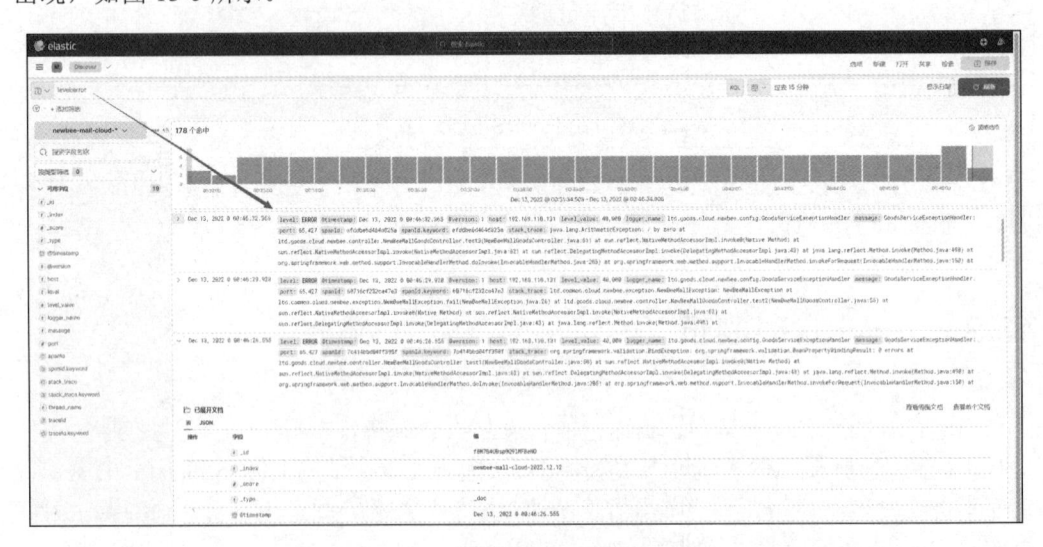

图 13-5 根据日志级别搜索日志

当然，如果想要更精确，那就加上一些条件。比如，搜索最近 15 分钟 GoodsService ExceptionHandler 输出的 error 日志，就可以在输入框中输入"level:error and logger_name: "ltd.goods.cloud.newbee.config.GoodsServiceExceptionHandler""，相关的信息就会显示出来，如图 13-6 所示。

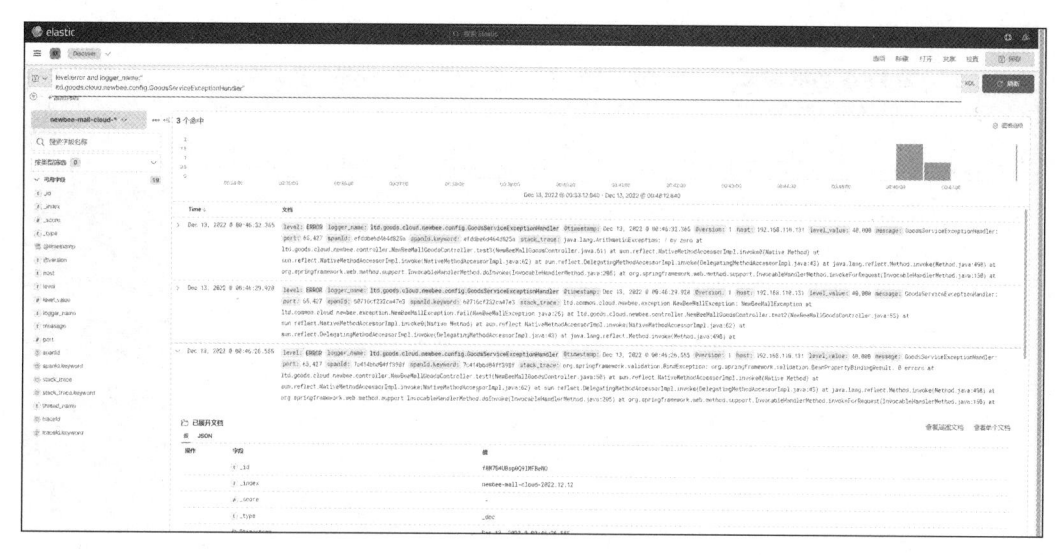

图 13-6　多条件搜索日志

除此之外，还可以使用在代码中自定义的一些字符，如"mamimamihong""zhimakaimen"，都是一些自定义的信息。如果觉得哪里可能会出问题，就打印一下日志，这样定位问题也更快一些。当然，也不要打印太多日志，没问题后就把一些没用的日志输出代码及时删掉。

笔者在平时上班的时候，到工位上的前两件事肯定是查看邮件和查看负责业务的错误日志，在浏览器中一直打开着 Kibana 页面，时不时地刷新一下，在遇到问题时赶紧定位，然后处理。

13.2.4　根据 traceId 搜索日志

使用 Sleuth 和 Zikpin 可以完成一套链路追踪系统的搭建，帮助开发人员串联调用链中的上下游访问链路，快速定位线上异常出现在哪个环节。不过，仅仅只是日志打标和追踪还不够，如果想要更加详细的信息，还要借助程序中输出的日志信息。这样，链路追踪加上日志中心，整个链路追踪就实现了闭环，有日志打标、日志收集、日志索引、日志精确搜索、链路可视化和日志的统计报表等。

简单举一个例子。看到 error 日志之后，想要查看上下游微服务实例的一些情况，就可以把这条 error 日志中的 traceId 或 spanId 作为搜索条件在 Kibana 中查询所有相关的日志信息，查询条件就输入"traceId:××××××××"或"spanId:××××××××"，或者直接搜索 traceId 或 spanId 的值，如图 13-7 所示。

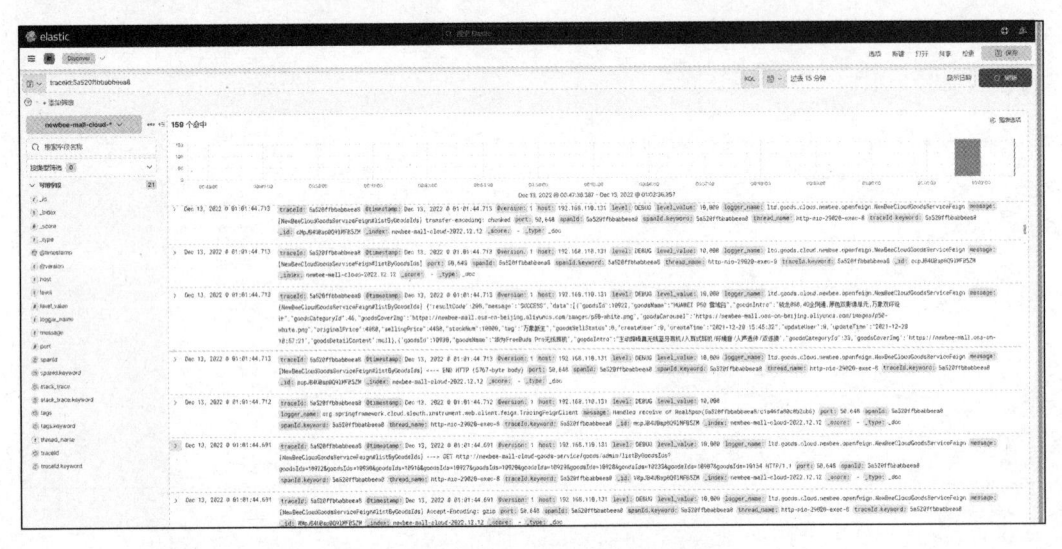

图 13-7　根据 traceId 搜索关联日志

拔出萝卜带出泥，与之相关的一些日志就都显示出来了。

为了代码演示需要及展现出更好的效果，笔者在代码里把一些日志级别调成了 debug 级别。在私下测试时可以这样做，在企业开发中千万不能这样做，因为 debug 级别的日志真的太多了。这可不是开玩笑的，把太多 debug 级别的日志弄到日志中心很麻烦。

好的，ELK 日志中心的整合及配置也讲解完毕了，整个链路追踪过程和 ELK 日志中心的搭建及整合的知识点也完成了闭环。

虽然新蜂商城项目微服务版本的功能模块已经全部讲解完毕，但是新蜂商城的优化和迭代工作不会停止，更新和优化的内容都会上传到开源仓库中供读者学习和使用。行文至此，笔者万般不舍。在本书的最后，诚心地祝愿读者能够在编程的道路上寻找到属于自己的精彩！